T0315054

DESERT NAVIGATOR

Desert Navigator

THE JOURNEY OF AN ANT

Rüdiger Wehner

THE BELKNAP PRESS OF HARVARD UNIVERSITY PRESS

Cambridge, Massachusetts, and London, England   2020

Copyright © 2020 by the President and Fellows of Harvard College

All rights reserved

Printed in Canada

First printing

Library of Congress Cataloging-in-Publication Data

Names: Wehner, Rüdiger, author.
Title: Desert navigator : the journey of an ant / Rüdiger Wehner.
Description: Cambridge, Massachusetts : The Belknap Press of Harvard
    University Press, 2020. | Includes bibliographical references and index.
Identifiers: LCCN 2019015093 | ISBN 9780674045880 (hardcover)
Subjects: LCSH: Ants—Behavior—Sahara. | Animal navigation. | Neural circuitry.
Classification: LCC QL568.F7 W427 2020 | DDC 595.79/6156—dc23
LC record available at https://lccn.loc.gov/2019015093

For Sibylle

Contents

Prologue

It is certain that there may be extraordinary mental activity with an extremely small absolute mass of nervous matter: thus the wonderfully diversified instincts, mental powers, and affections of ants are notorious, yet their cerebral ganglia are not so large as the quarter of a small pin's head. Under this point of view, the brain of an ant is one of the most marvelous atoms of matter in the world, perhaps more so than the brain of a man.

Charles Darwin, 1871

This book is about an oriental beauty, an elegant, long-legged, extremely speedy ant. It strikes the eye immediately when one wanders about in the Sahara or in any other Old World desert. I became fascinated by this remarkable ant—*Cataglyphis* by name—half a century ago, and my fascination has never ceased. Besides its conspicuous stature, running style, and extraordinary love for the blazing desert heat, it is the superb means of finding its way on far-ranging foraging journeys across large expanses of barren land that continues to astound me.

Over several decades dozens of highly motivated and talented graduate students have shared my fascination in unraveling the navigational repertoire of these little desert dwellers. Moreover, species of some other genera of ant, *Ocymyrmex* and *Melophorus,* which inhabit the deserts of southern Africa and central Australia, respectively, and resemble *Cataglyphis* in foraging style, heat resistance, and navigational skills, have joined us, as have colleagues from various research schools all over the world. The Danish physiologist and Nobel laureate August Krogh once remarked that "for a large number of problems there will be some animal of choice, or a few such animals, on which it can be most conveniently studied."[1] For various reasons, which will become apparent in this book, *Cataglyphis* is an ideal model organism for studying problems of animal navigation.

A foraging cataglyph may leave its underground nest for distances of several hundred meters, i.e., for more than 10,000 times its body length. It does so completely on its own. How can such a lone ant endowed with a miniature brain that is "not so large as the quarter of a small pin's head" and weighs less than 0.1 of a milligram,[2] accomplish its amazingly complex navigational tasks? This book aims to answer this question. As we shall see, the ant combines for its guidance all the sensory data at its disposal—visual information from sky and earth, geomagnetic, olfactory, tactile, and proprioceptive information—and feeds it into compasses, integrators, imaging devices, and various kinds of memory stores. We shall discuss these tools, study their modes of operation, and finally ask how the ant combines all this information to compute the courses to steer. What emerges is a network of highly advanced navigational strategies, which extend far beyond instinctual and simple association-based behaviors, but which are mediated by miniaturized neural circuits densely packed within the insect brain. These are challenging topics, which are tackled first and foremost by behavioral and neuro-computational scientists, but are increasingly stirring excitement among evolutionary biologists, cognitive scientists, and roboticists as well.

This book is a discovery story rather than a comprehensive review. I focus on key experimental paradigms that have been instrumental in further developing the field, and sketch out how the experiments have been devised and performed. As we shall see, focus on detail of experimental design and implementation is often crucial for avoiding hasty and vague conclusions. This effort adds some historical and epistemological touch to the stories told in the book, and may often imply that the researchers had to embark on a long odyssey. The biophysicist and molecular biologist Francis Crick once remarked that when a new scientific concept has been established, it is difficult, if not impossible, to see how it was before[3]—and, as we may add, how tortuous the ways have been that led to the concept. This remark might well apply to the current research agendas of 'spatial cognition' in animals.

Besides its emphasis on the ants' navigational strategies, this book is also an excursion into the natural history and general biology of a social insect,[4] which makes its living within the particular ecological setting of harsh arid environments, where food is scarce and foraging distances are large. Hence, after we have set the scene by introducing the navigator and its navigational needs (Chapter 1), and before we start to cover the ants' navigational tools (Chapters 3–6) and their interactions (Chapter 7), we take a special look at the cataglyphs' thermophilic way of life, and consider the suite of structural, physiological, behavioral, and social traits that enable the ants to efficiently perform their wide-ranging foraging journeys (Chapter 2). After all, it is on the hot ecological stage of the desert that the cataglyphs act their navigational play. In this play we shall get to know the actors from different angles and perspectives, and develop a feeling for the organism—the desert navigator—and the world in which it is continually faced with the decision when and where to go.

It is also important to say at the outset what the book is not. It is not about the inner workings of *Cataglyphis* colonies, and hence does not cover the high degree of physiological and behavioral organization pertaining in these 'superorganisms'.[5] It is about the individual organism. However, as the individual's behavioral repertoire— in the present account, its navigational tool set—has been shaped by the needs of the higher-level system, the colony, the potentialities and constraints of this repertoire must always be seen in the light of this higher-level evolutionary perspective. Surely, in the superorganismal 'society of minds' collective intelligence emerges from the self-organized interactions of 'simple minds', but as I like to convince the reader, these individual minds themselves are by no means simple.

I have written this book for the proverbial general reader rather than the specialists in the respective fields of research. With this in mind, I use a narrative style throughout and avoid technical jargon and considerations as much as possible. Literature references are omitted in the text, but indicated by numerical superscripts and presented in annotated form as Notes at the end of the book. All figures, which often contain information combined and adapted from several publications, have been newly designed. Credits and full references are provided after the Notes section. May the narrative style help to pique the interest of those who at the outset might wonder why at all a scientist is inclined to investigate the cockpit machinery of an insect navigator, and who are curious enough and open to surprises in unusual or exotic areas. Over the years many colleagues and friends of mine from both the sciences and the humanities have kept asking me what I was actually doing while studying my favored desert navigators, and why these little creatures have captivated me. In this book I try to provide an answer, while inviting the reader to participate in the challenges and joy of discovery.

"Monet is only an eye," Cézanne once said of his friend, "but my God what an eye."[6] I hope that at the end you will agree that *Cataglyphis* is only an ant, but truly what an ant.

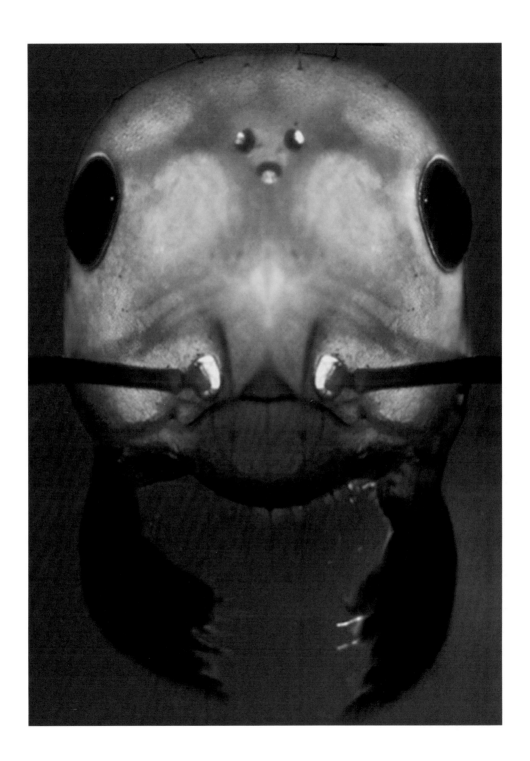

1 Setting the Scene

One hot summer day 50 years ago I went out into the vast expanse of a Saharan salt pan, the Chott el Djerid in southern Tunisia. The shimmering heat conjured up mirages in the far distance. Down on the ground there was a shiny black ant winding its way in a frantic search for food across this largely featureless desert plain. The ant—*Cataglyphis fortis,* as it later turned out—had put the hind part of its body, its gaster, into an upright position while it dashed and darted across the desert floor. It did so completely on its own. No other ant was in sight. After I had followed this solitary forager for quite some while, it suddenly detected the corpse of a large fly, grasped it, and loaded with quite heavy prey hurried off straight back until it vanished into an inconspicuous hole on the desert floor, the entrance to its underground colony. When I looked back to the site where the ant had found the fly, and which I had marked with a stick, I was amazed to see how far the ant had ventured out from its nest into the flat and extremely hostile terrain (Figure 1.2). In fact, it had left its nest for a distance of more than 100 meters, and had done so by meandering to and fro on the salt-pan floor. What amazed me even more than the sheer distance was the way the swift cataglyph had returned home: not by retracing its tortuous outward path but by following an almost straight line—the proverbial beeline—between the starting point and the end point of its foraging journey. It somehow must have been able to integrate all the directions it had steered and all the distances it had covered during its outward journey into a mean home vector (for an

1.1 Head of *Cataglyphis bicolor* illuminated from the inside

The head is glued on a tiny light-guide fiber and viewed through a stereo microscope. The bright spots in the left and right compound eye correspond to those visual units (ommatidia) that have their optical axes aligned with the optical axis of the microscope and thus with the viewing direction of the observer.

1.2 The salt-pan habitat of
Cataglyphis fortis. Chott el Djerid
in southern Tunisia

example, see Figure 1.3). For humans this task would be difficult to
accomplish without the help of instruments and some fairly detailed
knowledge in mathematical calculus or, of course, a global positioning
system (GPS) device. Here was a small-size animal less than a centi-
meter in length, endowed with a tiny brain, which I hoped—maybe
naively though—would be small enough to be amenable to neurobiological analyses, and there was a fascinating behavioral trait, a computational task complex enough to captivate the interest of neuroethologists and—as it should turn out later—even of cognitive scientists
and the neuroinformatics community.

Since this first sighting we have recorded hundreds of such foraging
trips, outbound and inbound alike, under a variety of experimental
conditions, but at the time of the first observation I did not even know
that the cataglyph, which I had followed on its long-distance journey,
was a member of a species that lived only in the salt-pan areas of

1.3 Foraging journey of a worker of *Cataglyphis fortis*

N, start of journey (nest site); *F*, prey finding site. The *thin blue* and *heavy blue lines* depict the ant's outbound and inbound path, respectively. Time marks *(black dots)* are set every 60 s. *Inset:* A forager with its gaster elevated.

Tunisia and Algeria.[1] The salty and crusty surfaces of the North African chotts and sebkhas inhabited by this halophilic desert ant can be smooth like the floor of a tennis court, but more often they are covered by polygonal mosaics of cracks and ridges. Even in the latter case, however, large salt-pan habitats such as the Chott el Djerid and Chott el Melrhir do not provide the cataglyphs with sufficiently reliable landmark cues, which could guide the animals on their long-distance journeys.

Other species of *Cataglyphis,* in fact most of them, inhabit more cluttered desert environments, e.g., the areas bordering the chotts, where the ground is covered by gravel, low scrub, or sparsely distributed tussocks of dry grass. In addition, loosely scattered xerophilous trees and bushes may provide a variety of terrestrial cues that the ants can and do use for guidance. How are these and a variety of other cues detected, processed, and integrated in the cataglyph's navigational tool kit? What does the spatial representation look like that the ant may finally generate of its outside world, its foraging terrain? In the light of such questions it seems worthwhile to embark on a short *tour d'horizon* about the ant's main navigational performances. In this tour I will draw upon some of the main issues, results, and conclusions to be described and discussed in detail later in this book, thus facing the risk of getting a bit ahead of my story. But I will do so, as otherwise it might be difficult to imagine the mega tasks that this mini navigator must accomplish.

The Safety Line

The cataglyphs are central place foragers.[2] They always start from and routinely return to their central place, the nest entrance. Whatever navigational tasks they must accomplish to achieve this goal, and whatever spatial knowledge they will finally acquire about their landmark environment, the ecological need of central place foraging has set the stage for the ant's navigational tool kit to evolve. An integral part of this tool kit—actually the ant's principal navigational routine—is path integration. In path integration the navigator acquires information about its current position (in relation to a given point of reference) only by recording its own movements. This means that positions are defined not by site-specific cues—"there is no there there," as one could metaphorically cite Gertrude Stein's quip about Oakland, California[3]—but only by how the animal has got there, by information acquired *en route* rather than *on-site*. Hence, path integration is an incremental, moment-by-moment process that starts when the animal departs from the nest and runs continuously while the animal is on its way. It is as if the animal dragged behind itself an invisible, tightly pulled Ariadne's thread, some kind of safety line connecting the animal with the point of departure and enabling it to return to this point without the aid of any landmark.

As a Greek myth tells us, King Minos of Crete engaged the famous inventor and craftsman Daedalus to build an almost impenetrable labyrinth, in which he imprisoned the bull-headed monster Minotaur.[4] This mazelike labyrinth had only one outward door and was so full of winding passageways and blind alleys that no one who had ever en-

1.4 View onto a test field provided with a grid of 1 m squares painted on the flat and barren desert floor

tered it could find a way out. All visitors fell victim to the monster. The navigational escape problem was only solved when Ariadne, the daughter of King Minos, was advised by Daedalus to provide her lover Theseus with a ball of thread and to fix the loose end of the thread at the door. While Theseus negotiated his way through the labyrinth and proceeded toward the center, he paid out the thread behind him. After having slayed the monster, he followed the thread back to the exit door.

Owing to the structure of the labyrinth with its walled alleys, Theseus could not pull the thread tight, but in the open terrain of a desert plain a cataglyph, while winding its way in search for food, can do so, at least metaphorically. In contrast to Theseus, who had to make all former turns in reverse direction, the ant can employ its path integrator and return straight to the start (Chapter 5). If it forages in cluttered environments, it is forced to negotiate a natural maze and to run obstacle courses. However, unlike Theseus it does not return by exactly retracing its outward path. It usually chooses a novel homeward route that is as close as possible to the direct vector route. When the ant happens to visit a goal repeatedly, it rapidly develops remarkable route fidelities by using landmark memories—a kind of Ariadne's thread as well—to stick to idiosyncratic outbound and inbound routes (Chapter 6).

1.5 The displacement paradigm

(a) Outbound run of an ant in the training (nest) area. The journey starts at the nest (N) and leads to an artificial feeder (F) at which the animal is captured shortly before it starts its homeward run (full-vector ant), and displaced to a distant test area. (b) Inbound run in the test area from the point of release (R, the fictive position of the feeder) to the fictive position of the nest (N*, open circle). The *red dot* indicates the end of the straight inbound run and the start of the area concentrated search. Time marks (black dots) are set every 10 s. (c) Inbound trajectory of an ant, which has performed a 10 m foraging run in the field and is subsequently mounted on top of a spherical treadmill (see Figure 1.7). *Cataglyphis fortis.*

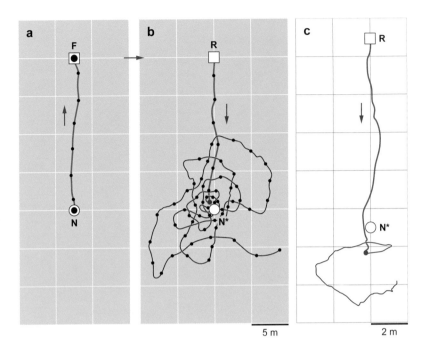

5 m

2 m

When pulled tight, the metaphorical thread is a vector that points either back from the cataglyph to the starting point or from the starting point to the animal's current position (depending on where the base of the vector is fixed). Hence, path integration can be called vector navigation.[5] In computing this path integration vector the cataglyphs use direction and distance sensing devices to monitor their angular and linear components of movement, and finally employ an integrator for combining these components in computing the home vector. A striking example of a round-trip journey, on which a cataglyph is guided exclusively by path integration, is shown in Figure 1.3. On its straight way home the ant reels off the path integration vector that it had rolled up during its outbound journey. When the animal has arrived at the nest, the vector is zeroed.

Of course, methodically one question remains. How can we be fully sure that the cataglyph really relies on path integration and does not take advantage of one or another familiar signpost unnoticed by the human observer? We can be sure only after having performed a critical displacement experiment, in which the ant is captured just after it has grasped its prey and is displaced to a nearby test field. This field is provided with a grid of white lines as a means of facilitating the recording of the ants' trajectories (Figure 1.4). The result of the displacement experiment is clear-cut. Upon release the ant first performs some turning movements to adjust its body orientation with the desired predisplacement homeward direction, then runs off in this direction rather than in the direction of the real nest, and travels for a distance equivalent to its predisplacement homeward distance. Having reached the fictive position of the nest, it starts to systematically search around the location where according to its home vector the nest is most likely to be (Figure 1.5b).

We have applied this experimental procedure in several modified versions. The ants are routinely trained to an artificial feeder, which is provided with a particular kind of bait such as biscuit crumbs or small pieces of meat, fruit, or cheese, and sunk into the ground, so that the ants cannot see it from a distance. After an ant has grasped a morsel, it is captured, marked individually (by applying a two-dot or three-dot color code to head, alitrunk, and gaster), and transferred to the unfamiliar test area, where its homeward run is recorded at reduced scale on squared paper or more recently by means of an electronic graphics tablet (Figure 1.6). At certain time intervals (usually every 10 seconds, as indicated by an electronic timer) time marks are

1.6 Recording an ant's walking trajectory

The inbound path of a cataglyph, which has been displaced from its habitual home range area to a distant test field, is recorded by means of a gridwork painted on the desert floor and a corresponding scaled-down gridwork on an electronic graphics tablet. The latter is powered by a solar panel mounted on the back of the experimenter. Graduate student David Andel recording an ant's path.

1.7 A cataglyph on a spherical treadmill

An individual *Cataglyphis fortis* is mounted
on top of an air-cushioned sphere. The ant is
glued to a filament of dental floss, which is
attached to a fine steel rod (diameter 0.08 mm)
rotating in a vertical glass tube. Thus the ant is
able to rotate about its yaw axis with almost no
friction, but when moving forward it is kept in
place and instead moves the sphere. A
circular screen (not shown) around the top of
the treadmill prevents any view of the
landscape. **(a, b)** Close-up views of a mounted
ant walking on the sphere *(a)* and detached
from it *(b)*. **(c)** Experimental setup at our field
site near Mahrès, Tunisia.

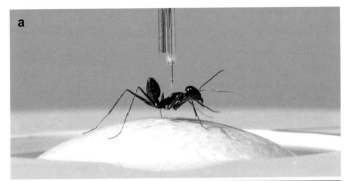

added to the ant's path. All recorded trajectories are later digitized and
computer analyzed. On sandy ground, where a test grid cannot easily
be applied, the spatial layout of the runs is recorded by placing num-
bered flags on the ants' trajectories at 10-second intervals. Afterward
the coordinates of all flagged positions are determined by using differential GPS and evaluating the data by means of geographic information systems (GIS) software. Moreover, for tracking entire foraging
journeys over large distances, sufficiently accurate information can be

obtained by just following the ants with a handheld GPS device. For analyzing the fine spatial detail of the ants' trajectories as well as the ants' head and body postures, high-speed video recordings are performed.

One might wonder whether the human observer, while recording the ant's trajectories, would disturb the experimental animal or deflect it in one way or another from its intended navigational course. Innumerable control experiments have shown that this is not the case. The observer is constantly changing his or her position relative to the ant, and hence provides the animal with an unreliable visual signpost. In general, if an animal must decide whether it should take advantage of a given landmark as a navigational aid, it should certainly apply the rule 'if it moves, don't use it.'[6] Finally, in many experiments the homing ant walks underneath a rolling optical laboratory (see Figures 3.8 and 3.18d), within which it can see neither the investigator nor the surrounding landmarks, but with which its view of the sky can be manipulated in particular ways.

Recently Hansjürgen Dahmen and Matthias Wittlinger from the Universities of Tübingen and Freiburg in Germany have succeeded in recording the cataglyphs' homeward runs by tethering the animals and letting them walk on a spherical treadmill under the open sky directly at our North African study site (Figure 1.7). In this device the ant is comfortably mounted on top of an extremely light air-suspended polystyrene foam ball. Even though it is fixed in its position, the ant is able to adapt its posture with respect to height, pitch, and roll, and to rotate about its vertical body axis with its own moment of inertia. In the special setup used here, the apparatus records changes in the ant's running direction. By monitoring this ant-induced motion of the sphere, the fictive path taken by the ant on top of the sphere is reconstructed.[7] The cataglyphs get so accustomed to walking and navigating under these constrained conditions that they pay out their home vector as accurately as they do when they freely walk on the desert floor. Compare the homeward runs performed by an ant displaced from the feeder either to the test field (Figure 1.5b) or to the top of the spherical treadmill (Figure 1.5c). Quite certainly, the trackball setup first developed for fruit flies in the laboratory and now well advanced for the cataglyphs in the field will open a Pandora's box of

1.8 Cluttered desert environment inhabited by several *Cataglyphis* species

The parched terrain in the Djerid area of southwestern Tunisia is the habitat of *C. savignyi, C. albicans,* and *C. rubra.*

new experimental paradigms, including recordings from the brain while the animal is simultaneously presented with a variety of natural or artificial sensory stimuli.

Panoramic Vision

The path integrator prevents the cataglyphs from getting lost when other navigational routines fail. However, if landmarks are available, as they usually are in most desert environments (Figure 1.8), the ants acquire, retain, and use visual information derived from the surrounding landmark scenes, in order to return to specific locations and to navigate visually guided familiar routes. All this information is based on views originally acquired while the ants rely on path integration.

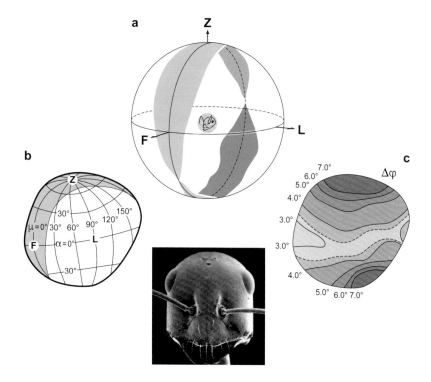

1.9 Holoptic vision in the cataglyphs: full panoramic view

(a) Three-dimensional representation of the visual fields of the left and right compound eye. The *light blue* and *dark blue shading* in the ant's visual sphere depicts the binocular range and the blind zone, respectively. (b) Circles of longitude and latitude of the outside world projected onto the left compound eye. The head is inclined relative to the horizontal by the angular amount that the animals, on average, adjust during walking. (c) Receptor spacing (interommatidial divergence angles, $\Delta\varphi$) of the left compound eye. Those parts of the eye that look at the horizon form a band-like 'visual streak' of slightly higher resolution. Data derived from optical pseudopupil measurements in live animals. *Inset:* Frontal view of head. *F,* frontal; *L,* lateral; *Z,* zenith. *Cataglyphis bicolor.*

With its omnidirectional vision—its large panoramic field of view—a cataglyph is able to receive visual information simultaneously from almost all parts of its celestial and terrestrial world. As optical measurements (Figure 1.1) show, the visual fields of the two compound eyes taken together sample the entire 360° arc of the horizon and cover no less than 93% of the unit sphere surrounding the animal. There is only a small 'blind zone' below and behind the ant's body. In front of the ant the visual fields of both compound eyes overlap and provide the animal with a narrow binocular zone (Figure 1.9).

Independently of body size, this wide-field, panoramic vision applies to all *Cataglyphis* species studied so far. In *Cataglyphis bicolor* eye size varies in relation to body size from 600 visual units (ommatidia, see Figure 3.3) in the smallest workers to 1,300 units in the largest ones, but the size of the total visual field remains constant. Consequently, smaller individuals have lower spatial resolution than their larger conspecifics, meaning that they forgo visual acuity for a whole-field view of the world through which they move.[8] The smallest angles between adjacent visual units, the so-called interommatidial divergence angles—and thus the highest levels of spatial resolution—occur in a band-like zone, which runs along the equator of the eye. Even in this 'visual streak' the divergence angles between the individual visual units are not smaller than about 3° (Figure 1.9c), compared to 0.02° in the center of the human fovea. Technically speaking, the cataglyphs' compound eyes—as compound eyes in general—act as spatial low-pass filters discarding fine spatial detail.[9] However, coarse-grain vision can even be advantageous when it comes to guidance in more cluttered environments, in which high spatial frequency cues are likely to provide ambiguous information. In this case, low-resolution combined with full-view vision leads to more robust, i.e., less error-prone, performances and suggests that whole scene recognition may play a major role in *Cataglyphis* navigation.[10] Note, however, that the highest acuity reached in the forward direction may also emphasize the importance of frontal vision.

This raises the question of what kind of navigationally relevant information the cataglyphs select from this large panoramic visual surround. Imagine a desert environment similar to the one shown in Figure 1.8. There are all kinds of landmarks—small and large, nearby and far-off—and high up in the sky there is a peculiar optical phenomenon caused by the scattering of sunlight in the earth's atmosphere (the 'pattern of polarized light'), which is invisible to humans but conspicuous to the cataglyphs. Further imagine that within this desert environment a cataglyph provided with full panoramic vision would move forward along a straight line. Then the images of nearby objects appear one after another in the frontal field of view, move backward and disappear in the rear. The velocity of this self-induced optic flow, which radiates out across the entire retina from a 'point of expansion' in the direction of travel, is very low in the frontal visual field, i.e., close to the direction of travel, and reaches its maximum to the side. Near features move faster than more distant ones, so that in the visual world of the running ant the foreground will move with respect to the relatively constant distant panorama. As the sun and other skylight cues, such as the aforementioned pattern of polarized light, are effectively at infinity, the cataglyph would not experience any image motion of these cues as long as it keeps its direction of travel constant, but such image motion would immediately occur when the animal rotates, i.e., turns in a new direction. Given this structured diversity of visual stimuli impinging on the ant's omnidirectional visual system, the question immediately arises how the cataglyph's brain handles and uses this kaleidoscopic wealth of information. Has it acquired, during the course of evolution, particular sensory and neural 'filters' matched to the predictable nature of the ant's visual world, so that it is able to pick up and process only those kinds of visual information that are relevant for one aspect of navigation or another?[11]

Irrespective of what the answers to such questions are, what the cataglyphs acquire in the first place are nothing but local views, 'snapshots' taken from particular vantage points and stored within the ant's own system of reference. The important question, then, is how they organize their spatial memories and retrieve the proper information at the right time and context. Are they able to link them in a way that a coherent spatial representation of their foraging terrain is formed? This would be a computationally demanding task, because the ants would arguably need to construct such a geocentric (landscape-centered) representation of space from merely egocentric (animal-centered) landmark views and path integration coordinates. If they succeeded in this task, they would at least in principle be able to acquire what since the middle of the previous century has been called a 'cognitive map,' a mental analogue of a topographic map. This concept, which was originally derived to account for the spatial behavior of rats and mice in laboratory arenas, has more recently also been applied to describe the navigational behavior of honeybees. We will discuss this overarching topic in Chapters 6 and 7.

Constructing a topographic representation of familiar terrain is indeed the way a modern human cartographer proceeds in linking positional information about places. But during the course of evolution the insect navigator might well have adopted different means of coping with its navigational mega task to find its way through the cluttered, mazelike desert environment. Hence, we must be cautious not to phrase our questions simply in the way a human navigator would go about these tasks. Human problem solving relies on first principles, i.e., on reasoning based on mathematics and logic. It usually aims at general-purpose solutions, which tend to be computationally complex.

However, the question we have to deal with here is how efficiently the cataglyphs accomplish the tasks with which *they* must contend in the most immediate ways, and how well *they* perform within the *Cataglyphis*-specific ecological setting rather than within the conceptual framework of the human investigator's rational mind.

Late in Life

Defining the navigational tasks in terms of the cataglyphs' particular ecological setting brings us to another decisive point, which is worth considering at the outset. It refers to the ant's life history. The *Cataglyphis* workers, like workers of almost all ant species, change tasks as they age. They become foragers only during the very last part of their lifetimes. How long they are then engaged in these outdoor duties can be read off their survival frequency curves (Figure 1.10). Before they reach this stage they are engaged in quite a number of other tasks.[12] After eclosion from the pupal case they first stay as rather inactive 'repletes' (storers of liquid food) close to the brood. They are then characterized by swollen gasters and stretched abdominal intersegmental membranes; they sit mostly still inside the nest chambers and, if disturbed, move around only sluggishly. Subsequently, the workers become nurses specialized in various brood-caring and nest-maintaining tasks. Near the end of this stage they move toward the periphery of the nest, where they perform excavation and dredging work, and finally commence foraging. This centrifugal pattern of locomotor activity and task allocation continues in the outside world, as foraging distance increases with age.[13] Hence, at any one time a colony of these desert ants contains a number of 'temporal worker castes' with each caste performing a certain set of behaviors and each individual accomplishing different tasks at different ages (temporal polyethism).[14]

While accomplishing these tasks, the cataglyphs are able to identify the species-specific, colony-specific, and even task-specific status of their nestmates. For example, when the cataglyphs display some quite sophisticated behavior to rescue nestmates caught under collapsing sand or trapped in other ways, the foragers administer their

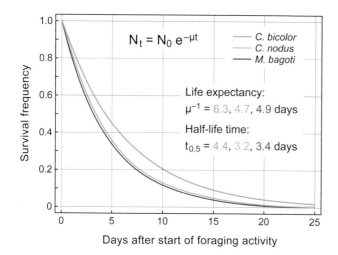

$$N_t = N_0 \, e^{-\mu t}$$

— C. bicolor
— C. nodus
— M. bagoti

Life expectancy:
μ^{-1} = 6.3, 4.7, 4.9 days

Half-life time:
$t_{0.5}$ = 4.4, 3.2, 3.4 days

Survival frequency

Days after start of foraging activity

1.10 Forager survival frequencies recorded in the ants' natural habitats

Cataglyphis bicolor in the lowland steppes of Tunisia, *C. nodus* in the Thessalian plains of Greece, and *Melophorus bagoti* in the semiarid desert habitats of central Australia. In all species the survival frequencies follow an exponential decay function.

aid especially to foragers, which they can distinguish from inside workers.[15] As in other ants, the compositions of cuticular hydrocarbons act as multicomponent recognition cues, which are highly variable among workers of different species, colonies, and developmental status. By constantly 'antennating,' i.e., contacting each other with the chemosensory batteries of their antennae, the members of a colony are supposed to learn these hydrocarbon profiles of their fellows and their status, and thus acquire a 'neural template' which they later compare with the cue profiles of the nestmates they encounter.[16] Similarly, as we shall see in Chapter 6, when the cataglyphs employ landmark views for navigation, they continually compare memorized panoramic images taken at particular vantage points ('templates') with the currently experienced images.

As mind-boggling as the ant's navigational performances might appear, the cataglyph's brain must take charge of a rich repertoire of other demanding behaviors as well. After all, it is for about 80% of their lifetimes that adult *Cataglyphis* workers are usually engaged in fulfilling their indoor duties exclusively within the darkness of the subterranean colony.[17] Even when they later become foragers and develop their navigational skills, they must also accomplish a variety of other tasks such as detecting and identifying prey, avoiding predators, or recognizing members of foreign conspecific colonies.

Recently, a colleague and friend of mine from the theoretical physics department approached me with a question, after I had told him about the mega tasks that our mini navigators accomplish at the end of their lives: "Why don't your ants, once they are out of the nest and adopt their foraging mode of behavior, take full advantage of the computational power of their entire brain to assume GPS function whenever needed?" My immediate answer was that the ant's brain, as any animal's brain, is composed of modular networks, which have evolved in a mosaic way. Driven by the vicissitudes of natural selection, the various guidance mechanisms and their underlying sensory and neural processing circuits, which now constitute the ant's navigational tool kit, might have been shaped in different ways and at different rates. More basic mechanisms have become modified and specialized, and new circuitries and interconnections between them have been established. How plastic, then, is the ant's brain in solving complex tasks? Can it recruit the computational capabilities of brain areas that are primarily involved in other tasks? Can the same task be solved in different ways? Questions like this will accompany us on our journey throughout the book and will take center stage when we will inquire in more detail about the architecture of the cataglyph's navigational tool set. At this juncture, they may lead to some general considerations and historical reflections about the design of the insect's mind.

Outlook

For more than a century a hotly debated issue has been whether insects—and especially social insects such as wasps, ants, and bees—are preprogrammed automatons or cognitive beings. Are they mere "mindless machines" capable of nothing but strictly stereotyped, reflex-like, "mechanistic" modes of behavior performed "with rigid mathematical precision," or must they be credited with some "higher psychological functions," certain "mental faculties," and even "rationality"?[18] Even though all these arguments are profoundly semantic and might conjure up nebulous realms, they nevertheless span the full range of views from considering insect behavior as the mere outcome of stimulus-response interactions to providing bees and ants with cognitive capacities similar to the ones observed in higher vertebrates.

"To know everything and to know nothing according as it acts under normal or exceptional conditions: that is the strange antithesis presented by the insect race," remarked the Midwestern biologist Roy Abbott in the 1930s. He referred to the complex, though apparently rigid, sequence of behavioral acts that *Sphex* digger wasps perform when they provision their underground burrows with paralyzed insect prey, but which they seemed unable to adapt to artificial changes in the cues that triggered the behavior. For decades, this *Sphex* story has been used by computer and cognitive scientists such as Dean Wooldridge, Douglas Hofstadter, and Daniel Dennett to characterize the mindlessness of seemingly thoughtful behavior.[19] 'Sphexishness' has become a philosophical metaphor increasingly separated from its biological underpinnings. In fact, later investigations have shown that within certain limits the wasp can adapt its presumed rigid sequence of actions to situational contingencies, meaning that on closer biological scrutiny the insect's behavior exhibits much more flexibility and adaptability than proclaimed by the sphecists.[20]

The catchy *Sphex* metaphor and its long-lasting narrow-minded view of insect behavior may serve as a warning against rigidly adhering to preconceived general views, be they good old sphexishness, on the one side, or the presently fashionable application of concepts derived from human ways of problem-solving, on the other side. Two great nineteenth-century naturalists, to whom we owe some of the most painstaking behavioral analyses of their times, Jean-Henri Fabre from the Vaucluse and Darwin's young friend John Lubbock, did not set out to pursue such overarching theoretical concepts.[21] They rather focused on the marvelous diversity and complexity of the behavior that insects display in various biological contexts, and tried to unravel in detail how the insects—these little 'aliens' to us humans—do it in their own idiosyncratic ways. Had the current tools of the neurosciences already been available to them, they might have become the first full-fledged insect neuroethologists.

The enduring debate about the insect's "prerational intelligence" highlights an important point: insect behavior is characterized by rigid, stereotypical, domain-specific routines as well as by experience-dependent adaptability, flexibility, learning, memory, and forms of decision making, which at first glance would be attributed only to higher and more cognitively minded vertebrates.[22] For more than a century, the divergent views mentioned above have fallen in and out of favor. After all, what we finally need to understand is a species' behavioral repertoire in its habitat, and the structure of the neural circuitries mediating that behavior. Then it may emerge that highly advanced behavioral capabilities can result from computations performed within relatively small neural networks. It is this miraculous emergence that neuroethologists try to unravel. The question asked in the beginning of this section is certainly ill-posed.

As the cataglyphs' navigational performances must be seen in the light of the animal's foraging ecology, let us now turn, at least for a while, to the very ecological niche in which these desert dwellers have evolved a suite of anatomical, physiological, and behavioral adaptations enabling them to forage and navigate successfully under the harsh conditions of their desert lives. In fact, the *Cataglyphis* navigators as well as their ecological equivalents in southern Africa *(Ocymyrmex)* and central Australia *(Melophorus)* inhabit the subtropical desert belts, where the combination of heat and aridity reaches its extremes.

2 The Thermophiles

The cataglyphs are thermophiles—heat-loving, heat-seeking thermal warriors. In restricting their foraging times to the hottest times of day and year, which are avoided by the majority of other ant species, they risk near-lethal temperatures to maximize their foraging success.[1] With their long legs, slender body, and high running speeds, they are conspicuous creatures, both morphologically and behaviorally, that can readily be dubbed the racehorses of the insect world.

When more than two centuries ago the young biologist Jules César de Savigny accompanied Napoléon Bonaparte on the French expedition to Egypt, where he was responsible for collecting invertebrate animals, he came across these elegant creatures and portrayed them—still nameless—in the magnificent edition of the *Description de l'Égypt* (Figure 2.2).[2] Even Greek and Roman writers might have referred to them, as Herodotus and Pliny mention large-size ants, which live in the desert "on the borders of the city of Caspatyros and the county of Pactyike," present-day Kabul or Peshawar. They describe them as running with high speed and making their dwelling in the ground, from which they carry out gold to the surface.[3]

If one wants to detect and observe a cataglyph, be it gold digging or not, one must be vigilant. These animals do not occur en masse. They do not march in groups along conspicuous trails, as the omnipresent harvester ants do. They are solitary foragers. One of them may run about here, another one there. Their widely dispersed colonies are

2.1 Thermal respite behavior of *Cataglyphis bombycina*, the silver ant

A forager radiates off excess heat on top of a stalk of dried vegetation. Sand-dune desert near Tozeur, Tunisia.

2.2 The first portraits of *Cataglyphis*

Plate 20 of the *Description de l'Égypt*
documenting the scientific results of
Napoléon Bonaparte's expedition to Egypt
in 1798–1799. The plate "Fourmis"
portrays various species of ants, but lacks a
detailed legend. Certainly, the upper row
depicts specimens of *Cataglyphis*. From left
to right: a male (enlarged in *inset below*), a
reproductive female, and a worker of *C.
savignyi* alongside a soldier of *C. bombycina* (accompanied by the isolated
mouthparts including a saber-like
mandible).

rather small, comprising only a few hundred or thousand members.
Nevertheless, if one happens to come across one of these foragers hur-
rying over some barren ground, this high-heeled creature can hardly
be overlooked.

Due to its extraordinary running speed and jerky movements, it will
be difficult to capture, but one should give it a try. Carefully grasp the
agile runner by its long legs—it does not sting—and look at it more
closely: at the usually dark red head with its oval-shaped compound

eyes on the side, its three punctate small ocelli on top of the head, and the triangular mandibles each endowed with a row of five teeth. Further note the slender, elongated middle part of the body (the alitrunk) with a conspicuous slit-like opening on either side (the so-called propodeal spiracles used for breathing); the extremely long and slender legs with their large and powerful first segments (the coxae); and the small often knob-like petiole linking the middle part of the body to the hind part (the gaster), which at its posterior tip has a striking brush, the so-called acidopore, a circular tuft of hairs surrounding the anus (for further morphological details and terms see Figures 2.3 and 2.4). Based on such characters, any ant taxonomist will easily recognize these ultrarunning ants as members of the genus *Cataglyphis.* However, at the species level taxonomic distinctions are not always easy and still a matter of debate. At present, about 100 valid species are recognized.[4]

Even though a look at a cataglyph may not raise the aesthetic excitement caused by an iridescent butterfly or a jewel beetle, the superb drawings of Eva Weber unveil the delicate beauty of these desert dwellers (see, e.g., Figures 2.3, 2.5, and 2.17). Some species are bright yellow *(C. livida),* others are pitch-black *(C. fortis),* but the majority is bicolored to various degrees, with head and alitrunk from light orange to dark red and the gaster brownish or black (e.g., *C. bicolor,* *C. viatica,* and *C. savignyi*).

Unique beauties are the silver ants (*C. bombycina,* Figure 2.5),[5] which inhabit the vast expanses of the Saharan sands. They strike the eye by their shiny silvery glare. When they dash about the surface of the sand in full sunshine, they can often be detected only by the shadow they cast on the bright surface of the sand. Their silvery appearance is caused by the remarkably long and dense pubescence, which—as we shall see later—provides these animals with an exquisite heat protection device. Another morphological trait, which characterizes all *Cataglyphis* species as well as some other desert ants, is particularly striking in *C. bombycina:* the ammochaetae, rows of long, slender hairs on the clypeus, the mandibles, the maxillary palps, and other parts of the underside of the head. Taken together, these hairs form a psammophore, a basketlike structure used in carrying sand or soil when the ants excavate their nests.[6]

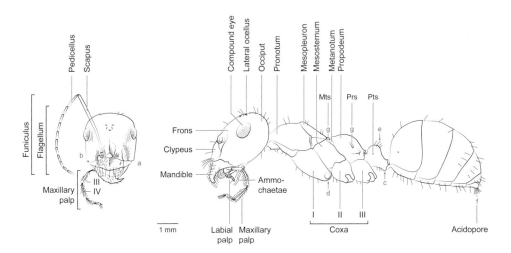

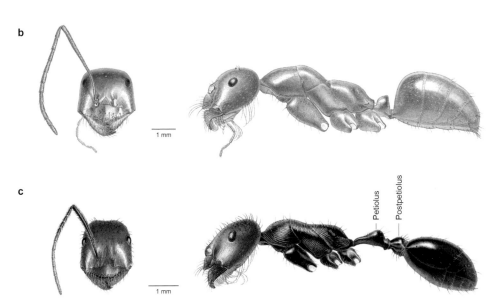

2.3 The three thermophiles: morphology

(**a**) *Cataglyphis bicolor;* (**b**) *Melophorus bagoti;* (**c**) *Ocymyrmex velox.* Note the set of long hairs (ammo-chaetae = macrochaetae) at the underside of the head and the inner margin of the mandibles. The workers of the formicine genera *Cataglyphis* and *Melophorus* are provided with long, extended maxillary palps. In *Ocymyrmex* as well as in all myrmicine workers the maxillary palps are much shorter and concealed underneath the mandibles. Spiracles: *Mts,* metathoracic spiracle; *Prs,* propodeal spiracle; *Pts,* petiolar spiracle. *Blue signatures* indicate the sites of the views taken in the scanning electron microscope (SEM) pictures of Figure 2.4 *(a–f)* and Figure 2.28a, b *(g).*

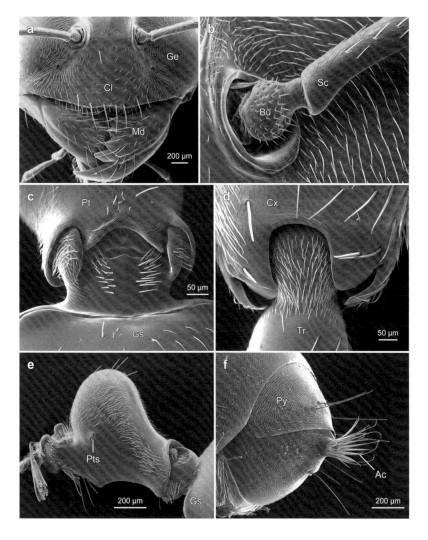

2.4 Morphological details

(a) Frontal view of head (lower part). (b) Insertion of antenna in head capsule. (c) Ventral view of petiole-gaster joint. (d) Coxa-trochanter joint of foreleg. (e) Petiole. (f) Terminal part of gaster. Note the fields of mechano-receptive hairs in b, c, and d. Ac, acidopore; Bu, bulbus of antenna; Cl, clypeus; Cx, coxa; Ge, gena; Gs, first segment of gaster; Md, mandible; Pt, petiole; Pts, petiolar spiracle; Py, pygidium; Sc, scapus of antenna; Tr, trochanter. *Cataglyphis fortis.*

Even though the cataglyph workers vary widely in body size, both within and across species, most of them are quite monomorphic—independent of body size all individuals of one species are more or less isometric copies of one another.[7] The most conspicuous exception to this rule is again the silver ant. This species possesses one of the most remarkable morphological soldier castes known in ants (Figure 2.5). Not only are these majors much larger than the minor workers, but they are also provided with bulky heads and huge saber-shaped mandibles, which are most probably used in nest defense, e.g.,

a

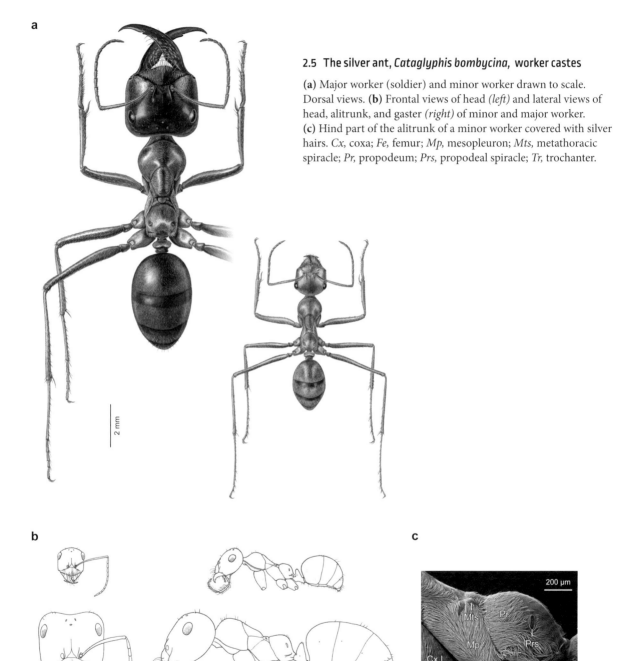

2 mm

2.5　The silver ant, *Cataglyphis bombycina,* worker castes

(**a**) Major worker (soldier) and minor worker drawn to scale. Dorsal views. (**b**) Frontal views of head *(left)* and lateral views of head, alitrunk, and gaster *(right)* of minor and major worker. (**c**) Hind part of the alitrunk of a minor worker covered with silver hairs. *Cx,* coxa; *Fe,* femur; *Mp,* mesopleuron; *Mts,* metathoracic spiracle; *Pr,* propodeum; *Prs,* propodeal spiracle; *Tr,* trochanter.

b

c

200 μm

Mts　Pr

Mp　　Prs

Cx I

Tr　　Cx II　　Cx III

Fe

against lizard intruders, but also in dissecting and handling large prey items.[8]

One morphological aspect that will become important when we later discuss the cataglyphs' running style is the shape of the petiole, the connecting link between the alitrunk and the gaster. The more original species such as the members of the *C. cursor* species group as well as *C. bombycina* have maintained the scalelike (squamiform) petiole, which is characteristic for the closely related wood ants, *Formica,* and many formicine workers in general, while the most agile cataglyphs, which are able to elevate the gaster, possess a nodular (nodiform) petiole. This derived trait is most conspicuous in members of the *C. bicolor* and *C. albicans* species groups (compare the squamiform petiole in Figures 2.1 and 2.5b with the nodiform one in Figures 2.3a and 2.4e).

Foraging Life

Ecologically the cataglyphs occupy a unique niche. They are strictly diurnal 'thermophilic scavengers' rather than predators or harvesters. Of course, 'scavenger' might be a bit inappropriate a term used to characterize these elegant, skilled, and vivacious little runners, but in ecological parlance scavenging is exactly what these ants do (Figure 2.6a, b). They forage for insects and other arthropods that have succumbed to the environmental stress of their desert habitats, while they themselves are able to withstand these harsh conditions well beyond the thermal limits fatal for their prey. In spring one can sometimes witness a fascinating spectacle, when myriads of painted lady butterflies migrate from their winter breeding areas in North Africa to Europe.[9] Many succumb on their way and become easy prey for the cataglyphs. In such bonanza situations even the soldiers of the silver ants engage in foraging, dissect the large butterflies into pieces, and join their smaller nestmates in securing these precious food items. Even though the cataglyphs are visually guided animals to the extreme, they detect food items by smell, using the odor plume that emanates from a food source as a guideline toward the source. Plume-following behavior is regularly elicited by so-called necromones, volatile decomposition products of arthropod corpses. In *Cataglyphis fortis* linoleic acid has

2.6 Food: animal corpses and plant exudates

(a) Two workers of *Cataglyphis bombycina* dragging a chelicer of a juvenile *Buthus* scorpion backward to the nest. (b) A soldier of *Cataglyphis bombycina* pushing the wings of a painted lady butterfly, *Vanessa cardui,* into the nest hole. (c) *Cataglyphis bicolor* nectar feeding in a bush of Christ's-thorn jujube, *Ziziphus spina-christi. Inset: Cataglyphis bellicosa* visiting a *Euphorbia* inflorescence.

been found to be the key necromone odorant released from dead insects and preferentially used in detecting food items.[10]

As far as carbohydrate sources are considered, the cataglyphs lick exudates and secretions from plants (Figure 2.6c), and collect petals of flowers and berries. In one species, an endemic of southern Spain, flower petals can dominate the ants' food spectrum to such an extent that this species even received its name—*Cataglyphis floricola*—from this habit. If seeds are occasionally taken by *Cataglyphis* foragers, they are all provided with ant-attractive elaiosomes, which due to their fatty acid content might be taken by the ants for insect corpses.[11]

The cataglyphs are central place foragers, fanning out widely from their colony, their central place. Each individual goes its own way. If one takes a look at a cataglyph's colony by assuming a bird's-eye view and using a time-lapse camera, an impressive picture emerges. A constant stream of particles radiates out from a center and spreads over an almost circular area. Within this field of flow each (individually labeled) particle follows a radial path, lingers at some distance from the start, and returns straight to the center (Figure 2.7). Later, the same particle reappears, follows a similar path or choses a new one, disappears in the center again, reappears, and continues to do so until at the end of the day the whole stream of particles ceases to flow. Now assume that the time window is expanded from one to several days and the time machine speeded up. One would then become aware of particles disappearing from the scene forever after having performed a number of moves, and new particles taking their place. It is as if one were observing a huge slime mold, in which cells continually moved in and out from an aggregation center and probed for the most suitable sites of the environment.[12]

This kind of 'diffuse foraging' is not a stochastic process.[13] Each individual ventures out in a particular direction. Depending on previous foraging success, it either keeps this direction during its subsequent foraging trip or switches to a new one. Directional fidelity ('sector fidelity') is the more pronounced, and develops the faster, the richer the environmental food supply is and thus the higher the success rates are that the individuals experience. Moreover, during a forager's lifetime, the duration and length as well as the efficiency of locating and retrieving food items (the success rates) continually increase.[14] Just to give some numbers: In their Mediterranean habitats the large *Cataglyphis* species *C. bicolor* and *C. nodus* perform 5–10 foraging journeys per day, during which they spend a daily total of about 200 minutes outside the nest. With a forager's life expectancy of 5–7 days, half of the forager force comprising 15%–20% of the population of a colony is replaced every 4 days.[15]

Pheromone recruitment, which is so obvious in the closely related *Formica* wood ants, has secondarily been lost. The rectal sac, the likely source of the *Formica* trail pheromone, is still present in *Cataglyphis* (Figure 2.8),[16] but trail laying has never been observed. Moreover, all attempts to elicit recruitment experimentally have led to negative results. Even if highly rewarding and renewable food sites are established

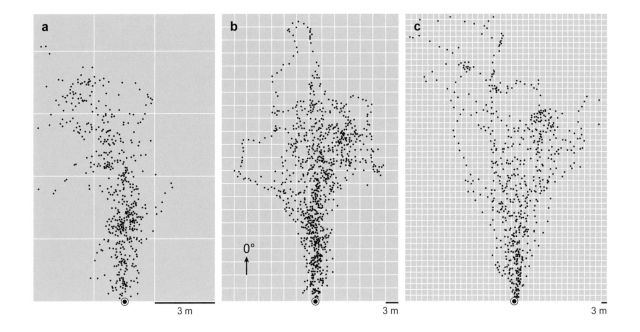

2.7 Spatial structure of foraging paths in three differently sized species of *Cataglyphis*

(a) *C. albicans,* (b) *C. bicolor,* (c) *C. fortis.* In each graph the digitized paths of 10 individuals are superimposed and rotated, so that their mean foraging directions (0°) coincide. Fixes of the ants' positions are taken every 10 s.

at some distance from the nest, individual foragers repeatedly return to that patch, but each forager remains an individual exploiter. The number of ants arriving at those sites for the first time (newcomers) increases linearly rather than exponentially, meaning that all workers appearing at a rich food patch have encountered that patch by themselves. It may happen that after an individual forager has returned from the patch, the general foraging activity of the colony temporarily increases, most likely due to the activation exerted by the successful forager underground. Moreover, when a large prey item has been encountered near the nest, some small-distance, nondirectional kind of recruitment may occur.[17] Are such limited communicative responses triggered by pheromone signals? We do not know yet, as we generally do not know much about the various functions performed by the cataglyphs' armory of exocrine glands (Figure 2.8).[18]

In line with their mode of diffuse foraging, the cataglyphs do not establish foraging territories. The foraging grounds of adjacent conspecific colonies largely overlap. This overlap is especially obvious in species that have a polydomous (multi-nest) colony structure. For example, in *Cataglyphis bicolor* the primary nest, which contains the

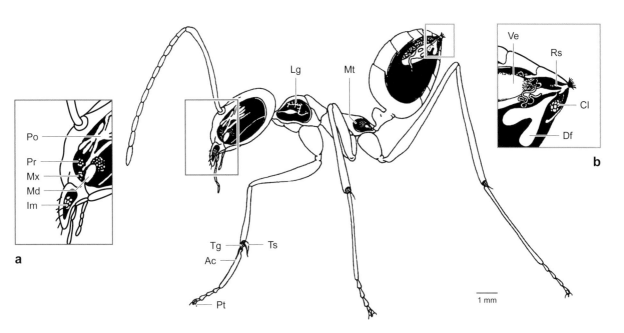

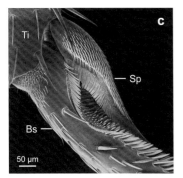

2.8 The cataglyph's battery of exocrine glands

Details of (**a**) head capsule and (**b**) terminal part of gaster. (**c**) Basitarsal brush, an antennal cleaning device. *Ac,* antenna cleaner gland; *Cl,* cloacal gland; *Bs,* basitarsus (tarsal segment 1); *Df,* Dufour's gland; *Im,* intramandibular gland; *Lg,* labial gland; *Md,* mandibular gland; *Mt,* metapleural gland; *Mx,* maxillary gland; *Po,* postpharyngeal gland; *Pr,* propharyngeal gland; *Pt,* pretarsal glands; *Rs,* posterior part of rectal sac; *Sp,* tibial spur; *Tg,* tibial gland; *Ti,* tibia; *Ts,* tibial spur gland; *Ve,* venom gland. The unicellular tectal and sternal glands in the gaster are not shown. *Cataglyphis bicolor.*

only queen, is associated with three to six satellite nests. All these nests are connected with each other not by subterranean tunnels, but by above-ground transport behavior. The multi-nest structure provides the colony with a number of foraging outposts and thus enables it to exploit its food environment more efficiently by shortening individual travel distances and thus minimizing foraging costs. It has indeed been shown that foraging activity and efficiency of the colony increase with the number of outposts.[19] Social coherence is ensured by colony-specific cuticular hydrocarbons, which serve as nestmate recognition cues and thus virtually bind the members of the colony together. As first described in the cataglyphs, but now accepted for ants in general, the postpharyngeal gland (Figure 2.8) acts as a store and flow-through system of these hydrocarbons, which are perpetually exchanged among the nestmates by grooming and food sharing.[20]

In the cataglyphs' natural habitats clumped food patches rarely occur. If one experimentally generates either aggregated or homogeneous food distributions within the ants' foraging area, *C. bicolor* is more efficient in retrieving food items from the homogeneous than the patchy distribution.[21] At artificially created food patches, fighting between members of different colonies may occur, e.g., in *C. fortis,* but usually aggressive behavior is restricted to the immediate nest surroundings, where it sharply declines in a few meters' distance from the nest entrance.[22]

Given their thermophilic foraging behavior, one may wonder whether the cataglyphs have to be afraid of predators at all. However, have a look at Figure 2.9. There are predators of all kinds lurking in ambush in a small shrub or in the shade of a stone, waiting to attack the cataglyphs by a rapid strike from the air (robber flies) or from the ground (jumping spiders). Some spiders, the zodariids, are even specialized to prey upon ants. They have evolved vague morphological and behavioral resemblances to their prey species, live close to the cataglyphs' nests, and attack the ants preferentially in the early morning and late afternoon hours, when due to the lower temperatures the ants exhibit a more sluggish behavior. Other stalk-and-ambush hunters, which lie in wait in the shade for cataglyphs passing by, are lizards. In the Saharan sands the Fringe-fingered Lizard, *Acanthodactylus dumerili,* preys heavily upon the silver ants, which it tries to catch in a rapid dart. Moreover, the cataglyphs can fall victim to airstrike attacks performed not only by robber flies, but also by small parasitoid wasps. These ichneumonids approach the cataglyphs from behind, alight on the ant's gaster, and deposit their eggs through an intersegmental membrane. The ants, which obviously sense the wasp's approach, try to escape by rapid whirligig movements. Finally, danger is imminent also from below. Carnivorous larvae of tiger beetles, *Cicindela* species, sit near the opening of self-constructed vertical burrows in such a way that their flattened head,

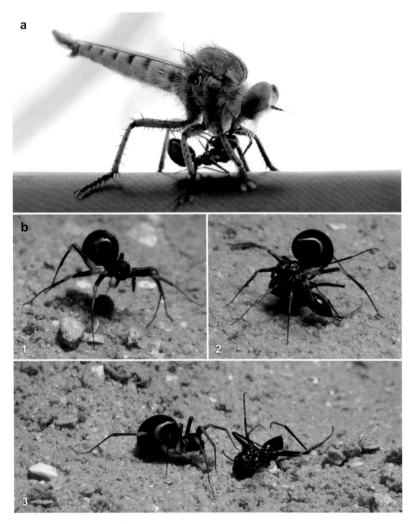

2.9 Predators: air and ground attacks

(a) A robber fly, *Promachus laciniosus*, has captured a *Cataglyphis fortis* worker. (b) A spider, a *Zodarion* species, approaches *(1)* and bites *(2)* a *Cataglyphis bicolor* ant. Having executed the rapid bite, it immediately retreats, but stays close to the paralyzed ant *(3)*, repeatedly touches it with its forelegs, and waits until the ant no longer shows any sign of movement. The motionless prey is then carried away to a retreat.

which bears large single-lens eyes, is level with the desert floor. When a cataglyph comes within reach, a predatory strike is released, and the unsuspecting ant is grasped from underneath with the larva's large mandibles.[23] Hence, danger is lurking around every twist and bend. As a result, the short life expectancies of the foragers are largely caused by the high environmental, especially predatory, pressure that the cataglyphs encounter in most of their habitats. In the laboratory the workers can live through several months and even a few years.

Species Diversity

The genus *Cataglyphis* was first described in 1850 by the eminent German entomologist Arnold Foerster. In the beginning of the twentieth century, Felix Santschi provided a taxonomic revision of the genus, but due to the immense splitting of species into subspecies and varieties, his species-level taxonomy must be treated with care. Subsequently, the field lay fallow for more than half a century, until Donat Agosti presented a thorough reclassification of the genus, a key to the species groups, and a catalog of the species names then recognized as valid.[24] When the operational tools that are now available in population genetics and phylogeography are fully applied in *Cataglyphis* taxonomy, we will certainly arrive at a much more diversified picture of the large species spectrum, which has evolved in the wide distributional area covered by the these desert ants. This area extends all the way from the Mediterranean Basin and the Sahara to the Middle East, the Arabian Peninsula, and as far as central Asia with the Takla Makan and Gobi Deserts (Figure 2.10). It thus covers all arid lands in the southern Palearctic region, including a variety of habitats from coastal sand flats and salt lakes to large fields of sand dunes, from macchia and garrigue shrubland to arid steppes and mountains. The cataglyphs' vertical distributional range extends from sea level up to about 3,000 meters, where the ants have been found in the Asir Mountains of Saudi Arabia and the gravel plains of the Hindu Kush and Pamirs area.[25]

As to their evolutionary history, the cataglyphs share a common ancestor with the wood ants *Formica,* which inhabit the temperate Palearctic region north of the *Cataglyphis* realm. Recent molecular phylogenetic studies tell us that the two lineages must have diverged more than 60 million years ago in early Paleogene times.[26] We do not know where this divergence occurred, but in a scenario supported by morphological and biogeographic evidence the first cataglyphs seem to have appeared in central Asia. It is there that the closely related *Alloformica* has its center of distribution, and it is also there that one of the most original *Cataglyphis* species occurs, *C. emeryi*. This species, which still resembles *Alloformica* in a suite of morphological traits such as relatively short legs, the structure of the alitrunk, and a scale-like petiole, inhabits the grassland deserts of central Asia.[27] It could

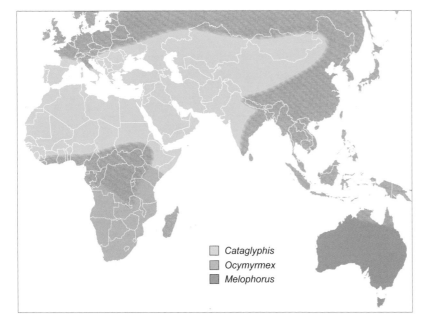

2.10 Distributional ranges of the three genera of thermophilic desert ants: *Cataglyphis, Ocymyrmex, Melophorus*

In East Africa the ranges of *Cataglyphis* and *Ocymyrmex* potentially overlap.

☐ Cataglyphis
▨ Ocymyrmex
▩ Melophorus

have been during the drier temperature regimes in Miocene times that *Cataglyphis* spread westward from central Asia.[28] A northern branch comprising members of the *C. cursor* species group proceeded further to the Balkan area and finally reached southern France and even northern Spain, while a southern branch, which gave rise to the most thermophilic desert dwellers (most prominently represented by the *C. altisquamis, C. bicolor,* and *C. albicans* species groups, but also by the *C. bombycina* species group), established a marked speciation center in the Middle East. Finally, the eastern and southern Mediterraneis and the Sahara were included into the *Cataglyphis* territory. The spread to North Africa could have been fostered by the final closure of the Tethys seaway and an arid desert climate that started to prevail across much of North Africa in the mid- and late Miocene. As documented in Figure 2.11 for Tunisia, the three most abundant and largest-size species of the *C. bicolor* group (*C. viatica, C. bicolor, C. savignyi*) exhibit an almost textbook-like parapatric geographical distribution, with *C. viatica* in the uppermost northern regions and *C. savignyi* in the south. This zonal distribution, which follows more or less the local orographic and climatic divisions, is overlaid by the distributional

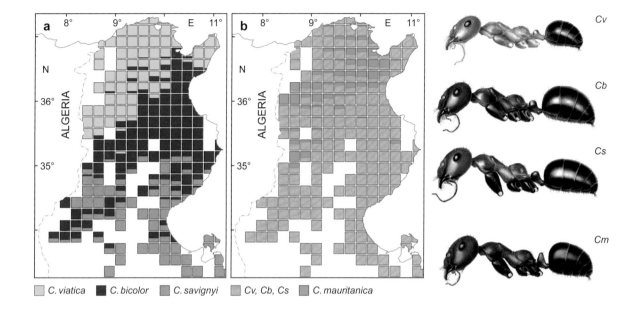

| ☐ *C. viatica* | ■ *C. bicolor* | ▨ *C. savignyi* | | ☐ *Cv, Cb, Cs* | ▨ *C. mauritanica* |

2.11 Zonal distribution of the four large-size *Cataglyphis* species in Tunisia

(a) The distributional ranges of the Tunisian members of the *C. bicolor* species group (*C. viatica, C. bicolor, C. savignyi*) and **(b)** the range of *C. mauritanica* (*C. altisquamis* species group) superimposed on the ranges of the three *C. bicolor*-group species. The spatial distributions are depicted by the species frequencies recorded in 20×20 km^2 sampling squares. *Right:* Side views of worker ant specimens. The workers of the four species differ in more morphological characteristics than the variable color differences shown in the figure.

range of *C. mauritanica,* a member of the *C. altisquamis* species group, which occurs especially in the highland steppes. It might be an attractive hypothesis to test whether North Africa has been reached by two *Cataglyphis* colonization waves, potentially first by members of the *C. altisquamis-* and later by those of the *C. bicolor* species group.[29]

In the far west of the Maghrebian area, members of some species groups even moved to the Iberian Peninsula and then became isolated after the Strait of Gibraltar had opened in the Pliocene. The latter scenario applies to members of the *C. altisquamis-* and *C. albicans* species groups, but not a single member of the otherwise widespread *C. bicolor* species group has made its way from North Africa to southern Spain. Nor has any member of the entire southern branch of *Cataglyphis* species crossed the Mediterranean to arrive at Sicily.[30] Last but not least, the many arid-humid fluctuations in the Pliocene and especially in the Pleistocene may have markedly influenced the diversification and speciation processes among the cataglyphs and shaped the present pattern of geographic distribution as well as the multitude of thermophilic lifestyles that have evolved. For example, *C. fortis,* the longest-distance navigator among the cataglyphs, which in the salt

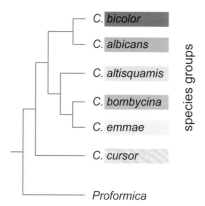

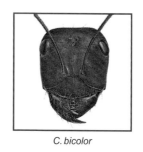

C. bicolor

C. albicans

C. fortis

C. bombycina

2.12 Phylogenetic relationships among the major *Cataglyphis* species groups based on a molecular (nc / mtDNA) phylogeny

pans of Algeria and Tunisia leads one of the most stressful thermophilic lives, could well have originated during a more humid Pleistocene period.[31]

By and large, this evolutionary scenario is in accord with the way of how systematists classify the about 100 *Cataglyphis* species and arrange them in species groups. The two molecular-based phylogenetic analyses performed in previous years consistently show that the major *Cataglyphis* species groups coincide with those that had previously been defined on the basis of morphological traits, and that these species groups are indeed monophyletic entities (clades).[32] The final and most comprehensive analysis defines the *C. cursor*-group—the northern branch in the biogeographic scenario outlined above—as the sister group to all other *Cataglyphis* species (Figure 2.12). Hence, these least thermophilic cataglyphs have gone their own way from early on.

Let us finally embark on an exciting journey to the mid-Eocene 'Baltic amber forests.' About 45 million years ago, cataglyphoid ants were running about in these subtropical savannah-like woods, which consisted of populations of oak, conifer, and palm trees associated with scanty ground vegetation. A first hint came from William Morton Wheeler's extensive survey of amber-embedded ants in the famous Royal Amber Collection of the Geological Institute of Königsberg, present-day Kaliningrad. Among the omnipresent fossils of *Formica flori* Wheeler found a few specimens that, by some *Cataglyphis*-like traits, struck his eye. He named them *Formica constricta* (Figure 2.13a). Nearly a century later the Russian myrmecologist Gennady Dlussky considered these traits decisive enough to establish a new genus: *Cataglyphoides*. Our recent micro-computed tomographic studies indeed show that the relative lengths of the remarkably long legs characterizing the *Cataglyphoides* workers match those measured in extant *Cataglyphis* workers of the *C. bicolor* species group and by far exceed those of *Formica flori* (Figures 2.13b, c, and 2.29a). In this and some other morphological characteristics *Cataglyphoides* resembles much more the members of the advanced *C. altisquamis*- and *C. bicolor* species groups than those of more ancestral groups.[33] How can this surprising result—the occurrence of an advanced cataglyphoid type of ant in the Baltic amber fauna—correspond with the evolutionary 'Out of Asia' scenario that we have hypothesized for the *Cataglyphis* radiation? In the mid-Eocene, the area where *Cataglyphoides* lived was still separated from Asia by the Turgai Strait, which might have constituted a severe barrier for the dispersal of cataglyphoid ants. Are we then left with the hypothesis that *Cataglyphoides* has evolved in parallel to the present highly advanced *Cataglyphis* species? As recent molecular phylogenies indicate, convergence seems to be much more widespread in ant evolution than previously suggested.[34]

Reproductive Strategies

Why have the cataglyphs been so successful in occupying such a particular ecological niche in such a vast geographical area? At present, this question is difficult to answer, as reproductive strategies and population structures—decisive factors in this context—are only beginning to be understood in the wide spectrum of *Cataglyphis* species. However, what has already emerged is something striking. While all species are thermophilic scavengers and thus share a rather uniform mode of foraging—of the way to maintain and grow their colonies— they are extremely diverse in their reproductive strategies by exhibiting a multitude of breeding systems and modes of sociogenetic organization—of ways to reproduce their colonies.[35] Some species are monogynous, others are polygynous (with single-queen or multiple-

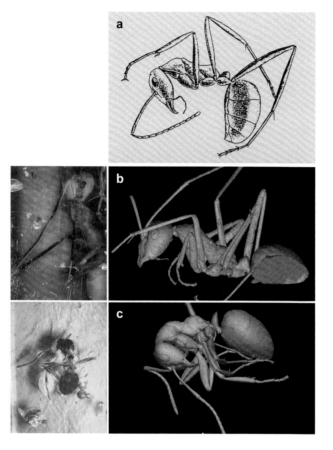

2.13 Fossil ants embedded in Baltic amber

(a) Wheeler's *Formica constricta* (1915) now transferred to the new genus *Cataglyphoides*. (b, c) Micro-computed tomographies, m-CTs, of (b) *Cataglyphoides constricta* and (c) *Formica flori*. In the micro-photographs *(left)* the two specimens shown in the m-CT images *(right)* are difficult to identify within their amber confinements, even though the photographs were taken with a high-resolution camera scanner and composed from image stacks. Note the rounded cuneiform *(b)* and edged squamiform *(c)* petioles.

queen colonies, respectively); some species are monandrous, but most species are polyandrous (with females mating with one or several males, respectively); in many species new queens and workers originate from sexual reproduction, but in others new queens are produced parthenogenetically, i.e., from unfertilized eggs, and this can occur in both workers and queens.[36] Even social hybridogenesis, a rather unusual mode of reproduction, has evolved in the cataglyphs. In this case different genetic lineages occur within a species. Reproductive females mate with males that belong to a different lineage than their own, and then produce genetically diverse workers as inter-lineage hybrids, while new queens and males originate asexually (by different kinds of parthenogenetic reproduction) and hence have a better chance to maintain their species-specific genetic background.

2.14 The male's sexual armature

(a) Side view of male (wings removed). (b) Genitalia in situ, hind view of male's gaster. (c) Genitalia isolated: *1,* subgenital plate; *2,* squamula with stipes *(st); 3,* volsella with cuspide (=lacinia, *cu*) and digitus *(di); 4,* sagitta. Genitalia 2–4 are paired structures. *Cataglyphis fortis.*

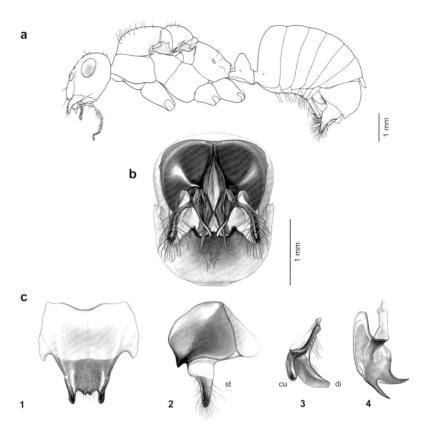

Given this variety of breeding systems, it is not surprising that the cataglyphs also display the two main mating and dispersal regimes found among ants. Some species adhere to the ancestral mode of ant reproduction, in which not only the males but also the virgin queens leave the natal colony, disperse, and mate with males of other colonies. Once mated, they found a new colony on their own. In the majority of *Cataglyphis* species this mode A of colony foundation has undergone a remarkable modification: Female dispersal occurs after rather than prior to mating. The virgin queens stay close to their natal nest, signal their presence to males by the release of sexual pheromones ('female calling') and mate—in most species with several males—within walking distance from their nest of origin. In *Cataglyphis* the male genitalia are especially prominent and diversified (Figure 2.14), so that structural details of the various parts of this sexual armature

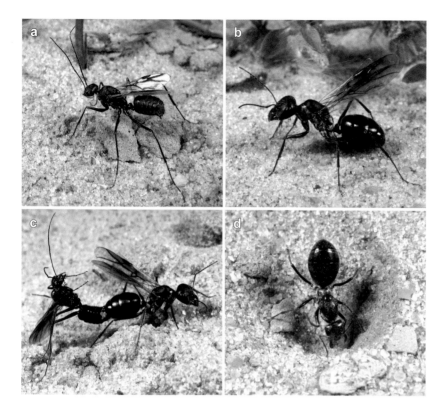

2.15 Mating behavior and nest excavation

(**a**) Male; (**b**) virgin queen; (**c**) copulation (*left,* male); (**d**) start of nest excavation by a dewinged female. *Cataglyphis savignyi.*

can readily be used as morphological traits in species identification. Functionally, this diversification of the male genitalia may be related to polyandry and the potential involvement of sexual selection.[37]

After mating, the queens perform their dispersal flights, land, shed their wings, dig a hole in the ground, and found a new colony (Figure 2.15). This polyandrous and monogynous reproductive strategy seems to be an ancestral *Cataglyphis* trait, but many species in different species groups have independently diverged from this pattern. They are typically polygynous and found new colonies by 'fission,' i.e., by the fragmentation of older colonies (mode B). In this case the mated queens, which are often short-winged or even permanently wingless ('ergatoid'), disperse by foot and depend on the support of a group of nestmate workers, which excavate the new nest and guide or even carry the mated queen to that site.

In the cataglyphs, members of both breeding systems often occur side by side in the same area. For example, in the steppes of the central Tunisian Sahel the monogynous mode-A species *C. bicolor* coexists sympatrically with the polygynous mode-B species *C. mauritanica*. The two species, whose nests are sometimes only 10 meters apart, share almost all foraging traits, such as diet preferences, foraging distances, and times, and do not show any sign of interference competition when they meet while wandering about in their foraging grounds, but their life histories differ markedly.[38] The two modes of reproduction also affect the way of how the ants' populations are genetically structured. The derived mode B leads to a genetically viscose population structure, in which the genetic differences between colonies gradually increase with the geographic distance of the colonies (e.g., in *C. mauritanica*). In contrast, in mode-A species with their relatively large distances of queen dispersal such a correlation between genetic and geographic distances does not occur, so that genetically unstructured populations result (e.g., in *C. bicolor*).[39] Ecologically speaking, *C. bicolor* could be considered an 'explorer' capable of colonizing new areas over quite large distances, whereas *C. mauritanica* might be a more efficient 'exploiter' of areas that it once has happened to occupy and where it can now successfully compete with other species.

In general, most ants adhere to the original type-A mode of mating and dispersal. Why, then, have so many cataglyphs opted for the derived mode B? This option could have been favored by their relatively small colony sizes and hence limited number of sexuals. In desert environments, in which favorable nesting areas are patchily distributed, dispersal of mated queens by flight may be a high-risk endeavor. In any way, the variety of reproductive systems may finally help us in understanding how the cataglyphs have managed to populate the various arid habitats in their vast distributional range with a rich spectrum of species. Given the recent advances in phylogeography, and the current interest of the scientific community in such questions, we may expect that when zooming in on the yellow area in Figure 2.10 in 10 years from now, we would discover a network of species boundaries more closely knit than hitherto imagined.[40]

Ecological Equivalents

So much for Old World *Cataglyphis*. But should one not expect that outside the Palearctic desert belt inhabited by the cataglyphs, i.e., in the other true deserts of the world, ants of different phylogenetic descent have evolved the same ecological lifestyle of a thermophilic scavenger, solitary forager, and visually guided navigator? Indeed, two genera, which are endemic to southern Africa and Australia—*Ocymyrmex* and *Melophorus,* respectively—occupy exactly that ecological niche.

Among these two genera a comparison with *Cataglyphis* is especially intriguing in *Ocymyrmex* (Figures 2.3c and 2.16c), because phylogenetically this genus is positioned furthest from *Cataglyphis*. While *Cataglyphis* is a formicine (a member of the subfamily Formicinae), *Ocymyrmex* belongs to the myrmicines (to the subfamily Myrmicinae) and thus renders this pair of genera an interesting example of parallel evolution of thermophilic foraging traits (Figure 2.17). In terms of geographic distribution, however, *Ocymyrmex* differs from *Cataglyphis* in being restricted to a much smaller part of our planet. It is confined to the dry areas of Africa south of the equator, where it ranges with fewer than 40 species from dry savannahs and shrubland to hyperarid deserts.[41] In contrast to *Cataglyphis,* in which the workers of each species come in a wide range of body sizes, in *Ocymyrmex* they are all of the same size, i.e., extremely monomorphic. They do not even differ very much from the reproductive females, which in *Ocymyrmex* are never winged. Even though there are many of these worker-like reproductive females ('ergatoids')—the virgin queens—in a colony, only one is inseminated. The others join the colony's forager force.[42] It would be interesting to see whether or to what extent their foraging behavior and navigational routines differ from those of the true workers.

This brings us directly back to foraging. In this respect *Ocymyrmex* differs from *Cataglyphis* in exhibiting a remarkable kind of group recruitment to food patches that occasionally arise here and there.[43] For example, when from time to time the slowly moving harvester termites appear above ground, one can frequently observe how a rapidly running recruiter is followed by a loose and often quite dispersed group of 5–10 equally speedy recruits. While the recruiter intermittently

2.16 The three thermophiles

One species of each of the three genera of thermophilic ants: (a) *Cataglyphis longipedem,* sandy gravel plain, Thar Desert, India. (b) *Melophorus bagoti,* sandy plain covered with grass tussocks and loosely scattered *Acacia* trees, central Australia. (c) *Ocymyrmex robustior,* sand flat, Namib Desert, Namibia.

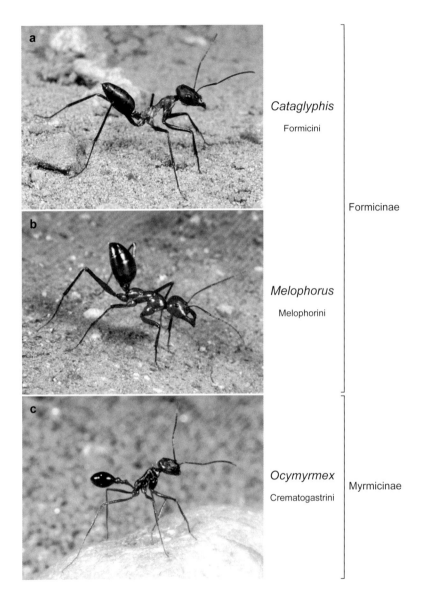

Cataglyphis

Formicini

Formicinae

Melophorus

Melophorini

Ocymyrmex

Crematogastrini

Myrmicinae

touches the surface of the ground with the tip of its gaster, it most likely deposits a volatile pheromone signal. Such recruitment events, which look like a mad rush for a distant food source, can occur during the entire diurnal activity period of the *Ocymyrmex* foragers and thus even at surface temperatures of more than 60°C.

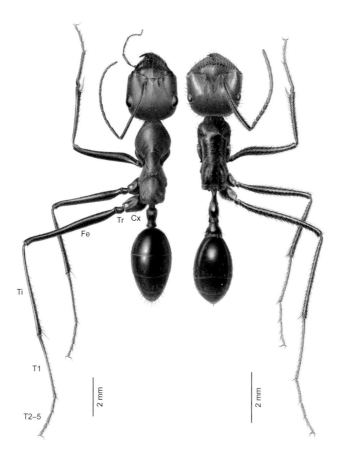

2.17 *Cataglyphis* and *Ocymyrmex*

Dorsal views of workers of *Cataglyphis savignyi (left)* and *Ocymyrmex velox (right).* Leg segments: *Cx,* coxa; *Fe,* femur; *T1,* basitarsus; *T2–5,* distal tarsomeres; *Ti,* tibia; *Tr,* trochanter. As already shown in Figure 2.3, the maxillary palps are large and extended in *Cataglyphis* species, but hidden (and hence not shown here) in *Ocymyrmex* species.

When we next turn to the other thermophilic ants to be considered here, to members of the genus *Melophorus* (Figures 2.3b and 2.16b), we encounter an even larger variety of foraging styles. Like *Cataglyphis* this genus belongs to the formicines, but to a different tribe. It is restricted to Australia, where it represents an abundant and diverse groups of ants. About as species rich as *Cataglyphis* in the Old World—according to a recent revision of the genus there are 93 *Melophorus* species—it reaches its highest diversity and abundance in arid and semiarid areas, but extends even into the tropical woodlands in the northernmost part of the country, where it is absent only from the high rainfall monsoonal zone.[44] Most species are—just as all *Cataglyphis* and *Ocymyrmex* species—thermophilic scavengers. However, *Melophorus* differs from the two other genera in including a wide variety of members with additional diet preferences, food usage patterns, and modes of foraging. While several species are general predators, some species have specialized on termite prey or on brood raiding of other ants. Even specialist seed harvesters occur.[45] According to these different lifestyles, *Melophorus* species have become morphologically modified in various ways. In many species the legs are relatively long, as is the case in all *Cataglyphis* and *Ocymyrmex* species, and in others they are stout and short; some species are bulky and stocky, while others are extremely slim or flattened. The largest and most heat-adapted species is *M. bagoti* in central Australia (Figure 2.16b).[46] The sandy, clayey soil of its habitat provides one of the hottest navigation platforms to which ants are exposed. For that reason we selected this 'red honey ant' for our navigation studies.

Another type of harsh desert environment—actually one of the most inhospitable desert habitats imaginable—are dried-out salt lakes, where the sun beats down relentlessly, where shade and hiding places are rare, and where food is scarce. In the chotts and sebkhas of Tunisia and Algeria *Cataglyphis fortis* is the salt-pan species par excellence. It is endemic to the salt pans of this particular geographical region. In the Middle East, e.g., in the sebkhas south of Aleppo and Palmyra in Syria, a species from another branch of the cataglyphs' phylogenetic tree, *Cataglyphis isis,* can play this hazardous ecological role.[47] Finally, it is also in the Australian *Melophorus* tribe that a salt-pan species has evolved. Drawing upon some early literature references that the Lake Eyre Dragon, the lizard *Ctenophorus maculosus* that is indigenous to the great salt lakes in South Australia, subsisted especially on ants, we undertook a small expedition to Lake Eyre, Lake Torrens, and Lake Hart (Figure 2.18), in order to search for this potential salt-pan species. Indeed, we found that at all three locations *Melophorus* ants existed that can probably be considered the Australian ecological equivalent of the North African *Cataglyphis fortis* (and have recently been described as *Melophorus oblongiceps* sp.n.).[48]

Thermal Tolerance

At the hottest times of a summer day the strictly diurnal thermophiles are often the only insects running about on the open desert floor. For

2.18 Salt-pan habitat of *Melophorus oblongiceps* sp.n, Lake Hart in South Australia

that reason they have been dubbed "furnace ants" *(Melophorus)* or, in Aboriginal language, "ituny, ituny," meaning "sun, sun." The Greek even call them "Englishmen," or "mad Englishmen," for their outdoor activity in the hot midday sun at siesta time *(Cataglyphis).*[49] Typical temperature profiles recorded in their habitats are shown in Figure 2.19 for a three-day summer period. The large fluctuations that occur between day and night on the sand surface contrast sharply with the rather constant soil temperature levels recorded already 20 centimeters below the surface.[50] Hence, the thermophiles experience a steep and sudden rise in temperature as they hurry out of their subterranean colonies and commence foraging.

At our North African study site, as well as in other parts of the Mediterranean zone, the cataglyphs' temporal foraging activities exhibit a

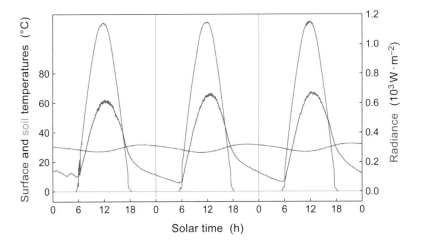

2.19 Solar radiation, surface and soil temperatures during three consecutive cloudless days in the Namib Desert

The measurements were taken during 23–25 November 2006 in the Kuiseb sand flats, the habitat of *Ocymyrmex robustior,* 23.6°S, 15.0°W. Surface temperature (*red,* recorded by a thermocouple covered by a monogranular layer of sand) and soil temperature (*green,* recorded by a thermocouple positioned 0.2 m below the sand surface) correspond to what the ants experience outside and inside the nest, respectively. Solar radiation is shown in *blue.*

distinct seasonal pattern with foraging activity in summer and hibernation in the cool winter months. During the winter torpidity period the members of a colony are strung together in dense clusters inside a single nest chamber. The same annual cycle is observed in *Melophorus bagoti* in central Australia.[51] However, when in winter surface temperatures exceed about 40°C, as it is regularly the case in the Sahara, the Arabian Peninsula, and southern Africa, *Cataglyphis* and *Ocymyrmex* species forage all year round. They then exhibit bimodal daily foraging profiles in summer and unimodal ones in winter (Figures 2.20 and 2.21). The depth and width of the midday trough in the bimodal summer profile is finely tuned to the currently prevailing ambient temperatures. Moreover, the bimodal summer profile can transform, from one day to another, into a unimodal one when temperatures are lowered experimentally or when they decrease naturally. Foragers respond quickly even to small, short-term changes in sur-

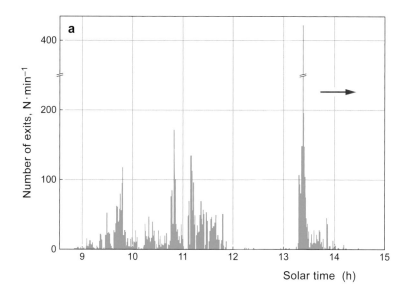
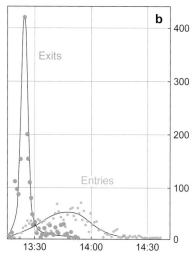

2.20 Diurnal foraging activity of the Saharan *Cataglyphis bombycina*

(a) Recordings of exits (nest-leaving events of individual ants) in 1 min bins. In **(b)** the afternoon activity (exits and entries) is shown in higher temporal resolution. Measurements and computations performed by Martin Müller at this very colony show that every day 1,600 foragers left the colony, each for 4 solitary foraging runs lasting for 24 minutes (mean values).

face temperature caused, e.g., by intermittent cloud cover.[52] In the silver ants several hundred foragers may leave the nest within a minute and frantically radiate out in all directions (Figure 2.20). Such mega outbursts, which often occur when the ambient temperature has reached a threshold level, may be triggered by a pheromone signal released by a few scout individuals.[53]

The temperature / activity correlations as shown in Figure 2.21 for *Ocymyrmex robustior* vary slightly but distinctly among different species of thermophilic ants and hence can be considered species-specific thermobiological signatures, which open the doors for comparative ecophysiological studies. Nevertheless, we should be careful not to ascribe all modulations of the thermophiles' foraging patterns to the effect of temperature alone. Ambient light intensity may play a role as well. As already the summary diagrams in Figure 2.21 show,

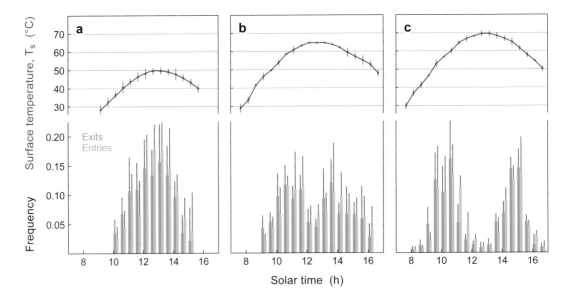

2.21 Diurnal foraging activity and its temperature dependency in the Namibian
Ocymyrmex robustior

Exits and entries depicted by *dark green* and *light green bars,* respectively, were
recorded in 5 min bins every 30 min during the months of **(a)** July and August
(winter) and **(b, c)** December (summer) in the sandflats near Gobabeb. Note that
from graphs *(a)* to *(c)* maximum surface temperatures increase from about 50°C to
70°C. Each histogram is based on data obtained during three 1-day recordings from
three different colonies.

the ants end their foraging in the evening at higher temperatures than
they start it in the morning, but the starts and ends often occur at about
the same light-intensity level.[54] Most likely, the ants' foraging activity
is primarily governed by endogenous activity rhythms, and depending
on species and ecological settings it is finally triggered by external visual and / or thermal cues.

The most striking feature of thermophilic ants is their ability to
boost their foraging activity and efficiency under conditions of exceedingly high temperatures, i.e., when surface temperatures reach 57°C–
63°C. These values have been measured in *Ocymyrmex robustior* and
Melophorus bagoti, and they correspond well with reports for *Cataglyphis bombycina.*[55] In contrast, the majority of other desert ants cease
foraging at soil temperatures in the range of 35°C–45°C, and even the

most heat-adapted species among the North American harvester ants such as *Pogonomyrmex californicus* and *P. pronotalis* stop foraging at temperatures at which *Cataglyphis* and *Ocymyrmex* have not even reached their peak activities.[56] What, then, are the body temperatures that these ants must and can endure?

To answer this question we move to the laboratory, confine an individual ant to a glass flask, expose it to a rising temperature gradient of, say, 1°C per minute, and determine the temperature at which the ant is no longer capable of proper locomotion, begins to exhibit muscle spasms, and shows signs of paralysis (heat coma). This threshold temperature is called the 'critical thermal maximum' (CT_{max}). When the ant is subsequently transferred back to lower ambient temperatures, it immediately recovers and starts to walk again in a coordinated manner,[57] but out in the fields, the general loss of locomotor coordination and controlled breathing would make it impossible for the animal to free itself from finally fatal conditions. Indeed, when the body temperature rises a couple of degrees further, the ant's upper lethal temperature (LT_{max}) is reached.

All *Cataglyphis* species exhibit critical thermal maxima and upper lethal temperatures that lie well above the corresponding values for harvester ants, *Messor,* and wood ants, *Formica.* This is illustrated in Figure 2.22a for *Cataglyphis bombycina* ($CT_{max} = 53.6$°C, $LT_{max} = 55.2$°C).[58] For these comparisons harvester ants are used, because they share with *Cataglyphis* the same habitat but forage at crepuscular times, as well as wood ants, which do not overlap with the cataglyphs geographically, but are closely related phylogenetically.

Another way of assessing an animal's thermal tolerance is to expose the animal to a fixed rather than a ramped temperature, and record the time until the CT_{max} and LT_{max} responses occur.[59] If this is done for a set of fixed temperatures, one can compute the critical temperature at which the animals display the characteristic CT_{max} or LT_{max} response after they have been exposed for a certain time—say, 10 minutes—to that temperature. Again, the results are clear-cut (Figure 2.22b). In conclusion, irrespective of whether the ants are tested in a temperature gradient (dynamic method) or in a constant temperature setting (static method), thermophilic ants stand out by their exceptionally high CT_{max} values. However, we might get even

2.22 Heat tolerance

(a) Critical thermal maxima determined by the dynamic method (CT$_{max}$, *closed circles*) and upper lethal temperatures (LT$_{max}$, *open circles*) of thermophilic desert ants *(Cataglyphis bombycina)*, harvester ants *(Messor arenarius)*, and wood ants *(Formica polyctena)*. For comparison, in *Ocymyrmex velox* and *Melophorus bagoti* the mean CT$_{max}$ values are 54.1°C and 56.8°C, respectively. (b) Critical thermal maxima determined by the static method. In the graph CT$_{max}$ is defined as the temperature at which 50% of the ants died after 10 min of exposure to that temperature. Comparison of *Cataglyphis velox* and *Messor capitatus*. (c) Operational environmental temperature at 4 mm above the sand surface and (d) respite frequency of *Ocymyrmex robustior* in relation to surface temperature. In both figures the temperature range T$_s$ = 40°C–60°C is *shaded and marked in red.*

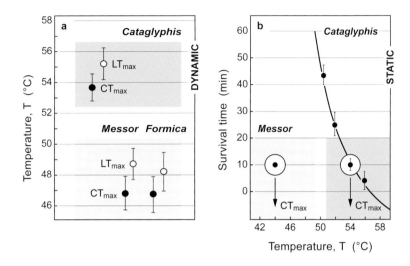

more excited when we embark on a quick detour to southern California, where we meet another thermophile. There, in the coastal sage scrub environment, tiny mites, *Paratarsotomus macropalpis*, 0.2 milligram light, are bustling about in the full midday sun. Most active when ground surface temperatures exceed 50°C, these mites reach a CT$_{max}$ value of 59.4°C, actually the highest value measured as yet in any animal.[60]

Under the high temperatures preferred by the thermophiles there is always the risk that neural and muscular coordination—in principle, cellular functions such as mitochondrial respiratory processes—are affected. To avoid such deleterious effects and still be able to forage on the extremely hot desert floor, the cataglyphs have evolved a number of morphological, physiological, and behavioral mechanisms. In the first place, there are some morphological traits that help the ants to reduce the amount of heat taken up from the environment. The particularly long and slender legs elevate the body several millimeters above the hot sand surface, minimize heat conduction from the ground, and let the ants exploit the steep temperature gradient above the desert floor.[61] Depending on ambient temperature and wind conditions, the temperature at ant height may be 10°C–15°C lower than on the sand surface.

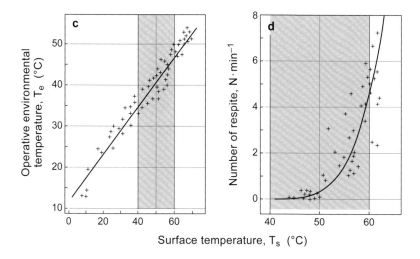

A quite spectacular heat-protection device is the 'thermal shield,' a coating of silvery hairs that covers the dorsal and lateral parts of the body of *Cataglyphis bombycina,* the silver ant, which received its name from this coating (Figure 2.23).[62] The silver hairs are straight and rigid and end in a sharply pointed tip, but their most remarkable feature is their triangular cross section. The lower of the three facets is completely smooth and oriented parallel to the surface of the body, whereas the two upper facets are fluted like Greek or Roman columns, but with the grooves running obliquely rather than parallel to the long axis of the hair. Functionally, the physical properties of these hairs endow the ants with two means of heat protection.

First, the prismatic hairs reflect solar radiation in the visible and near-infrared part of the electromagnetic spectrum (wavelengths 0.4–2.0 μm), where the radiation from the sun reaches its maximal values (Figure 2.24 bottom, left). As computations and modeling show, the total internal reflection occurring on the flat bottom surface of the hairs for a wide range of angles of incidence enhances the overall light reflection from the ant's body surface by almost an order of magnitude. Second, at longer wavelengths, i.e., in the mid-infrared part of the spectrum (wavelengths 4–16 μm), where solar radiation is negligible, the hair coating conversely acts as an antireflection

2.23 The thermal shield of the silver ant, *Cataglyphis bombycina*

(a) Frontal view of head. The head as well as dorsal and lateral parts of the body are covered by dense arrays of prismatic silver hairs seen (b) from above and (c) in cross-section.

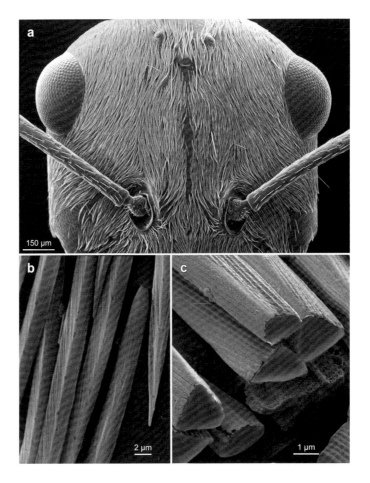

(gradient refractive index) layer: for wavelengths larger than 8 μm reflectivity of the hairs is reduced (Figure 2.24 bottom, right) and thus its emissivity enhanced. The 'thermal shield' becomes leaky and turns into a 'thermal window'. The leakiness enables the ants to offload heat by blackbody radiation from the body to cooler surroundings. This effect works all the time, when under full daylight conditions the foraging animals are exposed to the cool sky, but it is especially effective when the ants move to elevated and thus cooler places. There the silver hair coat enables the ants to efficiently emanate excess heat. Moreover, on the ant's bottom surface the silver hairs are absent: the mid-infrared thermal window, which has been opened by the

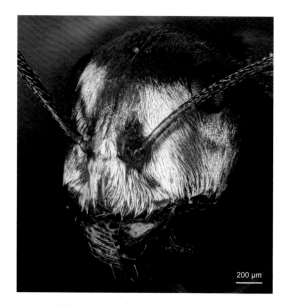

2.24 Heat protection: the silver hairs offer a twofold benefit

Top: Fronto-lateral view of the head of *Cataglyphis bombycina*. *Bottom:* Reflection of solar radiation in the visible range of the spectrum including the near infrared *(VIS-NIR)* and off-loading of heat by blackbody radiation in the mid-infrared range *(MIR)*. The *red and blue curves* depict the reflectivity of the ant's body surface with hair cover intact and with hairs removed, respectively. Solar radiation is indicated in *yellow*.

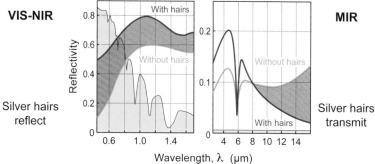

hair cover on the upper and lateral parts of the body, is closed again, so that radiative energy transfer from the hot sand surface to the ant's body is reduced.

Note, however, that other ant species such as *Cataglyphis fortis, Ocymyrmex robustior,* or *Melophorus bagoti,* which inhabit similarly hot desert environments, lack the silver hairs and must nevertheless forage under pretty much the same thermally stressful conditions. Certainly there are other means to cope with these conditions than enhancing radiative cooling via morphological specializations. Forced convection caused by the ants' running style and speed will be an important factor. The silver ants with their relatively short legs and inability to raise the gaster might have instead evolved their protective silver hair coat.

Physiologically, special expression patterns of heat shock genes play a protective role. If temperatures are elevated above the normal physiological temperature range, in all organisms, from bacteria to humans, a heat shock response is elicited. A suite of highly conserved genes related to heat tolerance, so-called heat shock genes, becomes transcribed, and heat shock proteins (HSPs) are synthesized. These proteins act as molecular chaperons by refolding denatured cellular proteins that have been damaged under heat stress conditions, and thus increase the organism's thermotolerance.[63] In the cataglyphs this heat shock response is much more pronounced than in the closely related *Formica* wood ants.[64] While in *Formica* any protein synthesis stops at 39°C, in *Cataglyphis bombycina* HSP synthesis continues up to 45°C. In particular, the basal—so-called constitutive—expression of HSPs is highly enhanced, meaning that these ants strongly express their molecular chaperons already under normal physiological conditions inside the nest. When all of a sudden the foragers exit the nest and thus rapidly expose themselves to extreme bursts of heat stress, they come already equipped with a properly high HSP content. On top of that, they exhibit a fast 'induced' activation of other sets of heat shock genes. In *C. mauritanica,* which experiences the relatively cooler and more variable temperature regimes of the Atlas mountains and the North African Sahel zone, this transient response is much stronger, and conversely the constitutive HSP expression is much weaker, than in *C. bombycina.* The wide spectrum of *Cataglyphis* species and their different thermal environments offer promising ways of inquiring into the species-specific fine-tuning of ecologically relevant stress responses, e.g., the trade-off between basal and transient HSP expression.[65] This is all the more rewarding as a suite of molecular heat shock response patterns seems to have evolved uniquely in the *Cataglyphis* lineage. Transcriptomics analyses in *C. bombycina* further show that the heat shock response especially preserves mitochondrial activity (energy production) and muscular structure and function, and thus supports the ants' advanced locomotor activity under high ambient temperatures.

Behaviorally, the thermophiles' extraordinary running speed and agility causes cooling by forced convection. Moreover, the ants are able to behaviorally exploit their thermal microhabitats (Figures 2.25 and 2.26) in effective ways. On the desert floor with its finely graded thermal mosaics, the ants' small body size and correspondingly small thermal time constants (low thermal inertia) is a double-edged sword rendering these little ectotherms, which cannot regulate their body temperature physiologically, extremely rapid heat exchangers in both directions. Depending on the temperature gradient between their bodies and the surroundings, they gain heat as rapidly as they lose it, and thus almost immediately equilibrate with the temperature at ant height. In this way, they can utilize thermal microenvironments that are not accessible to larger animals. Human investigators must think small enough to appreciate—and monitor—the micrometeorological conditions prevailing within the ants' physical environment close to the heated desert surface.[66]

Recording the ant's body temperature under these conditions causes technical problems. In a four-milligram *Ocymyrmex robustior* and a nine-milligram *Cataglyphis fortis* body temperature would be difficult to measure directly by the 'grab and stab' method, i.e., by capturing an animal, quickly inserting a small thermocouple into its body, and reading the temperature from a digital device. Instead, the so-called operative environmental temperature (T_e) can provide an estimate of the ant's body temperature (Figure 2.22c).[67] Defined as the equilibrium temperature attained by the animal in its microenvironment, it is usually measured by mounting a dead ant, or a model of it, on the tip of a thermocouple and exposing this device to the very environmental condition at which the ant's body temperature shall be determined. Measurements of this temperature show that on hot summer days *Cataglyphis bombycina* and *Ocymyrmex robustior* are most active when their body temperatures fluctuate, at least for short periods of time, in the range of 44°C–50°C.[68] Under these thermally stressful conditions, every now and then they retreat to cooler thermal refuges (Figures 2.1 and 2.27) by climbing stones or small sticks of dry vegetation, pause there, radiate off excess heat, jump down to the ground again, and continue foraging. The frequency and duration of this

2.25 Gravel plain desert inhabited by *Ocymyrmex velox*

Even the rather uniform gravel plains of the Namib Naukluft Desert offer fine-grain thermal mosaics, which are exploited by *Ocymyrmex* ants for behavioral thermoregulation.

2.26 Thermal microhabitats

The four panels depict thermovision recordings of surface temperature taken in 90 min intervals (*numbers,* solar time) within an area of gravel plain similar to the one shown in Figure 2.25.

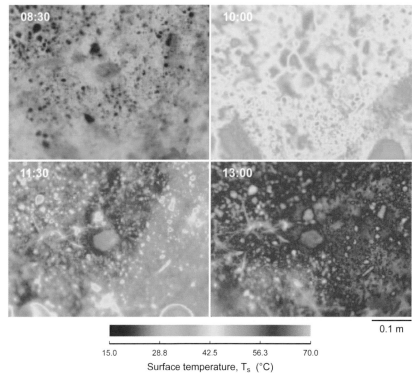

Surface temperature, T_s (°C)

thermal-respite behavior steeply increase as surface temperatures rise above $T_s = 50°C$ (Figure 2.22d), so that finally the ants may spend up to 70%–80% of their entire extranidal time in thermal refuges. While constantly shuttling up and down between such refuge sites and the hot desert floor, *Ocymyrmex robustior* ceases foraging only when surface temperatures exceed 67°C.

Finally, what happens when under thermally stressful conditions refuge sites are not available, so that the thermophilic ants have no way of loading off excess heat and thus cannot escape from their self-imposed 'thermal trap'? Then, inevitably they die. In the laboratory these fatal no-retreat conditions can be studied by exposing the ants to constant temperature regimes close to their CT_{max} values. For example, when *Cataglyphis bicolor* is subjected to such extreme heat-

2.27 Thermal respite behavior in *Ocymyrmex*

(a) *O. velox* and (b) *O. robustior* elevating their bodies above the hot sand surface by stretching their legs on top of a stone or a measuring rod, respectively. Namib Naukluft Desert, Namibia. The *inset* depicts an example of the daily temperature course above the desert sand surface.

stress conditions and its body temperature is recorded by thermovision methods, at a given temperature of, e.g., 51°C the ant's gaster and alitrunk soon equilibrate with the environmental temperature, but for several minutes the ants manage to keep their head at 44°C– 46°C by regurgitating a droplet of fluid from the mouth or excreting one at the anus and spreading it with their legs across the head. In such cases of emergency the ants can finally resort to evaporative cooling, in order to reduce the thermal load of their brain—or, in other words, to keep their head cool.[69] Out in the fields, however, cooling by evaporation is not a preferable option. This brings us directly to the next section.

Dehydration Risk

Thermophilic ants must contend not only with high temperatures, but also with low relative humidities. Taken together, these environmental constraints produce high deficits in water vapor saturation, exposing the thermophiles to the risk of rapid dehydration. Due to their small body size and hence relatively large body surface ants are especially susceptible to dehydration by the passive loss of water—a problem that is exacerbated in desert environments. Drawing upon the extensive literature on how insects can survive such water stress situations,[70] we have no reason to assume that desert insects either are able to store larger quantities of water or are physiologically more tolerant to water deficits, than insects of mesic habitats. Hence, for thermophilic ants the only effective way to avoid continuous water deficits is to lose water more slowly than other ants, in particular to make any effort to reduce the passive water losses that inevitably occur through the cuticle (cuticular water loss) and through the spiracles when the animals breathe (respiratory water loss). The relative contribution of these two avenues of water loss to total water loss is a much debated issue, but we can at least state that in the majority of insect species examined so far—including desert ants—cuticular water loss is by far the major route.[71]

Under these conditions, one would expect that in thermophilic ants a strong selection pressure has acted upon strengthening this cuticular water barrier, the waxy lipid layer of the epicuticle. Indeed, in desert ants including *Cataglyphis* this barrier formed by cuticular hydrocarbons, especially unbranched long-chain saturated alkanes, tends to be the stronger, the more xeric the environment is that a species inhabits.[72] It might even be that the proportion of linear alkanes is higher in foragers than in indoor maintenance workers, and that this change in cuticular chemistry is triggered by the warm-and-dry conditions of the ants' outdoor desert habitat.[73] As mentioned in Chapter 1, cuticular hydrocarbons can also serve as nestmate recognition cues. In this case, however, branched alkanes with their high potential of molecular diversity play the major role.[74]

In order to maintain their water barrier, desert insects may face an additional problem. At a critical 'transition temperature' water loss increases rapidly, indicating that a sudden change in the permeability characteristics of the cuticle has occurred. At this temperature the lipids in the epicuticle undergo a transition from a solid, waterproof state to a fluid, liquid-crystalline state, which no longer holds water. This 'melting temperature' may be especially high in desert insects. In Californian grasshoppers, *Melanoplus sanguinipes,* which exhibit population-specific rates of water loss, lower rates are strongly correlated not only with larger amounts but also with higher melting points of cuticular lipids. Under laboratory conditions, *Cataglyphis bicolor* water loss rates increase more rapidly when ambient temperatures exceed 45°C.[75] Under the extreme conditions of a Saharan summer day one can occasionally observe how a foraging cataglyph succumbs to its stressful environmental conditions. The animal crumples and loses its firm grip on the ground, then is blown away by the slightest gust of wind. Is it at this moment, when the sudden curling and crumpling occurs, that the melting point of the cuticular lipids has been reached and has caused a fatal loss of water? Or have such losses of water resulted from the constantly high respiration rate exhibited by the ants during their long-lasting walks? Or has the thermal stress been the dominating factor? Most likely it has been the combination of these factors that has created the ultimately lethal conditions.

This brings us to the second way by which a cataglyph inevitably loses water: by respiration. At rest, the cataglyphs as well as many other arthropod species breathe discontinuously. During the so-called discontinuous gas exchange cycles (DGCs) the spiracles are tightly closed for much of the time, so that any gas exchange is essentially shut off. Only when the spiracles fully open for a small fraction of the DGC cycle respiratory gases are exchanged with the atmosphere and CO_2 and H_2O are expelled in a burst (Figure 2.28c). Not surprisingly, this discontinuous mode of respiration has been regarded as a functional adaptation to reduce respiratory water loss, because the very same tracheole characteristics that favor gas exchange also facilitate water loss. In the past decades, in which DGCs have been intensively studied in various species under various conditions and with various results, other adaptive functions of discontinuous respiration have come to the fore as well, so that—as Natalie Schimpf from the University of Queensland at Brisbane, Australia, has aptly put it—the water conservation ('hygrig') hypothesis of DGC respiration must still be considered an ongoing "source of intrigue and great debate." Some authors have even proposed that the DGC pattern does not serve any adaptive function at all, but directly results from a down-regulation of brain activity during quiescent, sleeplike states of the brain. Then, in the absence of control from higher nervous centers, the thoracic and abdominal neural circuits would spontaneously display the energy-saving DGC mode of respiration. The jury is still out.[76]

Of course, what interests us the most is how the cataglyphs maintain their water balance not when they are at rest but when they are rapidly running. Then, like all other active insects, they breathe continuously. The switch in the ant's gas exchange pattern from discontinuous respiration during the resting state to continuous respiration during locomotor activity is clearly expressed in Figure 2.28c. A high rate of respiration is associated with a massive loss of water, so that an effective trade-off between gas exchange and water economy might become important. As the rate of spiracular water loss increases with metabolic rate, and as metabolic rate increases with speed of locomotion, which again increases with ambient temperature, a cataglyph running continuously at high speed over the hot desert surface will certainly experience substantial spiracular water losses. In this respect it is worth mentioning that all thermophiles—all *Cataglyphis, Ocymyrmex,* and *Melophorus* species—have especially large and slit-like

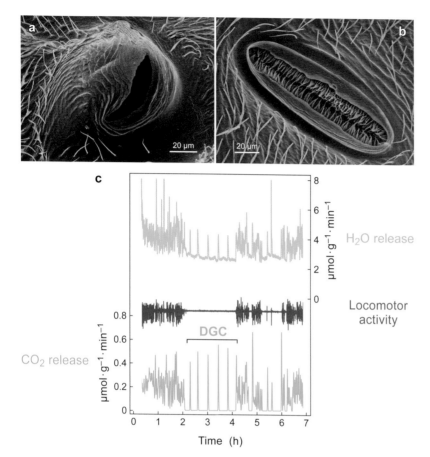

2.28 Respiration and water loss

(a) Metathoracic and (b) propodeal spiracle. Simultaneous recordings from these thoracic spiracles and the abdominal spiracles show that the former ones account for 80%–90% of the overall gas exchange. (c) Time course of the release of CO_2 *(blue)* and H_2O *(green)* in correlation with locomotor activity *(red)* at 25°C. *DGC,* discontinuous gas exchange cycles. *Cataglyphis bicolor.*

propodeal spiracles (Figure 2.28b) endowed with effective closing and opening mechanisms.[77] In the thermophiles' relatives that inhabit more mesic environments, e.g., in *Formica* and *Messor* ants, these spiracles are smaller and more circular in shape.

To keep respiratory water losses as small as possible, desert ants have been assumed to have especially low standard metabolic rates, i.e., low metabolic rates when at rest. In addition, in coping with their high energy demands during fast running, the cataglyphs are expected to exhibit high metabolic rates when foraging. To investigate these matters we need to take a close look at the ants' running style and speed.

Ultrarunners

Perhaps the most conspicuous behavioral trait of thermophilic desert ants is their extraordinary locomotor celerity and agility. Natural selection might have favored the cataglyphs' high running speeds for several interconnected reasons. Because of the low food density in the desert a solitary forager, which searches for isolated prey items, must cover large travel distances, but due to the environmental stress imposed on the desert dweller, travel times should be as short as possible. Moreover, high running speeds facilitate convective cooling, and the long legs associated with the high running speeds elevate the ant's body favorably above the hot sand surface.

It was already in 1787, when Johann Christian Fabricius, a famous Danish entomologist and former student of Carl Linné, described the first *Cataglyphis* species (then still included in the genus *Formica*), which he characterized as "velocissime cursitans," as running extremely fast. This was an exceptional remark, because in those days taxonomic species definitions written in Latin and consisting of some brief descriptions of morphological characters hardly ever included a behavioral trait. However, 15 years later another famous entomologist, Pierre André Latreille from France, again referred to *Cataglyphis* as "la fourmi coureuse, qui court avec une très-grande vitesse," as the high-speed cursorial ant. Finally, in the beginning of the twentieth century Victor Cornetz, working in Algeria, noted that "cet insecte ne marche pas, il court," that this insect does not walk or run, but races along; "it moves like a flash" continued Vladimir Karavaiev, a myrmecologist heading the Zoological Museum of the Ukraine Academy of Sciences. Finally, George Arnold, a polyglot English scientist and director of the Natural History Museum of Southern Rhodesia (now Zimbabwe), even claimed that *Ocymyrmex* ants "are endowed with the most marvellous celerity, far excelling in this respect all other ants . . . so much that they appear almost to fly over the surface of the ground."[78]

When surface temperatures rise well above 50°C–55°C, i.e., to levels at which peak foraging activities occur, the ants' running speeds reach one meter per second (as recorded in the silver ant *Cataglyphis bombycina*).[79] This might render our desert navigators the swiftest runners among all ants. Other insects known to be fast runners—Namibian darkling beetles *(Onymacris plana)*, American cockroaches *(Periplaneta americana)*, and Australian tiger beetles *(Cicindela hudsoni)*—reach maximal speeds of 1.0, 1.5, and 2.5 meters per second, respectively.[80] However, to put these numbers into perspective, one must note that the beetles and roaches run in a different size class. With body masses in the 1-gram range they are heavyweights as compared to the cataglyphs, which perform in the 10-milligram weight class.

If running speed is related to body length, as usually done in comparative studies, the silver ants' maximal speed value mentioned above transforms into 110 body lengths per second. In a small-sized species of tiger beetle, this number is even larger and amounts to 170. Seen in this light, the tiger beetles are the fastest runners—were there not the tiny Californian mites, *Paratarsotomus macropalpis*, which had already amazed us by their high critical thermal maxima. These mites can reach relative running speeds of up to 320 body lengths per second.[81] Again, however, in interpreting these data, body mass is important. Theoretical considerations supported by empirical data tell us that with decreasing body mass an animal's relative running speed should increase (and its absolute speed should decrease) in a regular way.[82] Given this interspecific scaling relation between relative speed and body mass for fast runners ranging widely in size, even the mite's exceedingly high relative running speeds are largely predicted by scaling. The champion would then be defined best as the one that deviates the most from the "norm"—that is to say, from the scaling prediction—and this might again be the tiger beetle. However, these predatory insects reach their high speeds only during short-lived chases, when they pursue their prey in a stop-and-go mode of locomotion.

Beyond the broad picture painted by such inter-taxa comparisons, we are interested the most in how the cataglyphs achieve their thermophilic high-speed, high-endurance style of locomotion, how they differ in this respect from related ants of more mesic habitats, and what species differences may exist within the genus. Even if one takes only a cursory glance at a cataglyph, one immediately recognizes a set of morphological traits that might be correlated with high running speeds: extremely long legs, a slender alitrunk, and a highly movable gaster (for morphological terminology, see Figure 2.3). Long-leggedness is indeed the most conspicuous morphological feature of

2.29 Leg allometry

(a) Lengths of the hindleg femur in thermophilic ants (*Cataglyphis* and *Ocymyrmex* species; *colored symbols*) and phylogenetically related non-thermophilic ants (*Formica* and *Messor* species; *black / gray symbols*). Data are shown on double logarithmic coordinates. Alitrunk length is taken as a proxy for body size. *Cataglyphis* species: ALB, *C. albicans*; BIC, *C. bicolor*; FOR, *C. fortis*. *Formica* species: LEF, *F. lefrançoisi*; POL, *F. polyctena*. *Ocymyrmex* species: ROB, *O. robustior*; TUR, *O. turneri*; VEL, *O. velox*. *Messor* species: ARE, *M. arenarius*; MIN, *M. minor*. The *open symbols* refer to the extinct species *Cataglyphoides constricta (squares)* and *Formica flori (circles)*, respectively, from Baltic amber. **(b)** Total leg lengths in four *Cataglyphis* species: ALB, BIC, FOR as in *(a)*, BOM, *C. bombycina*.

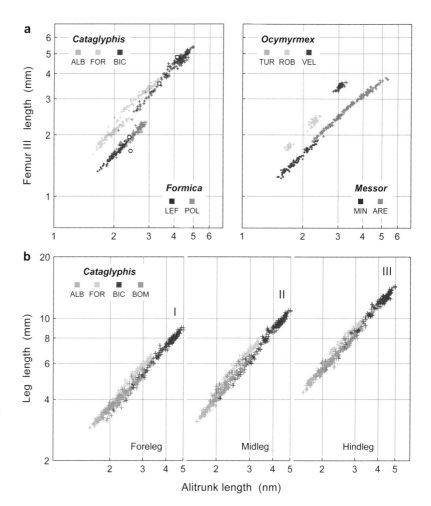

a thermophilic ant. Irrespective of whether we consider small-size, medium-size, or large-size species of either *Cataglyphis* or *Ocymyrmex,* the relative lengths of their legs highly significantly exceed those of non-thermophilic species belonging to closely related genera, in this case *Formica* and *Messor,* respectively (Figure 2.29a).[83]

Beyond this general long-leg trait there are interesting differences between the various species. For example, among the workers of the phylogenetically related *Cataglyphis* species considered in Figure 2.29a, the salt-pan species *C. fortis* inhabits the nutritionally most impoverished habitat, exhibits the largest foraging distances (Figure 2.7), and has the relatively longest legs. This supports the hypothesis mentioned in the beginning of this section that the evolution of long-leggedness, and with it of high running speed, might have been driven by the selection pressure to cover the largest possible search area per unit period of time. In *Ocymyrmex,* it is the largest species, *O. velox,* that strikes by the relatively longest legs. It forages on more rugose ground than the two other *Ocymyrmex* species included in our studies. Further analyses of such interspecific differences in the relative lengths of the legs will certainly shed more light on the subtle functional implications that relative leg length might have beyond that on mere running speed.[84]

There is, however, one telling example of the tight correlation between leg length and lifestyle. We only have to consider the ants' sexual morphs, discussed earlier, and have a look at Figure 2.30. As previously described, the reproductive females often employ a sit-and-wait strategy, when they try to attract males by pheromone calling. The males, which after having completed their dispersal flights search for females on foot by performing rapid 'mating walks,' have considerably longer legs—again relative to body size—than their reproductive female counterparts, but significantly shorter ones than the conspecific workers.

Moreover, in increasing running speed, natural selection not only has caused the legs to increase in length, but has concomitantly affected other parts of the body as well. For example, longer legs require longer leg muscles, which cause a shift in the endoskeletal elements, at which these muscles attach. As a consequence, the alitrunk of *Cataglyphis* has adopted a more elongated and slender shape than in related formicines with shorter legs. The most agile runners—members of the *C. bicolor* and *C. albicans* species groups—further improve maneuverability by their ability of elevating the gaster until it assumes a fully upright position (Figure 2.31) or is even drawn beyond the vertical, so that its posterior tip points forward (see inset in Figure 1.3). Due to this straight upright posture of the hind part of the body the French settlers in Algeria had dubbed *Cataglyphis bicolor,* one of the largest and most conspicuous members of the genus, *la fourmi gendarme.*[85] Morphologically, the ability of raising the gaster

**2.30 The three castes of
*Cataglyphis fortis***

In order to emphasize the
caste-specific differences in the
relative lengths of the legs, body
lengths have been equalized.
Compare scale bars.

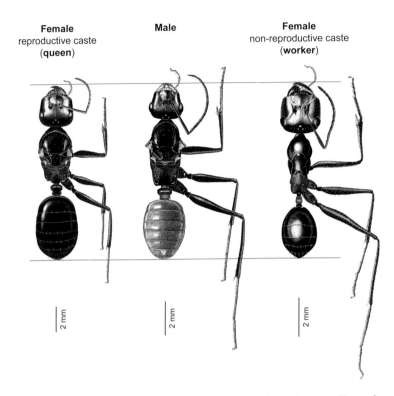

| Female
reproductive caste
(**queen**) | **Male** | Female
non-reproductive caste
(**worker**) |

is correlated with a nodular (nodiform) rather than the usually scale-
like (squamiform) shape of the petiole.

Functionally, the transformation of the original scale into a knob-
like joint between alitrunk and gaster, and the resulting ability of
raising the gaster, reduces the ants' moment of inertia when the animals perform fast turns—just as ballet dancers executing pirouettes
brings their arms closer to the axis of rotation. In trying to understand
how much this ability to lift the hind part of the body actually contributes to the ant's agility, we have modeled the moment of inertia
for a cataglyph, which has its gaster held at various inclinations above
the horizontal. In the model the ant's major body parts (head, alitrunk,
and gaster) are transformed into ellipsoids and specified by their actual
dimensions and masses (see inset in Figure 2.31). The analyses show
that by raising the gaster from the horizontal to the vertical, *Cataglyphis bicolor* and *C. fortis* shift the center of mass forward and upward, and reduce their moment of inertia about the center of mass by
more than one-half.[86] This enables them to increase their angular ac-

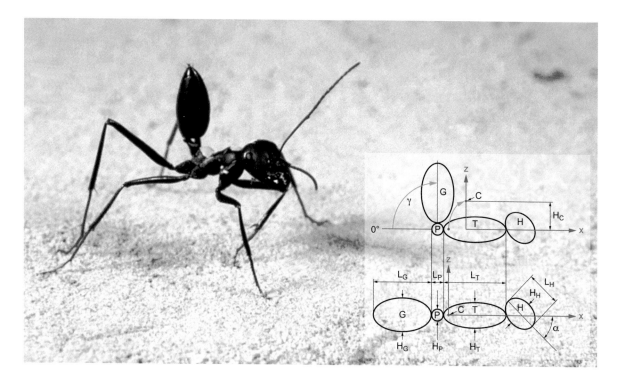

celerations substantially and to tackle tighter turns. Hence, it is a likely hypothesis that the behavioral trait of elevating the gaster has evolved for increasing locomotor agility in an energetically efficient way. As an elevated gaster experiences a cooler microenvironment, there will also be a thermoregulatory effect potentially riding as some kind of coadaptation on the enhanced-agility trait.[87] In this context it is worth noting that the silver ants, which inhabit the hottest desert areas and reach the highest forward speeds measured so far, have not evolved the ability to elevate the gaster. For considerable distances, they walk along almost straight lines.

Let us now return from turning to straight forward running, and have a look at the ants' basic running style and speed. At almost all speeds recorded so far the cataglyphs display an alternating tripod gait, as it has been observed in many insect species walking at moderate and high speeds.[88] In this tripod stepping pattern the front and hind legs of one body side and the contralateral middle leg are simultaneously on the ground (stance phase), while the other three legs

2.31 Elevating the gaster

Cataglyphis bicolor in full speed. The *white* and *blue dots* on head and alitrunk are marks applied for individual identification. *Inset:* Sagittal views of a model cataglyph with the gaster fully elevated *(top)* and in horizontal position *(bottom)*. The letters *H, T, P,* and *G* refer to head, thorax (alitrunk), petiole, and gaster, respectively. *C* marks the center of mass. The letters *H, L,* and *W* denote height, length, and width, respectively, of head, thorax, and gaster (see indexes). α, inclination of head; γ, elevation of gaster; *x* and *z* define the longitudinal and vertical body axis, respectively.

2.32 Patterns of leg movement during tripod locomotion

(a) Stride width (stride amplitude) as defined within an egocentric system of reference. Movement ranges of forelegs *(1)*, midlegs *(2)*, and hindlegs *(3)*. The left-hand side depicts the most extreme touch-down points of the legs. *AEP* and *PEP,* anterior and posterior extreme position, respectively (shown for the lower range of walking speeds). (b) Stride length as defined within a geocentric system of reference. Tripod gate performed by a cataglyph moving in a straight line forward. The tripods (synchronous substrate contacts of three legs) formed by the left foreleg and hindleg (L1, L3) and the right midleg (R2) are shown in *blue,* the alternate tripod R1L2R3 in *orange.* Stride length is defined as the distance between two successive touch-down points of L2.

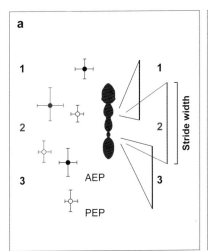
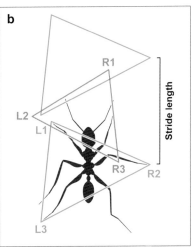

swing forward (swing phase). In the next step the two tripods change their roles. In other words, the legs of each tripod move in phase to each other and in antiphase to the legs of the other tripod. Figure 2.32b provides a schematic representation of this tripod stepping pattern. Note that in this figure stride length is measured within a space-centered (geocentric) system of reference, and hence does not reflect the stride width or amplitude performed by the legs within an animal-centered (egocentric) system of reference (Figure 2.32a).

The cataglyphs' walking behavior was first studied by Christoph Zollikofer, then a doctoral student and now a professor of physical anthropology at the University of Zürich, Switzerland. He filmed ants walking on smoked-glass plates, and then analyzed the geometry of the footfall patterns (tarsal imprints) in conjunction with the video-recorded temporal sequences of the ants' body positions. Twenty years after the original research, these laboratory findings were corroborated in the field, when Matthias Wittlinger and Tobias Seidl performed their doctoral theses at our North African field site and employed high-speed video analyses.[89] All of these recordings show that the cataglyphs maintain their temporal tripod coordination pattern over almost the entire range of walking speeds and under a variety of conditions: when they walk on straight or curved paths, on flat or rugose ground, and up and down on inclines. Cataglyphs even try to display this pattern when they are strongly handicapped by partial amputation. For example, with two legs removed (a hindleg and the contra-lateral midleg), one tripod is transformed into a 'monopod' consisting only of one foreleg. Even these four-legged ants, which must take half of their steps with one leg and thus clumsily stumble across the ground, try hard to restore their tripod coordination pattern.

In general, one could regard the hexapod ants as bipedal animals, in which each tripod represents a 'functional foot'. However, as a closer inspection of the footfall patterns in Figure 2.33a shows, the legs of one tripod do not touch the ground exactly at the same time. The ants actually unroll their tripod feet just as we humans do with our monopod feet. For example, in the swing phase, the hindlegs lift off first followed by the forelegs and finally the midlegs. It is only in the silver ants running on soft, sandy ground that all legs of a tripod take off (and touch down) simultaneously and thus effectively push the animal away from the ground. Taking all these results into account, the cataglyphs are very robust tripod walkers and refrain from this regular coordination pattern only under certain conditions, e.g., when moving very slowly at low temperatures in the morning or when they drag heavy and bulky loads backward to the nest. Then they do not simply reverse their normal forward pattern. Rather, the rigidly fixed inter-leg coupling gets relaxed, and the individual legs seem to act as loosely coupled separate units.

The most challenging question is how the cataglyphs reach their high running speeds. As speed is the product of stride length and stride frequency, maximal running speeds finally depend on how large a stride the ant can take, and how fast it can swing its legs back and forth. With increasing walking speed, both traits—stride length and stride frequency—increase concurrently. While stride length increases almost linearly up to the maximum walking speed measured so far (0.86 meter per second in *C. bombycina*), stride frequencies start to level off at plateaus of 35–40 strides per second (Figure 2.33b). These correlations, though rather strict within a species, may differ across species. The silver ants, which have retained a suite of ancestral traits such as relatively short legs, reach a given walking speed by ex-hibiting smaller stride lengths and higher stride frequencies—exceeding even 40 strides per second—than is the case in the longest-legged *C. fortis*.

Irrespective of these differences in walking style, from a certain walking speed onward aerial phases are included into the stepping pattern of either species. The swing phases of the legs get larger than the stance phases—in technical terms, the duty factor (the ratio of stance phase to cycle period) gets smaller than 0.5—and the cataglyphs start to gallop. In *C. bombycina* and *C. fortis* this happens already for walking speeds higher than 120 and 370 millimeters per second, respectively. The ants now jump from step to step, and George Arnold's early impression that they appear "to fly over the surface of the ground" finally comes true! This is especially conspicuous in the silver ants, in which running speed, stride frequency, and the relative amount of 'flying phases' reach maximal values. In contrast to the gallop of quadruped mammals, e.g., horses, the cataglyphs retain their common (tripod) gate over the entire range of walking speeds.[90]

All data and observations mentioned so far have been recorded in animals running on flat sand surfaces, but the cataglyphs can move swiftly in their alternating tripod gate also on rugose ground. Even under these conditions, when the animals negotiate tricky, irregular terrain (Figure 2.34a), one never gets the impression that they have difficulties in maintaining their locomotor stability. The flexible chain of tarsomeres with its distal claws and attachment pads may help in clinging to the surface, and elastic structures of the legs may provide stabilization through mechanical feedback.[91] However, when the ants are forced to run across corrugated surfaces (Figure 2.34b, c), in which the spatial corrugation period and amplitude are in the range of their stride lengths, their walking behavior is severely disturbed. The animals stumble, frequently step into troughs and bump into hills, or even get their stepping pattern entrained by the distance between the cor-

Figure 2.33 The cataglyphs in motion

(**a**) Tripod gate of a *C. fortis* worker at low *(left)* and high *(right)* running speeds. The upper figures (podograms) depict the temporal sequences of the stance phases *(colored)* and the swing phases *(white)* of the left and right legs (*L1–3* and *R1–3*). The stance phases of the tripods L1R2L3 and R1L2R3 are shown in *blue* and *orange,* respectively. The same color convention is used in the lower figures, which represent the spatial succession of the tripods. Stance phase overlap (all six legs on the ground) and swing phase overlap (all six legs off the ground) are indicated by *st* and *sw,* respectively. (**b**) Relationships between stride length *l,* stride frequency *f,* and walking speed *v.* For clarity, the data sets for *C. bombycina* are depicted only by regression lines *(red).*

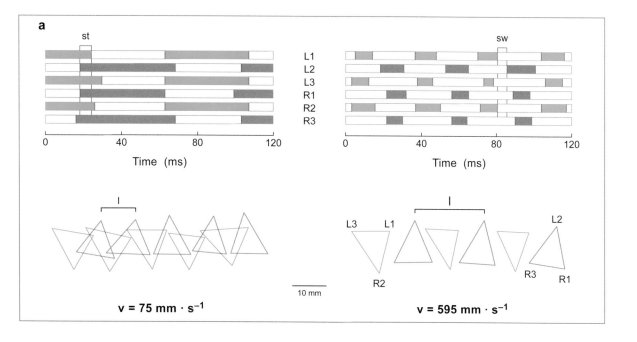

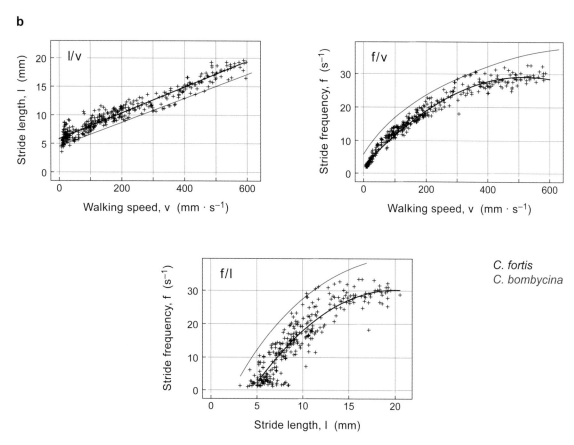

C. fortis
C. bombycina

2.34 Cataglyphs walking on rugose substrates

(a) *Cataglyphis nodus* dashing over stony ground. Peloponnese (Greece). (b) *Cataglyphis fortis* walking in a channel across a sinusoidally corrugated floor. (c) Three superimposed side-view frames of high-speed video recordings (taken through a transparent Perspex window in the side walls of the channel). The *red dots* mark successive touch-down positions of the left middle leg. The *blue lines* indicate body (alitrunk) orientation, with the *blue dots* marking the head-alitrunk and the petiole-gaster joints.

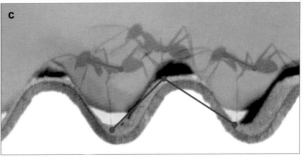

rugation peaks. Nevertheless, they keep going, and as we shall see in Chapter 4, their odometric navigational capabilities are not impaired at all when their inter-leg coordination pattern breaks down either by walking on such corrugated surfaces or by dragging heavy prey backward.

With their wide range of species differing in body size, relative leg length, foraging distance, habitat structure, and rugosity of the ground over which they move, the cataglyphs offer a large potential for studying functional adaptations to particular locomotor needs. In any way, investigating the kinematics and dynamics of six-legged locomotion in these ultrarunning ants, and inquiring into potential species-specific adaptive deviations from a general pattern, might be a fascinating project, all the more as ultra-miniature force-plate sensors have been developed, which allow one to measure the instantaneous ground reaction forces exerted by single legs of 10-milligram ants in the micro-Newton range.[92] Such force platforms have already been used to investigate the mechanical functions of individual legs in the cataglyphs, e.g., how pulling and pushing forces are distributed among the different legs when the animals walk on level ground or climb upward or downward.[93]

This also raises the question of how the legs touch the substrate. In *Cataglyphis,* as in ants in general, the 'toe' of each leg is provided with two movable claws and an adhesive pad, the arolium, which can be unfolded by increasing hemolymph pressure (Figure 2.35). However, the cataglyphs rarely walk on tiptoes. Ground contact is mainly made by the third and fourth tarsal segments, the ants' 'heels,' which are provided with rather long distally oriented hairs and lack the dense arrays of fine hairs that are characteristic for *Formica* wood ants. How these attachment structures vary across the cataglyphs, and how they contribute to generating the proper pushing and pulling forces of the legs on various substrates, again opens up a promising field of research in comparative functional morphology.[94]

For the sheer sake of curiosity let us finally compare an ultrarunning cataglyph with an ultrarunning human. On 4 October 1981 my colleague and friend Bernd Heinrich, a professor of biology at the University of Vermont, ran a world record in the Chicago 100-kilometer ultra-marathon race. He covered the 100 kilometers in about 6.5 hours, to be

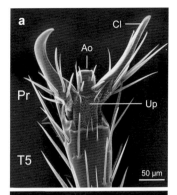

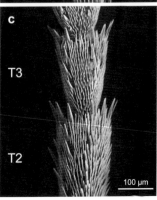

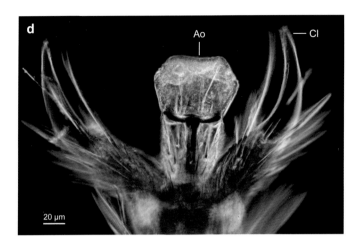

2.35 Toes and heels

Ventral views of **(a, d)** 'toe' and **(b)** 'heel' (*T3* and *T4*) of *Cataglyphis fortis*.
(c) Ventral view of tarsal segments of the wood ant, *Formica polyctena*. The ants usually touch the ground with tarsomeres *T3* and *T4*. *Ao*, arolium; *Cl*, claw; *Pr*, pretarsus; *T2–5*, tarsomeres; *Up*, unguitractor plate (part of a mechanical device involved in moving claws and arolium).

precise in 6:38:21. As Bernd kindly provided me with details of his running style and performance, I could compute that during his 100-kilometer run he must have made about 66,700 steps, each 1.5 meters wide (which translates into 33,350 strides, each 3.0 meters wide),[95] at a stride frequency of 1.4 strides per second. Comparatively, in the longest run recorded as yet in a cataglyph,[96] a specimen of *C. fortis* covered 1.2 kilometers in 68.7 minutes, and thus walked at a mean speed of 300 millimeters per second. During this far-reaching round-trip, the ant moved away from the nest for a maximal distance of 356 meters. While running steadily at a speed nearly 15 times lower than Bernd Heinrich's 15 kilometers per hour, it moved its legs with a stride frequency that was more than one order of magnitude higher (24 strides per second) than that of the human champion. This observation is in accord with the general relationship between stride frequency and body size: the smaller the animal, the higher its stride frequency.[97]

For physiological and finally cell biological reasons, locomotor muscles always face a trade-off between force generation and contraction frequency. This constraint may have exerted a selection pressure on the cataglyphs to reach their fastest running speeds more by increasing stride length—and thus increasing relative leg length up to the maximum reached in *C. fortis*—than stride frequency. *C. bombycina*, which has about 15% shorter legs than *C. fortis*, more readily includes aerial phases in its runs. As running speed increases, stride frequency levels off, though at rather high values (Figure 2.33b). The need to cycle their legs at high rates might make us wonder whether the cataglyphs exhibit higher rates of power output than one would predict from a 'general' running insect of their body size.

Power Output

To explore the energy demands during fast running let us first have a look at the world's premier ultrarunning animal, the pronghorn antelope of the American Great Plains. Running at maximum sustainable speeds of 65 kilometers per hour over distances of more than 10 kilometers, and reaching top speeds of up to 100 kilometers per hour, the pronghorn, *Antilocapra americana,* is certainly the best endurance runner that has evolved in the animal world. When Stan Lindstedt let these antelopes gallop on a treadmill and recorded their rate of oxygen uptake per unit of time and unit of body mass,[98] he arrived at a maximum metabolic rate—the pronghorn's 'aerobic capacity'—of $5 \ mlO_2 \cdot kg^{-1} \cdot s^{-1}$ (in words, 5 milliliters of oxygen per kilogram body mass per second, or 4.2 watts per kilogram). This is 25 times higher than the antelope's standard metabolic rate. However, we can consider this extraordinary 'metabolic scope' of 25 exceptional only if we put it into perspective by comparing it with some default value, i.e., with the metabolic scope expected for a 'general' mammal of the pronghorn's body size. Such a comparison indeed shows that the pronghorn is special. Its metabolic scope is 3.3 times larger than predicted for its body size.

How does the performance of this ultrarunning mammal compare with that of the ultrarunning cataglyph? By using miniaturized respirometric devices and applying gas analysis techniques, one is indeed able to measure the amount of CO_2 exhaled and / or O_2 inhaled by an individual ant, while it is confined to a small glass tube.[99] When tested this way, a resting *Cataglyphis bicolor* exhibits a standard metabolic rate of 0.04 $mlO_2 \cdot kg^{-1} \cdot s^{-1}$. For a 35-milligram cataglyph, this amounts to $1.4 \cdot 10^{-6} \, mlO_2 \cdot s^{-1}$. By converting this value to watts, we conclude that an individual cataglyph has a standard power output of 29.4 µW, i.e., roughly 30 microwatts.

This standard metabolic rate of a cataglyph is in accord with the values determined in some other ant species from various habitats and taxonomic groups (Figure 2.36a), and hence does not support the previous hypothesis that desert ants have adopted especially low standard metabolic rates.[100] It does not support another hypothesis either. This other hypothesis—the 'aerobic capacity hypothesis'—states that species which spend more energy on activity should also have a higher standard metabolic rate. The latter is the case in flying insects such as honeybees and vespine wasps, which engage in energetically extremely costly flight behaviors and concomitantly exhibit elevated standard metabolic rates (Figure 2.36b),[101] but there is no indication that the high-speed cataglyphs follow this pattern.

What do we know at all about the metabolic rates maintained by the cataglyphs during fast locomotion? Are the thermophilic ants special in the way of increasing their aerobic capacity when they are in full speed at high ambient temperatures? In the rare cases in which aerobic capacities have been measured in walking ants such as leaf-cutter and carpenter ants, the metabolic rates were four- to sevenfold higher than the resting metabolic rate. In the long-legged Namibian desert beetle *Onymacris plana,* whose fast running speed we had already mentioned before, the metabolic scope is about 12, while cockroaches—when in their fast running mode of locomotion—reach scope values of up to 23.[102] In *Cataglyphis bicolor* Stefan Hetz from the Humboldt University in Berlin, Germany, has measured the maximal metabolic rate at walking. It was 15 times larger than the standard metabolic rate. Most probably this value determined at 25°C with animals walking slowly in a test tube will certainly increase when we consider cataglyphs running at high speed in their natural habitat, so that the cataglyph's metabolic scope might well approach that of the pronghorn antelope. Nevertheless, the final question we have answered in the antelope—to what extent the ultrarunner's metabolic scope is larger than predicted for its body

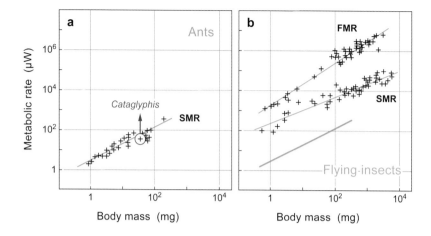

2.36 Metabolic rates

(**a**) Standard metabolic rates *(SMR)* of workers from various species of ponerine, myrmicine, and formicine ants plotted as a function of body mass. The SMR of *Cataglyphis bicolor* is highlighted by the *open red circle*. The *red arrow* points at the current best estimate of the metabolic scope of a running cataglyph. (**b**) Standard metabolic rates *(SMR)* and the corresponding flight metabolic rates *(FMR)* of a wide range of flying insects including odonates, lepidopterans, dipterans, and bees. For comparison, the regression line pertaining to the SMR of worker ants *(orange line)* is copied from *(a)*.

size—remains a fascinating, though hitherto unresolved, issue in the cataglyphs.

Why Thermophiles?

The reader will have noticed that the further we proceeded in this chapter, the more questions have been raised than answered. Nevertheless, there are at least two general conclusions that we may derive from this glimpse into the physiological ecology of our little desert navigators. First, the physiological adaptations, which the thermophilic desert ants have evolved, differ from those of non-thermophilic ants only quantitatively rather than qualitatively. As is the case with the ultrarunning mammal, the pronghorn antelope, the cataglyphs do not employ any grand trick, or novel mechanism, to accomplish their extraordinary running and foraging tasks under environmentally stressful conditions. Rather, they succeed by enhancing a number of individual features and traits that are generally normal to ants. For example, they have longer legs and can elevate their bodies higher above the ground and reach higher running speeds than predicted; they use some often slight deviations from an insect's standard morphological and physiological outfit for achieving higher thermal tolerances and lower rates of cuticular transpiration than predicted—with the predictions derived from large samples of ants. Second, however, these adaptive deviations would not suffice in a small ectotherm endowed with a large surface-to-volume ratio, if they were not associated with certain behavioral traits, especially with the ability to exploit the microclimatic mosaic available on the desert floor.[103] Hence, it is a whole suite of integrated morphological, physiological, and behavioral adaptations that defines the 'thermophilia syndrome.'

If this syndrome actually consists in the enhancement of quite a number of individual traits, one may wonder whether there are species of ants in which such traits have evolved at least partially or to a lesser degree. In the arid and semiarid areas of western North America the honeypot ants *Myrmecocystus* may come to mind. Their colloquial name derives from a special caste of workers, so-called repletes, which store liquid food in their crops and thus serve as living food containers. Apart from this specialization, their workers resemble those of *Cataglyphis* in so many morphological characters that once in the nineteenth century both genera had even been united to a single one.[104] In contrast to *Cataglyphis*, however, many species of *Myrmecocystus* are nocturnal, and the diurnal ones have substantially lower thermal tolerances ($CT_{max} = 43°C–48°C$) than the cataglyphs ($CT_{max} = 52°C–55°C$).[105] They do forage on dead insects, but concentrate to a large extent on termites, which constitute one of their major food sources, and collect nectar from flowers and even honeydew from aphids. Most remarkably, they can get engaged in spectacular mass recruitment events. We owe it to the thorough and elegant studies of Bert Hölldobler, then at Harvard University, that the ants' strategy of establishing flexible, spatially, and temporarily varying foraging territories, which are defended against workers of adjacent colonies by ritualized combat tournaments, has been unraveled in detail.[106] Nothing of this kind of behavior occurs in the individually foraging cataglyphs.

Quite a different ecological niche is occupied by the desert harvester ants belonging to the genus *Pogonomyrmex,* the most abundant desert ants in North America, which "nearly blanket the arid regions of Mexico and the western United States."[107] Most species are strictly diurnal foragers, which reach their peak activities at surface temperatures of about $T_s = 45°C$. Due to their rather high thermal tolerances

(CT_{max} around 52°C) they terminate foraging only at $T_s = 53°C–60°C$, yet these are the temperatures at which *Cataglyphis bombycina, Ocymyrmex robustior,* and *Melophorus bagoti* reach their peak activities. Furthermore, besides being predominantly herbivores, these seed harvesters differ from the cataglyphs and their thermophilic equivalents by employing elaborate trunk trail systems for mass recruitment, and by strictly partitioning their foraging grounds for avoiding intraspecific and interspecific confrontations.

Given these differences, we may well ask whether in the North American deserts—in the Great Basin, the Mojave Desert, and the Chihuahuan and Sonoran Deserts—there is any ant at all that resembles a thermophilic cataglyph. As Robert Johnson, one of the authorities of North American desert ant ecology, assures me, the dolichoderine genus *Forelius* provides the most thermophilic ants in this part of the world. In lifestyle, however, *Forelius* species markedly differ from the cataglyphs. They forage as mass recruiters with hundreds of tiny workers in large columns, and are often the first to detect, occupy, and exploit a food source before other species arrive. In conclusion, various groups of ants have evolved thermotolerant behaviors to various degrees and in various ecological contexts, but *Cataglyphis, Ocymyrmex,* and *Melophorus*—the paradigmatic thermophiles—stand out by the very suite of common characteristics that define the thermophilia syndrome.

We certainly cannot end this section without asking what happens in the Atacama Desert of South America. With a mean annual precipitation of mostly less than 50 millimeters, and at some places less than 5 millimeters, the Atacama contains the most arid areas on our planet. In some parts of this barren land there have been only four years with more than 2 millimeters annual precipitation within nearly half a century. "Not even a single cactus is growing here," remarked the botanist Rudolph Philippi when he traveled through the Atacama Desert in the middle of the nineteenth century.[108] In his early studies on the Chilean ant fauna, Wilhelm Goetsch considered *Dorymyrmex* the South American analogue of the cataglyphs, and so did Gennady Dlussky in his treatise on desert ants. However, more recent work clearly shows that this is not the case. *Dorymyrmex goetschi,* a species that lives at the fringe of the most arid parts of the

Atacama Desert, forages at surface temperatures in the range from $T_s = 15°C–50°C$ and thus avoids the very temperatures at which thermophilic ants are maximally active. It rather seems to be adapted to the lower temperatures prevailing in the relatively cool Atacama Desert.[109] In general, however, in the Atacama many barren and rocky parts of this virtually rainless desert seem to be devoid of ants altogether (Figure 2.37).[110]

As this look into the foraging ecology of desert ants has shown, the thermophiles are unique insofar as they assume an extreme position within a wide spectrum of foraging strategies adopted by desert ants. Throughout this chapter we have addressed this issue by asking *how*— by what morphological, physiological, and behavioral adaptations—the thermophiles operate under their stressful environmental conditions. In addition, we may wonder *why* they have acquired these adaptations in the first place. What selection pressures have caused them to occupy the unique ecological niche of a thermophilic scavenger? First, recall that in terms of abundance and biomass the thermophiles comprise only a minor fraction of all ants in desert habitats.[111] For example, in the gravel plains of the Namib Desert harvester ants account for more than 95% of the total forager biomass. Second, in aggressive interactions between thermophilic and non-thermophilic species the latter are usually the successful (behaviorally dominant) ones and chase off their thermophilic competitors. Based on quantitative assessments of such interspecific interactions at artificial baiting sites, ecologists use to define dominance hierarchies among the members of ant communities. Where such dominance hierarchies have been established, e.g., in the subtropical semiarid habitats of southeastern Spain and northern Argentina, the most subordinate species turned out to be the most thermophilic ones (several *Cataglyphis* species in Spain and *Forelius nigriventris* in Argentina). Obviously, there is a trade-off between thermal tolerance and behavioral dominance. This could suggest that subordinate species have evolved their thermophilia syndrome in order to avoid competition with the behaviorally dominant and much more abundant species, which occupy the desert foraging grounds at more temperate times of the day (Figure 2.38). Even though the dominant species (e.g., *Pheidole, Tetramorium, Tapinoma,* and

Messor species) are mainly seed harvesters, most of them exhibit quite some dietary flexibility and also include dead arthropod matter in their diet. Given their numerical dominance, their nutritional niche does indeed overlap with that of the thermal warriors. Consequently, it is a likely evolutionary scenario that the thermophiles have been able to become real desert dwellers, because their ancestors have pushed their thermal limits upward beyond the level that can be tolerated by an 'average' ant, and thus have opened up a new ecological niche.[112]

As a case in point let us have a look at the magnificent spectacle that occurs each summer day in the central Australian desert. When surface temperatures have risen above 45°C, the omnipresent

2.37 'No-ant land' in the Atacama Desert

Hyperarid parts of the Atacama Desert. The rocky area located about 70 km east of Copiapó, Chile, is virtually devoid of vegetation—and of ants.

2.38 In ecological terms: subordinate versus dominant species

Daily foraging patterns of subordinate (*Cataglyphis bicolor, Cb*) and dominant (*Messor arenarius, Ma*) ants in North Africa in summer (June, *JUN*) and winter (January, *JAN*). During the winter months *C. bicolor* hibernates. Representative recordings at one nest of either species. In the *Messor* histograms the *dark green* and *light green columns* depict the frequency of exits and entries, respectively, recorded at one nest in June and January. The *Cataglyphis* data refer to the frequency of exits. The *open arrowheads* mark the times of sunrise and sunset.

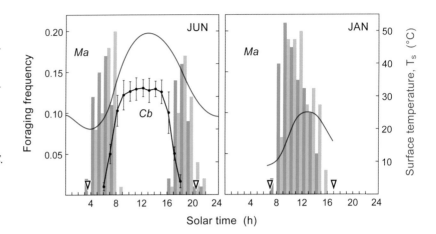

meat ants—these aggressive omnivorous scavengers, *Iridomyrmex purpureus*—retreat underground, and the thermophilic *Melophorus bagoti* appears on the scene.[113] Then, high-density mass recruitment gives way to diffuse thermophilic foraging—and the curtain opens for the navigation drama that will now unfold in front of our eyes.

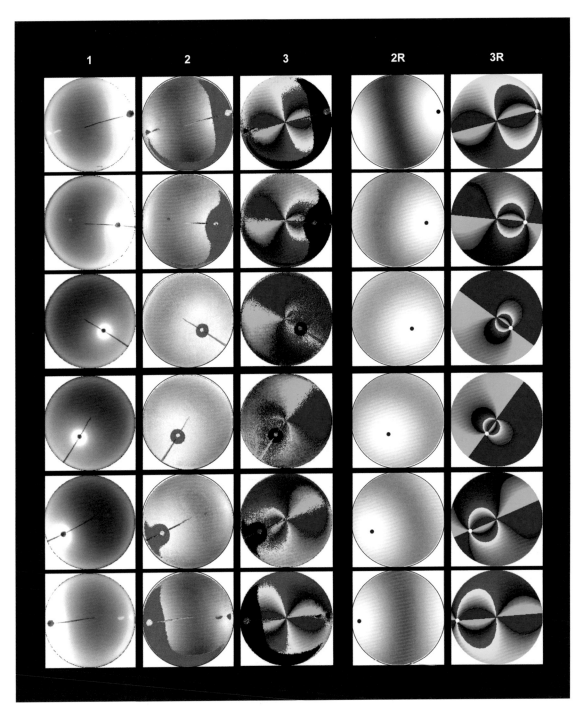

3 Finding Directions

Since early medieval times, at least since word had spread from China in the late twelfth century, human sailors have used a magnetic compass to steer a ship to its destination.[1] Foraging cataglyphs, however, rely mainly on a visual compass, a sky compass, which derives its information from the celestial hemisphere: from direct sunlight and scattered skylight with its marked polarization and intensity / spectral patterns. As during the course of the day the sun and with it all skylight patterns move across the sky, the cataglyphs must somehow integrate information about the changing position of the sun relative to an earthbound reference, e.g., the earth's magnetic field or the distant skyline, with information provided by their internal clock. Let us first investigate what kind of visual information the cataglyphs are able to derive from the sky. Later we shall see that the distant skyline, the earth's magnetic field, and the direction of the prevailing wind may provide the ants with additional—and in this case directly earthbound—compass information.

Polarized Skylight

In 1809 the French astronomer Dominique Arago looked through a double-refracting calcite crystal at the sky and observed that scattered light reaching the earth from the sunlit sky is polarized. This was only one year after polarized light had been discovered as a physical phenomenon. Some decades later the Scottish physicist Sir David

3.1 Full-view skylight images of celestial e-vector patterns

The polarimetric sky images depict the spatial distribution of the degree (*d*, columns 2 and 2R) and angle (φ, columns 3 and 3R) of polarization measured by full-sky imaging polarimetry. The sun is occluded by a *small black disk*. Column 1: Color photographs of the natural sky. In columns 2, 2R, and 3, 3R the distributions of *d* and φ are either recorded from the natural sky (*2* and *3*) or theoretically computed by using a single-scatter radiation model (Rayleigh skies, *2R* and *3R*). In columns 2 and 3 *red* and *black*, respectively, indicate overexposure. The angles of polarization (*3* and *3R*) are transformed into false colors: $\varphi = 0°–45°$ *(red)*; 45°–90° *(dark green)*; 90°–135° *(blue)*; and 135°–180° *(light green to yellow)*. *Black* indicates overexposure. Pictures were taken under clear sky conditions on 26 August 1999 from near sunrise *(top row)* to near sunset *(bottom row)* in the Chott el Djerid, Tunisia. North is up.

Brewster was so impressed by Arago's observation "that the blue atmosphere which overhangs us exhibits in the light which it polarizes phenomena somewhat analogous to those of crystals with two axes of double refraction" that he referred to this phenomenon as one of "the wonders of terrestrial physics." In August 1869, John Tyndall, while standing on top of the Aletschhorn in the Swiss Alps, used exactly such a crystal to scan the entire sky and to study the details of this impressive pattern. Soon after Tyndall's observations John William Strutt, the later Lord Rayleigh, was able to deduce the characteristics of polarized light from the scattering of electromagnetic waves by molecular-size particles (Rayleigh scattering), and to show that the pattern of polarized light in the sky originated from the scattering of sunlight by the molecules of nitrogen and oxygen within the earth's atmosphere—from the very same physical phenomenon that renders the daytime sky blue.[2]

Light is the kind of electromagnetic radiation that can be perceived by animal eyes. In an electromagnetic wave, in which the electric and magnetic fields oscillate at right angles to each other and to the direction of travel, it is only the electric component shown in Figure 3.2 that is responsible for vision. When the axis of the electric field, the so-called e-vector, oscillates within a plane so that its endpoints trace out a line, we speak of linearly polarized light, and call it vertically polarized, horizontally polarized, and so forth, depending on whether the e-vector is oriented vertically, horizontally, or at any other angle. Humans are largely blind to the polarization of light, because the light-absorbing visual pigment molecules (rhodopsin) are randomly oriented within our photoreceptor membranes,[3] but in insects and many other invertebrates they are more orderly arranged, and thus render a given photoreceptor cell sensitive to polarized light of a particular e-vector orientation (Figure 3.3). Note that a rhodopsin molecule absorbs light maximally when the e-vector is oriented parallel to the long axis of its chromophore (retinal).

Our natural polarization blindness might have been the reason why the discovery of polarization vision in animals had to await the middle of the previous century, when Karl von Frisch made his striking discovery in honeybees. He let the bees perform their recruitment ('waggle') dances on a horizontal comb, where he presented them

3.2 Skylight polarization

The wave traces shown in *white* depict the oscillations of the electric field vector (e-vector) of light propagating in the directions of the *orange arrows*. The *double-headed arrows* inside the panels mark the orientation of the e-vectors as seen by an earthbound observer. The light radiated by the sun **(a)** is unpolarized. In the extremely rapidly fluctuating electric field (much more rapidly fluctuating than visual systems and even modern measuring devices can detect it) all e-vector orientations are evenly distributed. Hence, direct sunlight perceived by an earthbound observer is unpolarized. **(b)** In contrast, light scattered by an air molecule *(yellow arrows)* is polarized. Maximal polarization occurs at a scattering angle of 90° **(c)**. At all other scattering angles light is polarized only to a certain degree (partially polarized light **(d)**). Background: Namib Desert between Mirabib and Gobabeb.

with nothing but a small part of the blue sky. Whenever he rotated a polarization filter in their field of view, the bees changed the direction of their dances. Several decades earlier the Swiss physician and myrmecologist Felix Santschi had already observed that ants could derive compass information not only from the sun, but also from small patches of unobscured sky, but as he was not aware of the phenomenon of polarized light, he could not draw the right conclusion from his startling observation.[4] After von Frisch's discovery the field lay fallow for more than two decades, until in the early 1970s, all of a sudden, and completely independently of one another, several Swiss, German, Russian, Dutch, and American research groups started to inquire about the bee's and ant's celestial polarization compass.[5] The way was thorny. Hypotheses abounded. A solution came within reach only after the discovery that in both the cataglyphs and the hon-

3.3 Polarization sensitivity in *Cataglyphis:* from eye to molecule

The light-sensitive rhodopsin molecule *(lower right)* gets maximally stimulated, when the e-vector oscillates parallel to the molecular axis of its chromophore (retinal). Furthermore, the absorption axes of the rhodopsin molecules are aligned with the axes of the microvilli. Like the bristles of a brush, the microvilli of a photoreceptor cell form the rhabdomere. Provided that all microvilli of a photoreceptor have the same orientation, this orientation defines the e-vector tuning axis of a photoreceptor. The rhabdomeres of each visual unit (ommatidium) are fused and form a light-guide structure, the rhabdom. *Ax,* axon; *Cn,* crystalline cone; *Co,* corneal lens; *E,* e-vector, here aligned with the long axis of the chomophore retinal; *Pp,* primary pigment cell; *Ps,* secondary pigment cell; *Rc,* photoreceptor cell; *Rd,* rhabdomere; *Rh,* rhabdom; *Rt,* retinal.

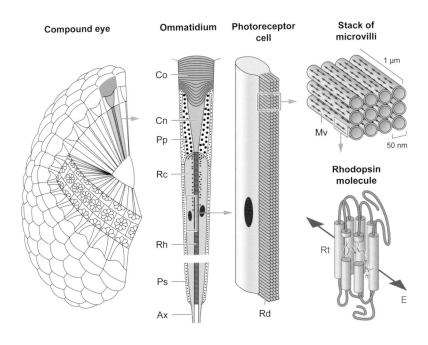

eybees a small part at the dorsal rim of the compound eye had all the properties to perform this task,[6] and that it actually did it.

To investigate this solution, let us begin with Figure 3.4a, which portrays the polarization patterns in the sky as seen by an earthbound observer. All e-vectors form concentric circles around the sun. In particular, let me draw your attention to two features. First, the solar meridian (the vertical from the zenith through the sun down to the horizon) and the antisolar meridian form the symmetry plane of all e-vector patterns. Along this symmetry plane light is invariably polarized parallel to the horizon. At all other positions in the sky the e-vector orientation changes as the sun changes its elevation. Second, as depicted by the size of the blue bars, the degree of polarization (d) varies across the celestial hemisphere from $d=0\%$ (direct, unscattered, and hence unpolarized light from the sun) to $d=100\%$ (maximally polarized light at an angular distance of 90° from the sun). Consequently, when the sun is at the horizon, a great circle of maximal polarization runs through the zenith. As the sun rises, this great circle of maximal polarization tilts down within the antisolar half of the sky.

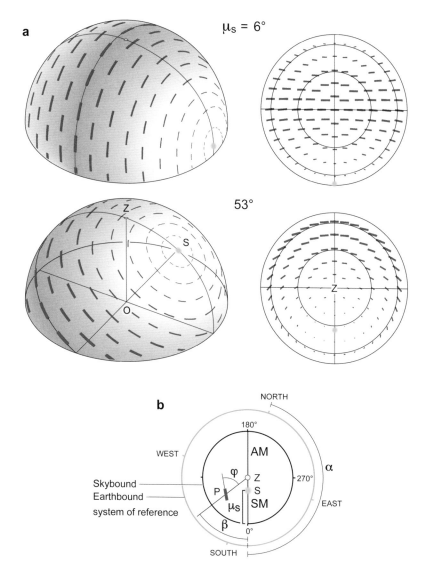

$\mu_s = 6°$

53°

3.4 Polarization gradients (e-vector patterns) extending across the celestial dome

(a) 3-D representations *(left)* and 2-D projections *(right)* for two elevations of the sun. The orientation and size of the *blue bars* (e-vectors) indicate the angle and degree of polarization, respectively. The symmetry plane of the patterns is marked in *red*. In the 2-D projections the two *inner circles* indicate elevations of 30° and 60° above the horizon. (b) Conventions and definitions. The solar meridian *(SM)* and the antisolar meridian *(AM)* form the symmetry plane of all skylight patterns (solar vertical). α and β denote azimuthal positions within an earth-based system of reference (α = 0°, north) and a sky-based system (β = 0°, solar azimuth), respectively. *Blue bar,* e-vector at celestial point P; φ, e-vector orientation (angle of polarization) relative to the vertical; μ_s, solar elevation; O, observer; S, sun; Z, zenith.

This implies that the antisolar half of the sky is always more strongly polarized than the solar one.

The above refers to an ideal atmosphere, within which each ray of sunlight is scattered only once—primary (Rayleigh) scattering[7]—and which is free of particles that are much larger in diameter than the wavelength of light. In the real atmosphere such particles almost always occur. Hence, in the natural sky the degree of polarization is markedly decreased by haze, dust, and clouds as well as by multiple scattering and reflections from the ground, so that the degree of polarization almost never exceeds 75%. This statement holds true even for the clear sky prevailing at our *Cataglyphis* research site in North Africa. Finally, there is one important aspect of skylight polarization that can easily be predicted from theory but was confirmed by measurement only recently, after we had performed wide-field polarimetric recordings in natural skies: Even though clouds, i.e., masses of water droplets, largely depolarize skylight, e-vector patterns are often undisturbed by patchy cloud cover as long as direct solar rays can illuminate the air space underneath the clouds, so that light scattering can occur within the volume of air between the clouds and the observer. As the polarization of light underneath the clouds follows the same geometrical rules as the polarization in the unobscured parts of the sky, the e-vector pattern can continue across the entire sky vault even under many cloudy sky conditions.[8]

There are various ways to visualize these celestial e-vector patterns. One method we applied in the early days is to take a picture of the entire sky vault through a wide-angle (180°) fish-eye lens provided with a linear polarizer or, better yet, with a Perspex dome that is equipped with a set of radial polarization analyzers, so-called polarization axis finders (Figure 3.5). Some years later, the physicist Gábor Horváth and his collaborator István Pomozi from Eötvös University in Budapest joined our endeavors and followed us to the same photographic site in the Tunisian Chott el Djerid, in order to perform computer-aided full-sky imaging polarimetry. The resulting high-resolution false-color images allow one to record the state of polarization in thousands of pixels of skylight simultaneously, and to sort out the effects of optical parameters such as the degree and angle of polarization (Figure 3.1). It also confirms that under certain conditions the e-vector pattern in the open sky can extend into the air space underneath patches of clouds, and that at night the e-vector patterns occurring in the moonlit sky are very similar indeed to the diurnal patterns in the sunlit sky, though lower in intensity by more than seven orders of magnitude.[9]

3.5 Visualizing celestial e-vector patterns

Looking with a 180° fish-eye lens at the sky. A Perspex dome provided with a set of radial polarization analyzers is vaulting the camera. *Left:* The axes of the dark hourglass-shaped figures appearing within the polarization filters mark the orientation of e-vectors in particular points in the sky. For polarimetric sky images taken at the same location, see Figure 3.1.

3.6 Providing the cataglyphs with eye caps

Prior to particular tests the compound eyes or parts of them are covered with light-tight lacquer sheets (as done here by graduate student Karl Fent). *Left:* Scanning electron micrographic (SEM) pictures of the right compound eye of *Cataglyphis fortis* with the eye cap applied *(top)* and, in preparation for the control test, removed *(bottom)*.

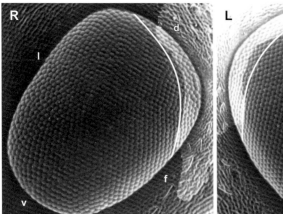
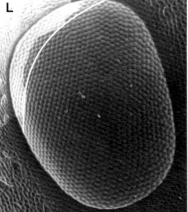

3.7 Leaving the dorsal rim areas (DRAs) open

In order to test for the exact position of the eye caps applied in the field (Figure 3.6), the ants' heads are prepared for SEM imaging by gold-coating them twice: once before and a second time after the eye caps have been removed. Hence, the eye regions that have remained uncovered receive double coating and thus appear brighter than the rest of the eye. The *white lines* demarcate the dorsal rim areas. *R* and *L,* right and left eye, respectively. *d,* dorsal; *f,* frontal; *l,* lateral; *v,* ventral. *Cataglyphis fortis.*

When at dawn the strongly polarizing upper layers of the atmosphere increasingly emerge from the earth's shadow, the scattering of sunlight creates a celestial pattern of polarization that becomes visible to the ants already about 30–35 minutes before sunrise, and remains visible to them until 30–35 minutes after sunset.[10] As the cataglyphs are strictly diurnal foragers, they usually do not experience these twilight polarization patterns. Yet properly designed experiments enable us to expose the ants to such patterns and test whether they are able to use them as compass cues. This leads us to the general question of how at all we can get the cataglyphs to disclose their compass secrets.

One way to analyze the ant's polarized-light-based sky compass, in short, its 'polarized-light compass,' is to provide the ants' eyes (Figure 3.6)—or parts of them (Figure 3.7)—with caps that either are light-tight or have different optical transmission functions and thus change the properties of the ants' visual input. Another way is to leave the eyes of the ants unimpaired and instead let the ants walk underneath a specially designed trolley, a rolling optical laboratory, that is moved along with an individual cataglyph as it performs its homeward run in the test field (Figure 3.8).[11] The trolley not only prevents the ant from seeing landmarks or detecting the direction of the wind, but also allows us to employ special filters, screens, and other optical gadgets for manipulating the polarization and spectral gradients in the sky, blocking out the sun, restricting skylight vision to particular parts of the celestial hemisphere, or providing the ants with artificially polarized beams of light. While one experimenter moves the trolley in such a way that the ant, which is continually viewed by the experimenter through cross-wires, remains centered within the optical devices mounted on the trolley, a second experimenter records the ant's trajectory.

These quite demanding experiments are considerably facilitated by the fact that the ant, surrounded by all this optical equipment, moves at a very slow pace. Nevertheless, one might argue that by moving the trolley, the experimenter could somehow influence the ant's directional choices. But this is completely impossible. The ant, not the experimenter decides where to go. Even if one tries intentionally to force the ant to move in a particular direction by steering the trolley accord-

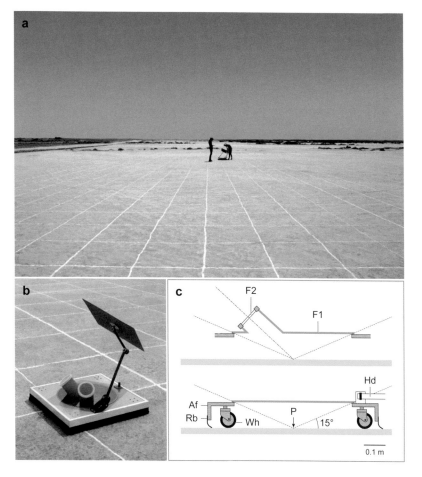

3.8 The filter trolley

(a) The trolley in action. The experimenter to the right follows the ant, which is walking underneath the moving trolley, while the other person records the ant's path. (b) The trolley equipped with a wide-field orange cutoff filter (for transmission function, see filter 560 in Figure 3.9h), two upward pointing cylinders loaded with linear polarizers, and a sun shield. (c) Vertical sections through two version of the trolley. *Af,* aluminum frame; *F1, F2,* optical filters; *Hd,* handle for moving the trolley; *P,* position of ant; *Rb,* rubber band; *Wh,* wheel.

ingly, the ant stubbornly sticks to its course, and finally gets rolled over by the vehicle. This assures us that the experiments designed this way do really work. The ants cooperate even when we have manipulated their eyes, e.g., by covering particular parts of their compound eyes, or their ocelli, with opaque lacquer sheets, and surrounded them with our optical trolley device. It is really amazing to see, and rewarding to learn, how well an animal as small as an ant performs in an open-field experiment, in which we have intervened so strongly in its visual world and physical integrity.

How, though, can we demonstrate that the ants really use a visual compass? This can be done most directly, though somewhat crudely,

3.9 Behavioral analysis: the ant's polarized-light compass receives its input exclusively from ultraviolet receptors of the dorsal rim area (DRA)

(a–f) Circular distributions of the cataglyphs' homeward courses recorded under different test conditions. +, start of homeward course; *0°*, goal (nest) direction. *Diamond symbols* indicate two data points. The trolley *(square icons on top of data sets)* is loaded with a sun shield *(vertical black bar)*, various spectral filters and, in *(e)* and *(f)*, in addition with a linear polarizer. Eye regions, which prior to the test have been covered with light-tight paint are shown in black (see *eye icons on top of data sets*). *DA*, dorsal area; *DRA*, dorsal rim area; *VA*, ventral area of compound eye; *triplet of small circles*, ocelli. The mean azimuthal positions of sun *(SA)* and skylight window (aperture *AP*) are indicated by *small yellow squares*. (g) Spectral sensitivity functions of the cataglyph's ultraviolet *(UV)* and green *(GR)* receptors. (h) Transmission functions of spectral filters. The numbers inside the figure denote the wavelengths at which the filters transmit 50% of incident light. The number 375 refers to the maximum transmission of a narrow-band ultraviolet filter. *POL*, linear polarizer. (a–d, g) *Cataglyphis bicolor*, (e, f) *C. fortis*.

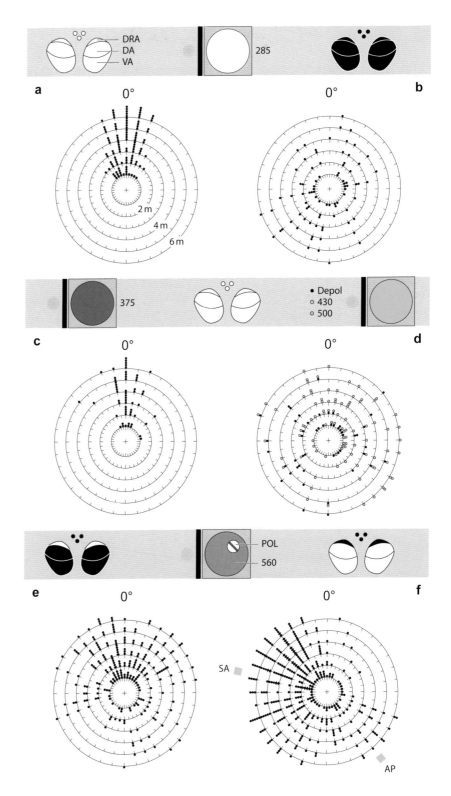

by abolishing vision altogether. While the control animals immediately select their proper home direction (Figure 3.9a), the blindfolded ones, which have their compound eyes and ocelli occluded, walk extremely slowly and search around in all directions (Figure 3.9b). In order to exclude the possibility that the mere procedure of applying the eye caps has a disturbing effect on the animals' behavior, the control animals are first provided with eye caps as well, but then the caps are removed before the ants are tested. Even after this treatment, the controls highly significantly select their homeward courses.

Next we leave the ants' eyes untreated but provide the trolley with special optical filters. One of these filters, a large-field 'compound filter', transmits all wavelengths visible to the ants, but depolarizes the sky and distorts its spectral gradient. Under these conditions, with all reliable skylight information gone, the ants behave as though blind, just as if their eyes had been completely occluded (Figure 3.9d, black symbols). Other filters change the spectral composition of the light in the sky. In this case the ants' ability to detect polarized skylight disappears completely at wavelengths larger than 410 nanometers, i.e., at wavelengths that include the entire range of wavelengths visible to humans (Figure 3.9d, colored symbols). The ants walk in random directions. However, when the spectral range is extended only a little bit into the ultraviolet, and even when filters are used that exclusively transmit ultraviolet light, the ants' navigational accuracy is completely restored (Figure 3.9c). Even though the cataglyphs possess two spectral types of photoreceptor in their compound eyes, ultraviolet and green receptors (Figure 3.9g), the behavioral experiments reveal that it is

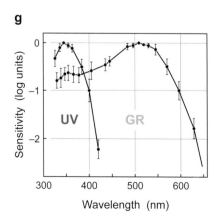

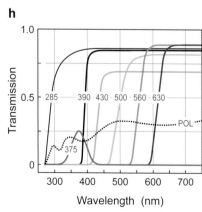

only the ultraviolet type of receptor that is necessary and sufficient for detecting the polarized light in the sky. The same has subsequently been shown for honeybees and a number of other insects including flies, dragonflies, dung beetles, and monarch butterflies.[12]

Why is it advantageous to have only one spectral type of receptor involved, and why, then, should one opt for the UV type? The use of only one spectral type of receptor (homochromacy) frees the system from the need to neurally disentangle information about the polarization and spectral content of the stimulus. The second question is especially intriguing, because in the clear, cloudless sky the radiant intensity and the degree of polarization are both lower in the ultraviolet than at longer wavelengths.[13] A direct answer to this puzzle—the UV-sky-pol paradox, as Gábor Horváth has aptly put it—comes from observations made by imaging polarimetry. The skylight polarization patterns are most stable, and the orientation of e-vectors can be determined most accurately, in the ultraviolet. Moreover, as mentioned above, the patterns can extend underneath clouds. It is especially under these conditions that the reliability of e-vector detection is the higher, the lower the wavelength of light.[14]

The use of ultraviolet light in polarization vision is just one secret the cataglyphs have disclosed. Another is that this sensory capacity resides in a small, sharply delineated region of the ants' compound eyes. In fact, only those UV receptors located at the uppermost dorsal rim of the eye—in what we have called the dorsal rim area (DRA)—are able to detect polarized skylight. Seen from the outside, the location of the DRA can be recognized by a small depression in the array of corneal facets, but otherwise there is nothing special or peculiar about this DRA, which comprises only 60–70 of the about 1,000 visual units (ommatidia) of the *Cataglyphis* compound eye. As we shall see later, hidden underneath the corneal facets there are special neural machineries perfectly tuned to detect and process polarized light in the most efficient way, but definite proof that the dorsal rim area—and only the dorsal rim area—is used for e-vector navigation must come from behavioral experiments. Such experiments indeed show that the cataglyphs can no longer perceive polarized skylight when the DRAs of their two compound eyes have been occluded (meaning that the DRA is *necessary* for e-vector detection), and that they are still able to do so when only the DRAs have been left open (meaning that the DRA is *sufficient* for e-vector detection).

Let us first focus on the criterion of sufficiency. In this case we must provide the cataglyphs with light-tight eye caps that cover their entire eyes except the DRAs, so that the ants while walking underneath the trolley are forced to steer their courses with only 6%–7% of their visual units functioning—and yet, they highly significantly select their home directions. I do not present these data here, because there is a snag. Do the ants tested under the conditions described above really rely on the e-vector pattern in the sky rather than on spectral or intensity gradients that they are still able to perceive? In order to answer this question, we let the ants look at the polarized light in the sky only through a small circular window (Figure 3.9e). Outside this skylight window all information about the polarization and spectral gradients in the sky is blocked by embedding the window into a large orange-colored, UV-absorbing cutoff filter. Even inside the window the ants are not presented with the natural sky. Instead, an artificial polarizer built into the window enables the experimenter to deliberately change the orientation of the e-vector in the patch of sky seen by the ant. The rationale is that if a particular e-vector is presented at an azimuthal distance of, say, 50° clockwise from where it occurs in the natural sky, the ants should deviate by 50° clockwise from their homeward direction. And this is what they actually do.[15] Given the extremely restricted stimulus situation—tiny parts of the eyes uncovered and a tiny skylight window available—the larger than usual scatter in the data is not astounding.

The same rationale can be applied in assessing the necessity of the DRA for e-vector detection. Now, however, we must occlude the DRA alone, so that more than 90% of the ants' compound eyes remain open and fully functional, but otherwise the very same experimental procedures are applied. Figure 3.9f shows the result. With their DRAs occluded, the ants no longer select the homeward courses that they are expected to take on the basis of the artificial e-vector presented to them. Nevertheless, they are not oriented at random. They exhibit two broad preferences: one in the direction of the skylight window and another more pronounced one in the direction of the sun—or, to be precise, in the direction of the sun's azimuthal position, because as in

3.10 With the DRA covered, the polarized-light compass ceases to operate

Exemplary path of an ant released at *R* under the conditions of Figure 3.9f. *0°,* goal (nest) direction; *AP,* direction of skylight window (aperture) provided with a polarization filter; *SA,* direction of solar azimuth. The sun itself is shielded off. Time marks *(black dots)* every 10 s. *Cataglyphis fortis.*

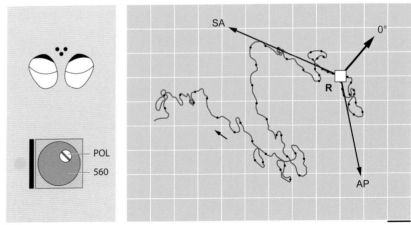

all experiments described so far the sun itself is shielded off. Nevertheless, the ants are able to determine the center of gravity of the light-intensity distribution caused by the sun and perceived through the orange-colored cutoff filter. With their DRA vision abolished and with no other reliable compass cue available, the ants seem to exhibit some kind of phototactic response toward the brightest points in their visual surroundings (for an exemplary path, see Figure 3.10).

We conclude that the ultraviolet receptors of the dorsal rim area are necessary and sufficient for steering compass courses by polarized skylight. Moreover, switching eye caps from one compound eye to the other shows that interocular transfer occurs. Compass information acquired during the outbound run by one eye can be retrieved during the subsequent inbound run by the other eye.[16] Finally, are there special parts of the visual field covered by the dorsal rim areas that the cataglyphs preferentially use for sky-compass navigation? The following experiment may help to answer this question. When the trolley is equipped with a shade that extends by different amounts from below or above into the ant's field of view, the cataglyphs perform remarkable head and body movements. Film recordings show that the ants tilt their head and alitrunk upward and downward when the boundary of the dark screen extends into the ant's field of view by more than 45° upward from the horizon and more than 30° down-

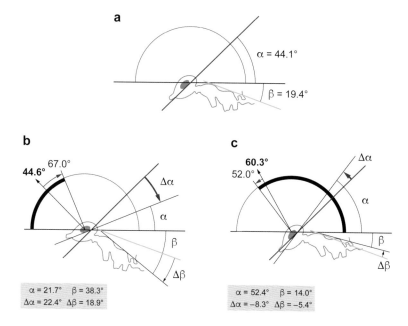

3.11 The cataglyphs perform compensatory head and body movements when their view of the sky is restricted from above or below

The ants walk underneath the trolley. The *heavy black arc* marks the part of the visual field that is shielded by a light-tight screen. (**a**) Normal body posture during walking. The *red* and *green lines* depict the tilt angles α and β of head and alitrunk, respectively. (**b, c**) Body postures after the screen is inserted from below or above in the ant's field of view. For example, when the screen extends up to 67° from the horizontal *(b)*, the ants tilt upward by a mean angle of $\Delta\alpha = 22.4°$, meaning that the lower margin of their preferred compass region in the eye is 44.6°. As a result, the preferred eye region involved in the task extends from 44.6° to 60.3° above the horizon. The postural changes ($\Delta\alpha$) are achieved mainly by raising and lowering the alitrunk ($\Delta\beta$) rather than tilting the head relative to the alitrunk. *Cataglyphis bicolor.*

ward from the zenith, respectively (Figure 3.11).[17] From these and other experiments we can conclude that the cataglyphs preferentially use areas of their compound eyes that are centered about 50°–55° above the horizon and 45° sideways of the median body plane and thus in the frontal parts of their dorsal rim areas.[18] When skylight windows are offered in these positions, the ants perform their straightest and thus fastest homeward runs. In other positions of the window, especially when the window is presented opposite to the goal direction, the ants regularly engage in turning movements so as to look at the patch of sky alternately with one or the other eye. When a zenith-centered patch is offered, they straighten up and raise their body as far as they can. Sometimes they even somersault backward.

Next we ask what strategy the ants may use in reading compass information from the celestial e-vector patterns. Let us start with some theoretical considerations about how one could proceed, if the sun were hidden by clouds and only small parts of the celestial e-vector pattern were available. Historically, the insect navigator, in this case the honeybee, was considered to solve the compass problem in a two-step way: first measure the orientation of the e-vector in individual

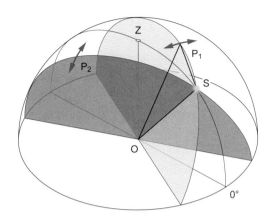

3.12 The basic rule of skylight polarization

As the theory of Rayleigh scattering predicts, in each point of the sky light is polarized perpendicularly to the plane of scattering, which contains the sun *(S)*, the observer *(O)*, and the observed point *(P)*. Hence, even if the sun is hidden by clouds, its position can be determined as the intersection of at least two such planes that pass perpendicularly through the e-vectors *(blue double-headed arrows)* in two pixels *(P₁, P₂)* of unobscured sky. Simply put, light is polarized at right angles to the triangle *POS*.

pixels of the celestial dome, and then apply some rules about the geometry of the celestial e-vector pattern to determine the position of the sun (Figure 3.12).[19] The cataglyphs, however, do not seem to follow such a computationally demanding two-step procedure. Neither do they apply a mere matching-to-memory strategy by remembering the pattern last seen and later trying to match the parts of the pattern currently available with the remembered one. Evidence for rejecting both of these hypotheses can be derived from experiments in which the cataglyphs are presented with the full e-vector pattern during training, but with only limited parts of the pattern in the test situation, or vice versa. Surprisingly, in this asymmetric training-and-test paradigm the ants make distinct and often quite substantial, but systematically varying, mistakes in selecting their homeward courses, and may thus give us a glimpse into their compass strategy.

In these tests the ants perform their outbound (training) runs within a narrow channel that restricts their skylight vision to a strip-like celestial window ('partial-sky condition'), and perform their inbound runs in an open test field ('full-sky condition'). In this situation the ants deviate systematically either to the right or to the left from the true homeward courses, with the sign and amount of these deviations depending on what part of the sky the ants have seen on their channel-based outbound runs (Figure 3.13a).[20] Yet, no errors occur when during the outbound runs the symmetry plane of the sky has been parallel to the axis of the channel. The ants' erroneous behavior is astonishing, as during the test the ants have access to the complete e-vector pattern spanning the celestial dome. Obviously, when trained under the partial-sky conditions the animals have set their compass in some 'wrong' way. If they followed the human theoretician's advice outlined above (Figure 3.12), navigational errors might not occur.

The cataglyphs' navigational errors leave us with at least two major puzzles. First, what do the experimentally induced errors actually mean in the ants' real-world behavior? Partial-skylight conditions as the ones created by our channel paradigm might easily occur in natural situations. If under such conditions the cataglyphs started their long-distance homebound runs with an initial error angle of, say, 20° to the left or right of their true homeward course, they would miss their goal, the nest, by quite some substantial amount. But in the thousands of natural foraging paths that we have recorded over the years in several *Cataglyphis* species, we have never observed such large compass errors. This is, of course, for a simple reason. Recall that in the experiments described so far the ants have been presented with a partial pattern only during one part (the outbound part) of their foraging journeys. During the other part (the inbound part) they could see the full pattern. Next imagine that we would perform an experiment in which the ants were provided with the same patch of polarized skylight during both their inbound and their outbound runs. Then no systematic errors should occur—and none are observed.[21] For example, when the ants are again trained within a channel but later tested with our trolley device that provides them with the same strip of sky which they have previously seen in the channel, the systematic errors disappear (see the data point depicted by an open circle in Figure 3.13a). Hence, on their foraging journeys the cataglyphs would be in trouble only if the atmospheric conditions—heavy clouding or haze—changed to such an extent that the ants' compass changed its zero setting while the ants are on their way. However, given the rather short durations of the ants' foraging endeavors, such dramatic changes in the skylight conditions would be a rather rare phenomenon. Further recall that under certain atmospheric conditions the e-vector pattern in the open sky even extends underneath the clouds.

The second question is much more challenging. Why do navigational errors occur at all when the ants are presented with this particular

3.13 Experimentally induced compass errors

(a) Strip-like sky windows. *Top:* Experimental setup. δ, angular deviation of the symmetry plane *(red)* of the e-vector pattern from the axis of the sky window. *Bottom:* Results. Error angles (ε), by which the ants deviate from their homeward course *(filled circles)*. The *sinusoidal green line* represents the prediction based on the results of *(b)*. *Open circle:* Ants tested with the very skylight window that they have experienced during training. *Cataglyphis fortis.* (b) Individual e-vectors. *Top:* Experimental setup. φ, e-vector orientation; β, azimuthal position of the e-vector with respect to the solar meridian. *Bottom:* Results. Azimuthal positions at which ants *(large open circles)* and bees *(small filled circles)* assume any e-vector to occur. The *light blue areas* denote the locations at which the e-vectors actually occurred in the sky during the course of the experiments. The *green lines* depict the e-vector distribution at sunrise and sunset. The azimuthal range *90°–270°* marks the antisolar half of the celestial hemisphere. *Cataglyphis bicolor* and *Apis mellifera.*

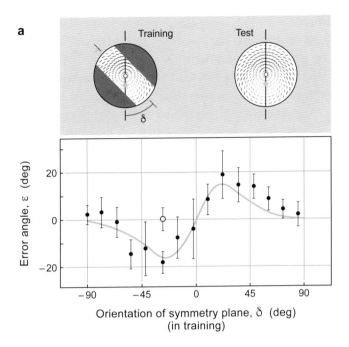

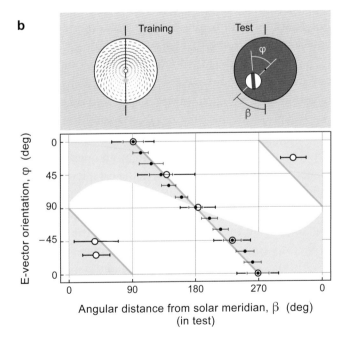

training-and-test situation, and what do these errors tell us about the compass strategy employed by the insect navigators? To come to grips with this question, at least from the behavioral side, we let the ants perform their outbound runs under the full sky, but restrict their vision during the inbound runs to only a single e-vector. From the navigational errors then observed, we can conclude where in the sky the animals expect any particular e-vector to occur (Figure 3.13b).

Before we interpret these results, let us look at studies with honeybees, in which we have applied the same experimental procedures. Of course, for obvious experimental reasons the bees are not followed, and their visual surroundings are not manipulated, while they are flying in the field, but while they perform their recruitment ('waggle') dances in the hive. As this indoor figure-eight waggle dance encodes the direction and distance to an outdoor food source, we can derive from it the same information that we would gain in recording a bee's actual flight path. The horizontal comb, on which the bees dance, is vaulted by a translucent Perspex dome, which allows us to present the bees, which have been trained under the full blue sky to visit a particular feeding site, with small 10°-wide pixels of skylight as well as with polarization filters. The bees' responses, i.e., the directions of their waggle dances, are recorded with a video camera mounted in the zenith of the Perspex dome.[22]

Such experiments elaborately performed by my collaborator Samuel Rossel are quite time-consuming, because celestial e-vectors must be presented to the animals at various positions in the sky and for various elevations of the sun, but the result is charmingly simple. Irrespective of the actual positions of the e-vectors in the sky at the time of the experiment, and irrespective of the elevation of the sun, bees and ants assume all e-vectors to occur at fixed positions relative to the symmetry plane of the global pattern. As in reality, the horizontal e-vectors define the symmetry plane of the pattern, but in contrast to reality, the vertical e-vectors are positioned invariably at right angles to that plane. All other e-vectors lie in fixed positions in between. The roughly linear φ/β relation derived from the experimental data (Figure 3.13b) represents the most uniform e-vector pattern that occurs in the sky when the sun is at the horizon (see Figure 3.4a). In descriptive terms, this pattern could provide a template used for comparisons with the actual e-vector information available in the sky. When we use this template for computing the theoretical error function depicted in Figure 3.13a, this function describes the actual experimental data surprisingly well.

Here I must add an important information: all individual e-vectors presented by means of polarization filters are invariably interpreted as lying in the antisolar half of the sky. However, if a circular aperture within the filter trolley is used to present them in the natural sky, in either the solar (lowly polarized) or antisolar (highly polarized) half, the cataglyphs expect them to lie in the correct half of the sky, but within this half not exactly in the positions where they actually occur at the time of the experiment. Again, as mentioned above, the ants behave as if they decided on the basis of the most uniform e-vector pattern. For disambiguating the two halves of the sky, they could rely on spectral and intensity gradients. Spanning the celestial hemisphere, these gradients provide additional compass cues. We shall turn to them in the next section.[23]

For the moment, let us ask a question that at the outset might appear a bit peculiar. Do the ocelli, this set of three small single-lens eyes on top of the head (see Figures 1.1 and 3.14a), play a role in this context? In the experiments about the function of the dorsal rim area we have always occluded the ocelli (Figure 3.9e, f), even though there was no immediate reason to assume that the cataglyphs would employ these 'simple eyes' as some kind of sky (or even polarized-light) compass. In a number of classical experiments, honeybees and wood ants did not exhibit any consistent visual behavior when their compound eyes had been blinded but their ocelli left intact.[24] As late as in the 1970s Martin Wilson, to whom we owe major contributions to our present understanding of the optical and physiological organization of insect ocelli, bluntly remarked that up to now "ocellar function is unknown," and still three decades later Eric Warrant and his colleagues started their in-depth study on the ocelli of bees and wasps with the remark that the "exact role [of the ocelli] is still a matter of conjecture."[25] Ocelli occur more likely in flying than in wingless insects, and it is in the flying ones, especially in dragonflies and locusts, that one important functional role of the ocelli has recently become apparent: the role of a horizon detector, which can con-

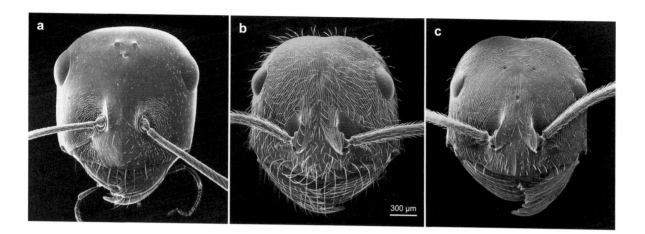

300 µm

veniently be used to maintain flight equilibrium by stabilizing body posture against inadvertent roll and pitch movements. The horizontal directions of view—the medial ocellus looks forward and the two lateral ones at right angles to the left and right—as well as the wide visual fields and defocused optics resulting in blurred vision perfectly support the function to keep the gaze level.[26]

In the cataglyphs ocelli occur not only in the winged sexual forms but also in the wingless workers (Figure 3.14a). Much to our surprise these cursorial foragers are still able to select their homeward courses when their compound eyes are fully occluded, and when they are tested under the filter trolley with the sun shielded off (Figure 3.15c, test 2). Even if they have access only to an individual e-vector presented by a polarizer within the filter trolley (Figure 3.15d, analogous to the DRA test paradigm depicted in Figure 3.9e), they select their homeward courses pertaining to the experimentally displayed e-vector.[27] However, in these situations, the ants left alone with their ocelli walk very slowly and continually turn to the left and right, as if they scanned the sky. Extremely tortuous walking paths result.

At present we can only speculate about how the cataglyphs accomplish this task. Surely, their ocellar ultraviolet receptors (Figure 3.15a)—the only spectral type of receptor that occurs in the median and lateral ocelli of *Cataglyphis*—are highly sensitive to polarized light,[28] but the orientation of the rhabdomeres within the retina does not exhibit an overt regular sampling matrix, at least not one that is as strict as

3.14 Presence and absence of ocelli in desert ants

Frontal views of the heads of (**a**) a worker of *Cataglyphis fortis* endowed with a triplet of ocelli, (**b**) a worker of *Ocymyrmex robustior* lacking ocelli, and (**c**) an ergatoid female of *O. robustior* exhibiting vestigial signs of ocelli.

3.15 The cataglyph's ocelli: visual properties and behavioral significance

(a) Spectral sensitivity of ocellar photoreceptors: polarization-sensitive ultraviolet *(UV)* receptors. (b) Fields of view *(purple)* of the three ocelli as defined by the half width of the angular sensitivity function (based on electroretino-gram recordings). *Red lines* mark the boundaries of the binocular field of view of the compound eyes. *Hatched pink area:* Dorsal rim area (DRA); *F,* frontal. (c) Homeward courses steered by animals in which either the ocelli *(1)* or the compound eyes *(2)* are occluded. (d) Ants with only their ocelli left open are presented with an individual (horizontal) e-vector in the position *AP*. In all experiments the sun is shielded off; for spectral characteristics of the filters, see Figure 3.9h; *0°,* goal (nest) direction; *AP,* mean azimuthal position of e-vector. *Cataglyphis bicolor.*

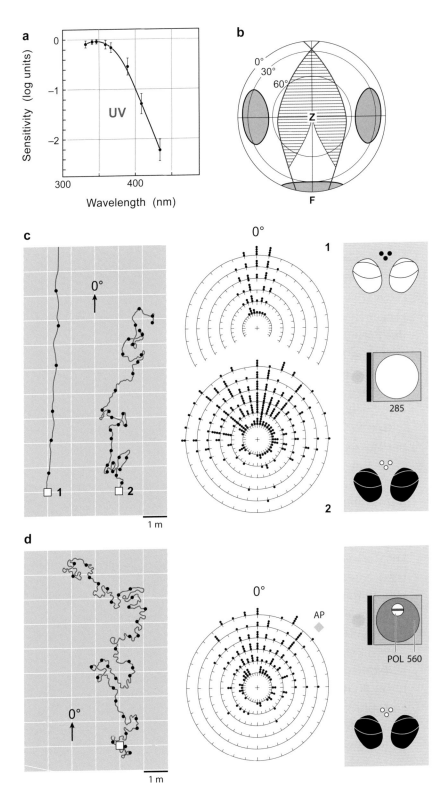

that in the dorsal rim area of the ant's compound eyes.[29] After all, providing sky-compass information might not be the primary function of the cataglyphs' ocelli. In some flying hymenopterans, and in the workers of at least one species of ant, the lenses and retinae are bipartite, with the ventral retina looking at the sky and the dorsal one toward the skyline.[30]

As to other potential functions, it should not go unheeded what Eugen Link remarked long ago: investigating the functional role of the ocelli in isolation by painting over the compound eyes will barely help in disclosing the full function of these *Stirnaugen* (frontal eyes), but "will do nothing but cast a shadow on them." Indeed, as recently found in flies, some ocellar interneurons enter the optic-lobe pathway, so that the high-speed ocelli may complement the high-acuity compound eyes in the controlling balance and gaze.[31] Hence, do the cataglyphs use this dual system for stabilizing the head, at least intermittently, about the roll and pitch axis, so that the skyline and the celestial hemisphere could be reliably projected onto the compound eyes?[32] Due to their large lenses and low f-numbers, ocelli are especially light sensitive and thus might be used in measuring ambient light intensity levels. Such a photometric function could enable them to control the timing of daily foraging activity or, more generally, to adjust the sensitivity of the compound eyes.[33] In any way, whatever function the ocelli might fulfill in *Cataglyphis,* we should not overlook that the *Ocymyrmex* navigators must get along without them (Figure 3.14b).[34]

At the end of this section, let us recall that Dominique Arago discovered polarized skylight two centuries ago. The discovery that certain groups of animals can detect this celestial phenomenon dates back only about half a century. When at all did animals discover polarized light in their environment and start to use it as a sensory cue? This question, by which we try to dig back into the evolutionary history of polarization vision, is difficult to answer. All we can say is that polarization vision did not evolve as a means of steering compass courses in the terrestrial world. It evolved underwater. Similar to air molecules in the atmosphere, water molecules in the hydrosphere scatter light and render it partially polarized. Even though these polarized underwater light fields exhibit rather small degrees of polarization (usually in the range of 20%–40%), they are omnipresent in the visual sceneries of aquatic animals. Recent studies show that they are actually exploited by polarization-sensitive visual systems in various groups of animals, generally for enhancing the contrast between objects and background and thus cutting through the 'veil of brightness' of the underwater space light.[35]

As rhabdomeric photoreceptors are intrinsically dichroic and thus polarization sensitive, the 'discovery' of polarized light by the predecessors of present-day arthropods may well date back for more than 500 million years to mid-Cambrian times. In those days the sunlit, well-oxygenated shallow seafloors on the margins of the barren, uninhabited continents were populated by a diverse variety of swimming, walking, or crawling proto-arthropods. These animals have been beautifully preserved in the Burgess Shale Formation of the Canadian Rocky Mountains, and also at several other places, so that they certainly had a worldwide distribution.[36] Many of them including the abundant trilobites—the 'butterflies of the seas'—possessed well-developed compound eyes, and even though retinal structures have not been preserved in the fossils, these eyes were most certainly equipped with some kind of rhabdomeric photoreceptors.[37] Hence, underwater polarization was probably exploited from early on in arthropod vision, millions of years before insects used the preexisting dichroic photoreceptors and built them into particularly designed polarization-sensitive visual pathways. However, there is more to the cataglyphs' sky compass than just using polarized light.

Spectral / Intensity Cues in the Sky

Once again, we did not expect at all what we observe now: Even when the dorsal rim areas of the cataglyphs' eyes and the ocelli are occluded, so that the ants become blind to polarized light, and even when the sun is shielded off, the ants are still able to derive compass information from the sky. In these experiments the homebound animals again walk underneath the experimental trolley, but now their view of the sky is restricted to a narrow, 10°-wide annular region in the lower part of the celestial hemisphere (Figure 3.16a). Under these conditions they can refer only to celestial intensity and spectral gradients, which are generated by the scattering of sunlight in the earth's atmosphere

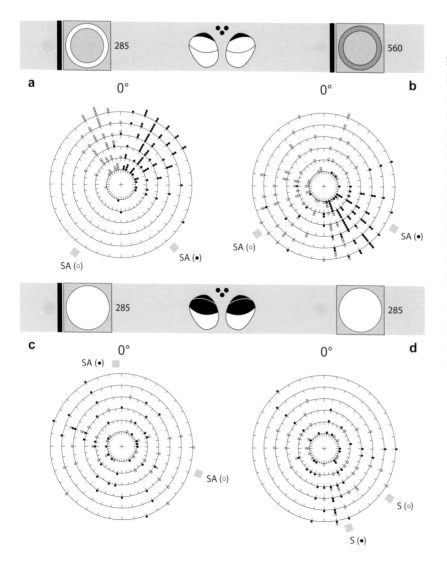

3.16 Orientation of ants with the polarization channel shut off

(**a, b**) DRA and ocelli painted over and sun occluded. Skylight view restricted to a narrow 360° wide angular window. Either the ants' UV and green receptors *(a)* or only the green receptors *(b)* are stimulated. (**c, d**) The entire dorsal halves of both compound eyes and the ocelli are occluded. The ants have visual access to nearly the entire celestial hemisphere, in which the sun is either shielded *(c)* or visible *(d)*. *0°*, goal (nest) direction. The *small yellow squares* mark the mean azimuthal positions of the solar azimuth *(SA)* or the sun *(S)*. For spectral characteristics of filters 285 and 560, see Figure 3.9h. (**a, b**) *Cataglyphis fortis,* (**c, d**) *C. bicolor.*

alongside the polarization gradients (the e-vector patters) considered so far.

Due to the physics of light scattering, skylight is the more dominated by short-wavelength radiation, the farther one moves away from the sun.[38] This can be easily read off Figure 3.17a. As the cataglyphs are equipped with short-wavelength (ultraviolet) and long-wavelength (green) receptors, they can exploit this source of information. Just as-

3.17 Spectral/intensity gradients in the sky provide the cataglyphs with compass information

(a) Radiant intensity (radiative transfer through the atmosphere) at wavelengths of 312 nm and 644 nm as a function of the azimuthal position in the sky. Data are given for an elevation of 31° (elevation of sun, 67°), but similar relations hold for other skylight conditions. (b) Response of a theoretical 'spectral opponent neuron' (S-neuron), which receives antagonistic input from the ant's polarization-insensitive ultraviolet and green receptors along eye meridians. The points in the sky used for the computation of the S-neuron responses are positioned on seven parallels of longitude from the solar meridian (0°) to the antisolar meridian (180°) and added up over four parallels of latitude.

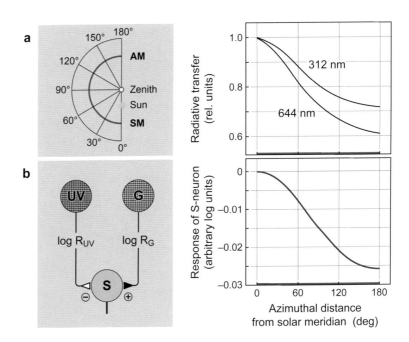

sume that the outputs of the two types of photoreceptor (Figure 3.9g) would interact antagonistically in the way shown in Figure 3.17b (left). Then, computations based on the spectral properties of the sky and the spectral sensitivities of the photoreceptors show that the responses of an underlying 'spectral opponent unit' sampling inputs along celestial meridians would vary systematically, when the animal shifts its direction of gaze through all points of the compass from the solar to the antisolar meridian (Figure 3.17b, right).[39] As in the behavioral experiments the dorsal rim areas of the eyes have been occluded, the polarization-insensitive parts of the eyes must be sufficient to perform that task. Yet there is something peculiar: The ants always deviate to a certain extent from their home direction toward the azimuthal position of the sun, the brightest part in their annular skylight window, the solar meridian (Figure 3.16a). Indeed, as a closer inspection of their walking trajectories shows, they seem to be in a constant conflict between exhibiting this kind of phototactic response toward the solar meridian and choosing their correct homebound course. They incessantly alternate between the two directions.[40]

Next, we go one step further and exclude not only the polarization contrast but also the spectral contrast. This is done by providing the annular skylight window with a short-wavelength cutoff filter, which stimulates exclusively the ants' green receptors. Under these extremely restricted stimulus conditions the phototactic response dominates (Figure 3.16b). Apparently, the ants can still determine the azimuthal position of the sun as the brightest point in their surroundings, but this information is not suitable to be used as a sun compass cue. However, when the sun itself becomes visible under such conditions, and especially when the ant's view of the sky is extended from the annular window to nearly the entire celestial hemisphere, the sun seen through the orange cutoff filter enables the ants to select their proper homeward course, but slight deviations toward the sun's azimuthal position always remain. What, then, is the role the sun plays in the ant's celestial system of navigation, and under what conditions does it play a role?

The Sun

The sun is a uniquely characterized point in the sky. It is unique in being not only the brightest point, but also the point that exhibits zero polarization and the highest ratio of long- to short-wavelength radiation. In fact, it is the pole of all skylight patterns. Why, then, does it come so late in our account on the ant's celestial compass system? Was it not the sun compass that had been discovered first as a means of deriving compass information from the sky, and was it not an ant—actually a harvester ant, *Messor barbarus*—in which this discovery had been made in the first place? Yes, both are true, but when more than a century ago Felix Santschi did his famous mirror experiment, in which he screened off the sun and reflected it with a mirror from the other side, he already noticed that the ants did not respond all the time.[41] The experiment worked much more easily in some genera of ants (e.g., in the trail-laying harvesters *Messor* and *Monomorium*) than in others (e.g., in the individual forager *Cataglyphis*), and it seemed to work better under some celestial and meteorological conditions than in others, e.g., when the sun was low in the sky and when the wind had stopped blowing. Even though these remarks were based only on a few behavioral observations, they already showed that the sun compass is only one part within an integrated system of navigation, and that this part plays different roles depending on the prevailing stimulus conditions and the foraging ecology of the species in question.

When it comes to the usefulness of the sun as a compass cue, there are at least two constraints that should be mentioned at the outset. First, large clouds may obscure the sun and shift the center of gravity of the light intensity distribution. The sky-wide polarization pattern offers spatially extended, more global information. Moreover, the polarized-light compass is, as we have seen, quite robust against local cloud covers and other atmospheric perturbations. Second, for obvious geometrical reasons, the precision with which the azimuthal position of the sun can be determined deteriorates the more, the higher the sun is in the sky (Figure 3.18a, b).[42] This is indeed what we observe in the cataglyphs, when they are tested under conditions under which they can rely only on their sun compass. Moreover, when in our experimental trolley the image of the sun is mirrored by only 20° upward or downward from the training situation (Figure 3.18d), the precision of the ants' compass readings decreases or increases, respectively.[43]

We may finally wonder how the ants behave when they are presented with conflicting information from the sun and polarized light in the sky. Such a conflict situation can be achieved in a rather tricky experiment, in which the naturally constant spatial relation between the solar azimuth and the e-vector in the zenith (which are always at right angles to each other) get experimentally decoupled. In these experiments the ants are trained to perform their outbound runs in a straight channel covered by a linear polarizer, which reduces the polarization pattern to a single e-vector overhead. During the course of the day the azimuthal position of the sun changes relative to this stationary e-vector through a wide range of angles, and provides the animals with constantly varying conflicting information. In their homeward runs in the open test field, the ants predominantly choose intermediate courses (Figure 3.19), but sometimes exhibit marked zigzagging behavior by alternately relying on one or the other com-

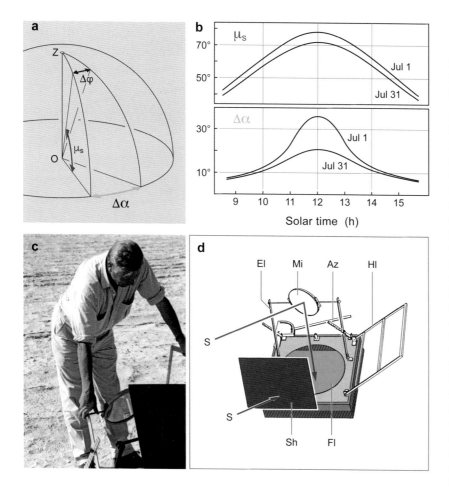

3.18 The sun compass becomes the less precise, the higher the sun is in the sky

(a) Relation between the sampling interval of the ant's compound eyes (divergence angle $\Delta\varphi$), the azimuthal difference $\Delta\alpha$ that can be resolved by $\Delta\varphi$, and the elevation of the sun (μ_s). O, observer; Z, zenith. (b) The daily course of the elevation of the sun (*top*) and the precision (reciprocal of $\Delta\alpha$) with which *Cataglyphis bicolor* can determine the azimuthal position of the sun (*bottom*) in the month of July at our North African study site at Mahrès, Tunisia. (c) Sir David Attenborough inspecting our experimental trolley during his work on the BBC nature documentary series *Trials of Life* at our Mahrès study site, July 1988. (d) Trolley designed for presenting the ants with various elevations of the sun. *Az, El,* devices for adjusting solar azimuth and elevation, respectively; *Fl,* spectral cutoff filter; *Hl,* handle for moving the trolley; *Mi,* mirror; *S,* direction of sunlight; *Sh* sun shield.

pass cue. Note that in these experiments the ants have been presented with a rather reduced 'polarization pattern,' and that their reliance on the sun or the polarized light in the sky may vary with the amount of e-vector information available in the experimental situations. Cue-conflict experiments performed in the Australian *Melophorus bagoti* also show that the weight ascribed to a celestial cue depends on its salience. When the polarization and/or spectral and intensity gradients are gradually weakened, the ants rely increasingly on the sun (reflected by 180° from its normal position). Furthermore, elaborate experiments in the cataglyphs provide clear

3.19 Cue conflict: sun compass versus polarized-light compass

Left: Training from nest *(N)* to feeder *(F)* in a channel covered by a polarization filter with its transmission axis *(blue double-headed arrow)* parallel to the channel axis, and with the sun *(S)* visible. α and β denote homeward directions indicated by the polarized-light compass and the sun compass, respectively. *Right:* Home directions in the test field for different training situations *(β–α)*. *Blue and orange lines* indicate the exclusive use of the polarization and the sun compass, respectively. *Cataglyphis fortis.*

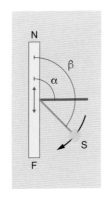

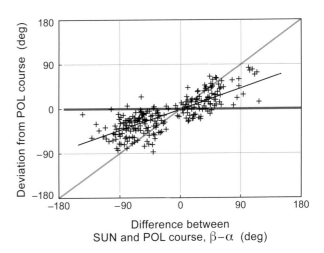

evidence that information obtained by the sun (in a polarization-free situation) can later be recalled by reference to polarized skylight (in a sun-free situation), and vice versa.[44]

We end this section by emphasizing that all celestial information is taken up and processed only by the dorsal part of the cataglyphs' visual system. When the entire dorsal halves of both eyes are experimentally occluded, the ants are no longer able to derive any compass information from the sky (Figure 3.16c, d). Even though all polarization, spectral, and intensity cues are still fully available, left alone with the ventral halves of their eyes the ants are unable to use these cues. This finding is all the more astounding as the ants tested under these conditions perform strong roll and pitch movements of their heads and thus, time and again, look at the sky with the ventral halves of their eyes, but this does not help. Obviously, the ventral part of the ant's visual system lacks the neural machinery for compass orientation by skylight cues. This conclusion is corroborated most impressively by the result shown in Figure 3.16d. In this experiment, in which the ants are presented with the entire sky including the sun, they tend to move toward the sun, which they perceive when they look at the sky with the ventral regions of their eyes, but they do not incorporate this information into their celestial compass system.

The Rotating Sky

Having considered the various compass cues that the cataglyphs derive from the celestial hemisphere, we now must return to the question asked in the very beginning of this chapter: What do the ants know about the rotation of the sky relative to an earthbound system of reference? The answer to this problem of time compensation is neither simple nor straightforward. Just have a look at the astronomical situation (Figure 3.20). Even though the sun moves along its arc through the sky with uniform velocity, with 360° per day, that is 15° per hour, its azimuthal position, which is used as the decisive compass cue, does not. Its rate of movement depends on three variables: time of day, time of year, and geographical latitude. These dependencies are expressed in the so-called solar ephemeris functions, which correlate the sun's azimuth with time of day. In Figure 3.20b they are depicted for the latitude of our experimental site at Mahrès, Tunisia. As a general rule, the azimuthal position of the sun changes slowly in the early morning and late afternoon, and relatively quickly near midday. After some early debate whether bees and ants can or cannot account for this celestial movement,[45] a suite of investigations in the 1950s convincingly showed that these insects as well as all other animal species studied in those days—crustaceans, spiders, and birds—were certainly able to do so.[46]

However, neither of these classical studies is detailed enough to allow for the conclusion that the animals have acquired fully time-compensated local ephemeris functions. In order to accomplish this task the animal must read time from an internal clock and correlate the time-linked positions of the sun with an earthbound system of reference. That an internal clock is at work has been demonstrated in bees, by time-shifting their clock or by training them at one geographical longitude and testing them at another one.[47] In either situation the bees did not orient in their correct home direction, and by this provided clear-cut evidence that they had not referred to any global earthbound system of reference based, for instance, on the earth's magnetic field. Nor had they used a global sky-bound reference such as the pole point around which the celestial sphere with all its visual cues rotates.[48] These and a number of companion studies provide ample

3.20 The daily movement of the sun across the celestial hemisphere at our Mahrès field site

(a) Three-dimensional representation: the sun's arc is given for its two extremes—the summer and winter solstice (21 June and 21 December)—as well as for the vernal and autumnal equinox (21 March and 23 September). $\Delta\alpha$, change in solar azimuth as the sun moves from position *1* to position *2*. (b) The solar ephemeris functions and (c) the solar azimuthal velocities for the times of the year indicated in *(a)*.

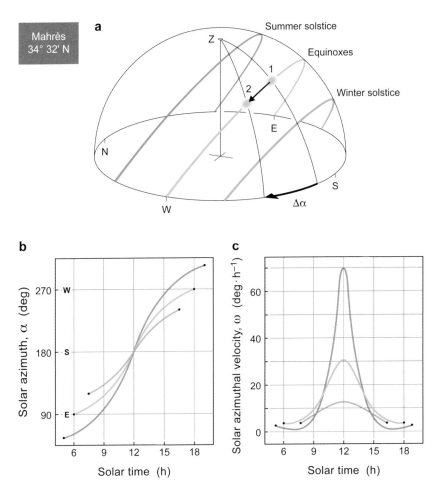

evidence that animals do account for the daily rotation of the sky, but what they know about the systematics of this rotation, and how they acquire this knowledge, has remained elusive.

On the one hand, it seems impossible to credit an ant or a bee with innate knowledge of all possible ephemeris functions that would provide the animal with a full set of data about the sun's daily movement for any particular time of year and any particular latitude at which the insect happens to start its foraging life. On the other hand, to acquire this knowledge by experience is by no means trivial either. Hence, it has even been assumed that bees and ants do not have any long-term knowledge about the sun's rate of movement at all, and just extrapolate linearly—by a running-average processing system—from the most recently observed rate.[49] This assumption based on an admittedly limited set of data could be refuted in the cataglyphs as well as in honeybees.[50] Drawing upon both evolutionary and individual experience, ants and bees seem to be provided with more detailed knowledge about the rotation of the sky than a linear extrapolation hypothesis suggests.

Assume that the ants, after having arrived at a feeder, are kept for some hours in the dark before they are released to perform their home run underneath the experimental trolley. If they are correctly informed about the azimuthal angle through which the sun has moved during the time interval between training and test, they should select their home direction accurately. Alternatively, if they deviate from their intended goal direction, say, clockwise by a certain error angle, a little geometrical consideration shows that they then assume the sun at an azimuthal position that deviates counterclockwise by this error angle from its actual position—or in other words, the ants have underestimated the movement of the sun by that very angle. By testing the ants under a variety of experimental conditions, e.g., by training them when the sun's azimuthal position moves slowly and testing them when it moves fast, and vice versa, one can clearly show that the cataglyphs compensate the movement of the sun fairly well (Figure 3.21a). However, there are still open questions. For example, why do some individuals tested as described in Figure 3.21b expect the sun to move in the counter-direction, i.e., through north rather than south across the sky. We shall return to this point later.

A central question bears on the earthbound system of reference used by the foragers to form a detailed memory of the sun's diurnal course. Likely candidates are the earth's magnetic field and the local landmark skyline. Direct evidence for the skyline hypothesis comes from an intriguing type of experiment, performed by Fred Dyer and James Gould at Princeton University, in honeybees.[51] The researchers 'shifted landmarks' in the bees' visual world, actually by displacing the hive to a new location, in which a line of natural landmarks appeared in a direction that differed from that in the training landscape. They then tested the bees in this rotated twin landscape under complete

3.21 Checking the ant's solar ephemeris function

(a) Ants trained either in the morning *(blue)* or at noon *(orange)* are tested in either the early or late afternoon, respectively. The symbols indicate at what azimuthal position they expect the sun to occur. *Solid black sigmoidal line:* Mean ephemeris function pertaining during the experimental period. *Colored lines:* Predictions according to the linear extrapolation hypothesis. *Cataglyphis bicolor* and *C. fortis.* (b) Derivative of ephemeris function *(solid line).* Data points indicate the ants' assumptions (under the condition of 1 h time intervals between training and test). *Cataglyphis fortis.*

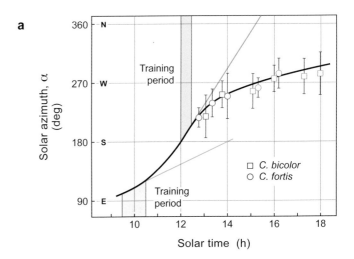

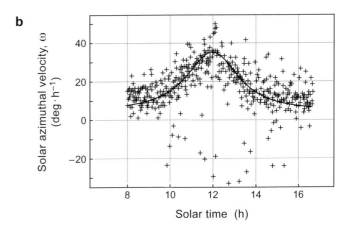

overcast conditions and inferred their knowledge about the position of the sun from their recruitment dances in the hive. This elegant experiment and a number of subsequent studies provide clear evidence that the bees detect and memorize the azimuthal position of the sun relative to the surrounding landmark panorama. More than two decades later William Towne and his collaborators at Kutztown University, Pennsylvania, took up this issue once again. They not only showed that the bees learn the sun's daily pattern of movement in relation to the entire landscape panorama, the skyline, rather than to individual landmarks, but also provided evidence that the bees are able to revise this relationship when they are transplanted to a new, panoramically dissimilar site.[52]

Next comes the question of how this spatiotemporal correlation occurs in the insect's mind. For example, bees and ants could construct the actual ephemeris function by fitting together a number of time-linked spatial data. However, without any preexisting knowledge about the general dynamics of the sun's daily course this would be a rather tedious enterprise. As the characteristics of the sun's movement conform to a general pattern, it is a more likely hypothesis that during the course of evolution the animals have incorporated the approximate shape of the solar ephemeris function as some kind of starting program into their neural compass system. In this respect, the most extensive studies have been performed by Fred Dyer and Jeffrey Dickinson in honeybees (Figure 3.22).[53] The researchers let incubator-reared bees forage at a feeder only in the late afternoon and prevented them from leaving the hive at other times of the day. After several days of training, they tested how these bees estimated the morning and midday course of the sun by allowing them to fly only when the sky was completely overcast. Even though the tested bees had not seen the sun, their recruitment dances in the hive indicated where they assumed the sun to be located in the sky.

The result is striking. The bees use an innate approximate ephemeris function that resembles a 180° step function. In this function the sun rises and sets at opposite points along the horizon, barely changes its azimuthal positions during the morning and afternoon hours, and 'jumps' at midday through 180° from the eastern to the western half of the sky. When later in foraging life the bees acquire more information at other times of day, this default function gets closer to the real ephemeris. Much of this learning process remains to be elucidated in both bees and ants. Questions remain, including to what extent are interpolation and extrapolation processes involved in shaping the actual ephemeris, or how and when do the foragers learn the direction of solar movement?

Note that in the Sahara the distributional ranges of several *Cataglyphis* species extend southward of the Tropic of Cancer, so that some populations live under conditions in which during parts of the year the sun moves counterclockwise rather than clockwise, i.e., from east

3.22 The 180° step function: an innate default representation of the solar ephemeris function

Solar positions estimated by incubator-raised honeybees that have seen the sky for several days only during 4 h training periods in the afternoon and are subsequently tested all day long under overcast conditions. The directions of their dances indicate their estimate of the sun's position *(crosses)* even at times of day under which they have never seen the sun before. A few bees assume the sun to move counterclockwise rather than clockwise *(open circles)*. The *colored lines* represent predictions of linear extrapolation and interpolation hypotheses: *1,* backward extrapolation from the sun's rate of movement at the beginning of the training period on previous days; *2,* interpolation between the sun's position at the beginning and end of the training period; *3,* forward extrapolation from the end of the training period on previous days. *Apis mellifera.*

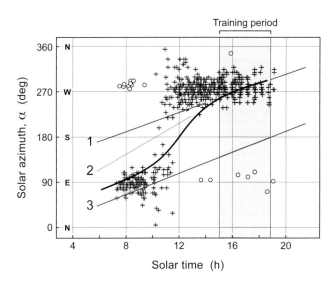

through north rather than south across the sky (Figure 3.23). We have not yet tested such populations, but we have been surprised to find that even in *Cataglyphis fortis,* which inhabits only areas north of the Tropic of Cancer, every now and then some individuals behave as if they expected a counterclockwise rotation of the sky (Figure 3.21b). This aberrant behavior, which occasionally has also been observed in honeybees starting their foraging lives (see Figure 3.22), indicates that it is also the direction of the sun's daily course that must be learned early in foraging life.[54]

A second look at Figure 3.23 provokes another and even more challenging hypothesis. The innate 180° step function seems to represent the mean of all possible ephemeris functions describing the sun's course in the geographical areas in which the cataglyphs occur. More globally, it is even the average of all these functions that can be observed at all latitudes and during all seasons.[55] Seen in this light, a cataglyph would start its foraging life with the best possible bet about the rotation of the sky. Whatever learning processes are later involved, they are built on an innate pattern that characterizes a general spatiotemporal correlation of the insect's environment.

This chapter began with the statement that skylight cues are ideally suited to serve as global compass cues for the simple reason that they

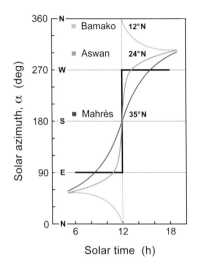

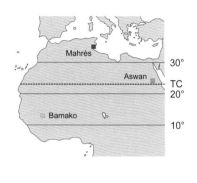

3.23 The daily movement of the sun at various North African sites inhabited by *Cataglyphis* species

Colored curves: The solar ephemeris functions as valid for Mahrès (Tunisia), Aswan (Egypt), and Bamako (Mali) on the summer solstice. Several *Cataglyphis* species occur as far south as Ivory Coast and southern Sudan. *Black curve:* Step function (solar ephemeris on the equinoxes at the equator); *TC*, Tropic of Cancer (23.4°N).

are virtually at infinity, and that this advantage must go along with the need for calibrating the 'rotation of the sky' against an earthbound system of reference such as the distant landmark panorama or the earth's magnetic field. Both compass cues could also be used later in foraging life, and so could the direction of the wind, if it were reliable over sufficiently long times. For that reason, we now turn to the geomagnetic field and regional wind regimes.

The Earth's Magnetic Field

When the cataglyphs switch from their indoor nursing to their outdoor foraging career, they perform short exploratory learning walks around their central place, the nest, in order to acquire the necessary landmark information that they later use when they return from their forays. As we will discuss in detail in Chapter 6, during these winding learning walks the ants often stop for 100–200 milliseconds and look back at the starting point, the nest entrance.[56] As the nest entrance itself is not visible to the ants, the animals must read out their path integrator to determine the goal direction. It was in this context that the earth's magnetic field came into play as a compass cue.

In the light of the previous findings, it was an obvious assumption that visual cues from the sky would provide the necessary compass information. However, this is not the case. When all relevant celestial cues—the sun and the polarization and spectral patterns—are blocked, the ants are still able to accurately adjust their gaze directions. They fail to do so only when Helmholtz coils are installed around the nest entrance (Figure 3.24a) and used to zero the horizontal component of the geomagnetic field. Whenever this occurs, the ants gaze in random directions, even though all celestial cues have remained unchanged. Moreover, when the horizontal field component is rotated through various angles relative to its natural orientation, e.g., through 90°, 180°, or −90°, the ants significantly gaze at the fictive position of the nest as defined by the altered magnetic field (Figures 3.24b and 3.25). From both sets of experiment we can conclude that the geomagnetic field provides the cataglyphs with the compass cue used in path integration during the ants' early learning walks.[57] Responses to manipulations of the earth's magnetic field have occasionally been observed in foragers of other species of ants and bees as well. For example, under certain conditions such manipulations can cause weaver ants and leaf-cutter ants when marching on their recruitment trails to deviate from their courses in a more or less reliable way, but these deviations largely vanish when celestial cues are available.[58] In contrast, the behavior described above for the learning walks of the cataglyphs is actually the first evidence that an insect can use the geomagnetic field as a compass cue that is both necessary and sufficient for accomplishing a particular—in this case path integration—task. Hence, the critical criteria of necessity and sufficiency, which we had found to be met by the dorsal rim area (DRA) of the cataglyphs' compound eyes for using polarized skylight as a compass cue, are also met in using the geomagnetic field as a compass cue during the cataglyphs' early learning walks.

This discovery raises some far-reaching questions. First, why do the ants employ a magnetic compass early in outdoor life? One may argue that the geomagnetic field could serve as a compass reference already during the ants' early indoor stages of life, when the animals must negotiate their ways through a labyrinth of subterranean gangways and tunnels, just as blind mole rats, which spend their entire lives underground, are proposed to do.[59] Upon first appearance above ground, the cataglyphs could maintain this compass reference for adjusting

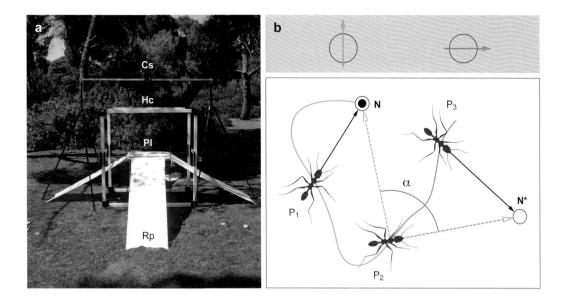

their turn-back behaviors to acquire goal-centered panoramic views, and they may even use it for calibrating their solar ephemeris function. As we had seen in the preceding section, honeybees most likely refer to the local landmark skyline as the geostable reference for calibrating the solar ephemeris, but in ants this is still an open question. In any case, the ants refrain from using their not yet time-compensated sky compass during their early learning walks. Second, why and exactly when do the cataglyphs switch from one to the other compass system? Does the sky compass outcompete the magnetic compass in terms of precision and speed of application? Moreover, do the ants continue to use the geomagnetic field as an additional compass cue employed in particular navigational tasks, e.g., in relearning walks performed later in foraging life (see Chapter 6)? Third and finally, how is the magnetic compass implemented in the cataglyph's sensory and neural system?[60] These exciting topics remain to be explored.

The Wind

In human navigation the wind compass preceded the magnetic compass. The various regional winds blowing across the Mediterranean

3.24 The cataglyph's magnetic compass

(a) Experimental setup within a pine forest of southern Greece. *Cs,* camera stand; *Hc,* Helmholtz coil system; *Pl,* experimental platform; *Rp,* ramp. (b) Sketch of a learning walk, in which the horizontal component of the geomagnetic field is rotated at point P_2 through $\alpha = 90°$. Path segments before and after the field manipulation are shown in *blue* and *orange,* respectively. Gaze directions are indicated by *ant icons* and *arrows.* The ant's gaze direction at P_2 is used to determine the position of N*. *N,* nest; *N*,* fictive position of nest after field manipulation. *Cataglyphis nodus.*

3.25 Rotation of the horizontal component of the magnetic field predictably changes the ants' gaze directions

The horizontal field component is rotated through various angles α *(orange)* relative to the natural situation *(blue)*. The circular diagrams depict the ants' gaze directions toward the real nest (direction 0°, *blue*) and the fictive position of the nest (directions 90°, 180°, and 270°, *orange*). The mean gaze directions are indicated by the *blue and orange arrows. Cataglyphis nodus.*

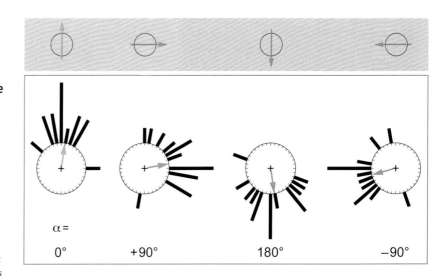

seas gave rise to the modern compass scale. At the Roman agora in Athens the octagonal Tower of the Winds, with its eight sides facing the points of the compass and its marble friezes portraying the eight gods of the winds, bears witness to the early importance of the wind compass. Until today, in French the compass rose is called *la rose des vents.*

In the coastline areas of our North African study site, where the ants' foraging grounds are extremely flat and largely bare of vegetation, the wind continually blows from a rather constant direction—at least during daytime hours, when the cataglyphs set out for their foraging journeys. This rather constant wind regime is due to the local land-and-sea breezes, which result in a periodical alteration between roughly southeasterly winds during daytime and north-northwesterly winds at night (Figure 3.26a). It allows us to expose the cataglyphs to different wind regimes in the training and test situations. When trained during daytime and later tested at night or in the early twilight hours at dawn, the ants deviate from their home direction by about the angular amount by which the direction of the wind has changed between day and night, but they immediately switch to the home direction, when 30–35 minutes before sunrise the polarization pattern in the sky becomes available as a compass cue.[61] The switch from wind-based to sky-based navigation occurs abruptly within a few minutes. It is not

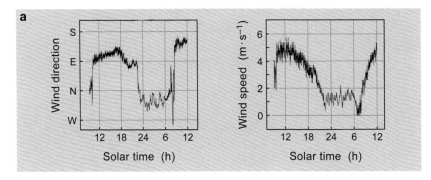

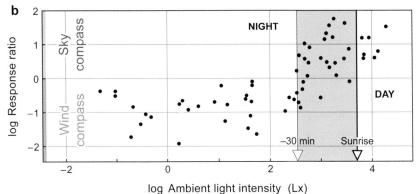

3.26 The cataglyphs' wind compass

(a) Daily course of wind direction and wind speed at the Mahrès study site. (b) The ants' use of wind-based and sky-based compass systems. The animals trained during daytime hours are tested at night, at dawn *(blue shading)*, and shortly after sunrise. The response ratio *(ordinate)* is a measure of how closely the ants' bearings correspond to the sky-based relative to the wind-based compass course. *Cataglyphis bicolor.*

influenced by the speed of the wind. During the short twilight transition period the ants repeatedly alternate between using their wind compass and their sky compass, between walking a few meters in the wind-defined direction and another few meters in the sky-defined direction (Figure 3.27). The corresponding behavior is observed at dusk. When the antennae are amputated or immobilized by applying tiny amounts of insect wax to their basal joints—the joints between head, scapus, and funiculus (see Figures 2.3a, 2.4b, and 7.7), where the mechanoreceptive wind detectors are located—all wind-based navigation is gone.

The twilight tests not only have led to the discovery of the cataglyphs' wind compass,[62] but also have provoked the question of how wind and sky-compass interact. In experiments performed during both twilight and diurnal times the polarized-light based sky compass outcompetes the wind compass, but when the cataglyphs are experi-

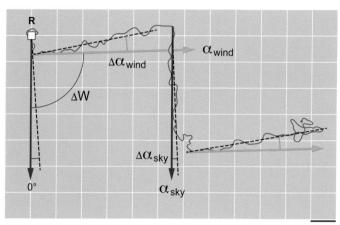

3.27 A test at dawn: wind and sky compass in competition

Example of an ant that switches repeatedly from using its wind-based and sky-based compass systems. This happens in a short period just when before sunrise the polarization pattern has appeared in the sky. *0°*, home direction during daytime; *R,* release point in the test field; *ΔW,* change of wind direction between day and night; $\Delta\alpha_{sky}$ and $\Delta\alpha_{wind}$, deviation of the ant's path *(blue)* from the sky-based *(red, α_{sky})* and the wind-based *(green, α_{wind})* compass course, respectively. *Cataglyphis bicolor.*

1 m

mentally prevented from perceiving polarized skylight and are left alone with the sun, the situation is different. Then the ants' behavior is dominated by the wind compass the more, the higher the sun is in the sky and hence the less precisely its azimuthal position can be defined (for the latter argument, see Figure 3.18a, b).[63] These results finally lead to questions about the relative significance and interdependencies of the various compass systems, how their information is weighted, and where they interact in the brain.

A First Look into the Cataglyph's Brain

This look is mainly concerned with the cataglyph's sky compass that is based on polarized light. We begin at the periphery, at the input stage within the compound eye. As the behavioral analysis has shown, the dorsal rim area (DRA) is necessary and sufficient for detecting polarized skylight. But what is so special about this DRA that it can accomplish this task?

In the cataglyphs there are at least five traits that render the DRA retina an efficient means of detecting polarized light in the sky.[64] First, in contrast to the remainder of the eye the microvilli of the DRA photoreceptors are strictly aligned for the entire length of the cell and thus ensure high polarization sensitivity (Figure 3.28b). Second, DRA rhabdoms are short (minimizing self-screening, which otherwise would jeopardize polarization sensitivity), but wide (securing high absolute sensitivity). Third, all polarization receptors are ultraviolet receptors and thus homochromatic (rendering the system 'color-blind'), and fourth, the ultraviolet receptors of each DRA ommatidium come in two sets, which have their microvilli arranged in mutually perpendicular ways. One set comprises receptors R2, R4, R6, and R8, which terminate with short visual fibers already in the lamina, while the receptors of the other set (R1 and R5) project with long visual fibers to the medulla (Figure 3.29a, b). Behavioral, physiological, and theoretical analyses consistently show how much the insect's polarization vision system benefits from this 'crossed analyzer' arrangement. By letting orthogonally tuned photoreceptors interact antagonistically, the polarization sensitivity of the detector gets enhanced and, given the right gain control of the inhibitory interactions, the polarization signal gets freed from responses to fluctuations in ambient light intensity. This means that each individual e-vector detector (consisting of two orthogonally arranged analyzer channels and an underlying comparator) responds exclusively to changes in e-vector orientation. Even though the very mini circuitry of this inhibitory interaction of DRA photoreceptors has not been uncovered yet, either in *Cataglyphis* or in any other insect, the polarization opponency resulting from this interaction is found in many neurons of the downstream polarization pathway (Figure 3.30c).[65] Further note that in *Cataglyphis* the two sets of polarized-light receptors contribute to the crossed-analyzer arrangement with remarkably different cross-sectional areas of their rhabdomeres. This asymmetric contribution of the two analyzer channels may enhance the e-vector tuning of the downstream polarization opponent neuron.[66]

Fifth and finally, within the DRA the e-vector detectors are arranged in a fanlike way (Figure 3.29d). This regular arrangement is an immediate consequence of the geometry of the compound eye and its ommatidial lattice. As the dorsal pole of the eye, from which the fan originates, does not look at the zenith, but deviates from it by about 30° toward the contralateral horizon, it draws, so to speak, the whole DRA receptor fan in the contralateral field of view. The DRA of the left eye looks in the ant's right visual field, and vice versa. As a result, each pixel of sky is viewed by both a polarization-sensitive visual unit (omma-

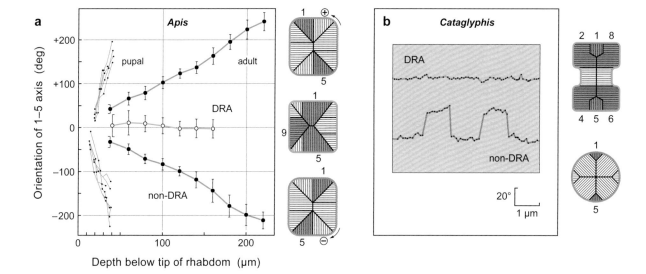

tidium) of the contralateral DRA and a polarization-insensitive unit of the ipsilateral retina outside the DRA. Our previous behavioral analyses had already shown that the cataglyphs can derive compass information from both the celestial polarization gradient (via the DRA) and the spectral/intensity gradients (via the non-DRA).

The information provided by the small-field polarized-light (e-vector) detectors in the retina gets spatially integrated in the optic lobes. It is already in the medulla—the largest and conspicuously multilayered visual neuropil—that wide-field polarizations-sensitive neurons sample the outputs of large parts of the DRA receptor fan. The prototypes of polarization-sensitive interneurons in the medulla, actually the first such neurons discovered in any insect, are the so-called POL1 neurons. They have been intensively studied in crickets (Figure 3.30), but similar ones have also been found in *Cataglyphis*.[67] When a linear polarizing filter is rotated within the visual field of these neurons, the spontaneous firing frequency is modulated sinusoidally with the activity peaks and troughs (φ_{max}-values and φ_{min}-values) lying 90° apart from each other (Figure 3.30b, c). The neurons do not respond either to variations in light intensity or to unpolarized light. They have wide visual fields and interconnect the ipsilateral and con-

3.28 Dorsal rim area (DRA) of the honeybee's and desert ant's compound eyes: photoreceptors have their microvilli strictly aligned.

Outside the dorsal rim area *(non-DRA)* the photoreceptors are twisted (**a**, in honeybees) or their microvilli are misaligned along the longitudinal axes of the cells (**b**, in *Cataglyphis bicolor*). *Insets:* Cross sections through rhabdoms with the UV receptors marked in *purple*. The UV receptors are provided with their conventional cell numbers. In the eyes of both bees and ants a short cell no. 9 occurs in the basal part of each ommatidium (not shown here), but in the DRA of the honeybee this cell no. 9 extends across the entire length of the ommatidium. In honeybees the receptor twist outside the DRA develops already in the *pupal* stage, before the rhabdoms are formed, and occurs either clockwise or counterclockwise, so that in the *adult* eye two mirror-image forms of rhabdoms occur.

tralateral medullae. Several types of similar medullar neurons have been described and characterized in locusts.[68]

Outside the DRA the first condition mentioned above—straight photoreceptors—is not met (Figure 3.28b). Along the lengths of the photoreceptor cells the microvilli are not strictly aligned, but irregularly change their direction from micron to micron, so that the photoreceptors largely lose their polarization sensitivity. In honeybees the same result is achieved by a regular corkscrew-like twist of the photoreceptor cells (Figure 3.28a). By the 'twist' and 'wobble' of their photoreceptors bees and ants abandon polarization sensitivity and thus render other visual performances such as color vision free of any disturbing contamination by unwanted polarized light cues of the terrestrial environment.[69]

First discovered in *Cataglyphis,* specialized dorsal rim areas have now been found in a large variety of insect species. All these DRAs share the crossed-analyzer arrangement of their polarization analyzers, but the spectral type of these analyzers varies among species. In most insects studied so far the polarization analyzers are ultraviolet receptors (in ants, bees, flies, dung beetles, and monarch butterflies), while in others they are either blue receptors (in crickets and locusts) or green receptors (in the cockchafer). Moreover, different DRAs come in quite different shapes and sizes of their visual fields. As we have seen in the cataglyphs, the polarization analyzers can form a regular, fanlike array overlooking a quite substantial part of the celestial hemisphere,

3.29 Dorsal rim area (DRA): polarization sensitive UV receptors, their interactions and viewing directions

(a) Rhabdom (electron microscopic cross section). Photoreceptor cells *R1* and *R5 (highlighted in yellow)* are marked by retrograde labeling. (b) Photoreceptor terminals (*S* and *L*, short and long visual fibers; *S9*, terminal of short basal receptor cell R9) and four types of monopolar cells *(Mp)*. *LA*, lamina; *ME*, medulla. (c) Polarization opponent *(P)* unit. Principle of operation *(left)* and response functions *(right)*; 'crossed analyzers' *(top)* and 'log-ratio amplifier' *(P unit, bottom)*. H, V, photoreceptors tuned to horizontal and vertical e-vectors, respectively. (d) How the DRA looks at the sky. Zenith projections. *Top:* Viewing directions of ommatidia. *Bottom:* E-vector tuning axes of R1 and R5, right DRA. *F*, frontal; *Z*, zenith. *Cataglyphis bicolor.*

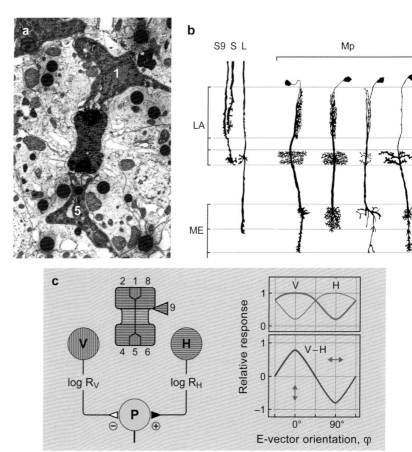

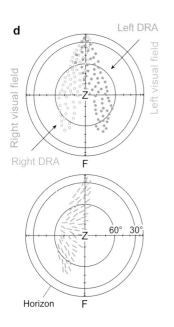

but in the cockchafer the DRA is small and looks with all its ommatidia more or less at the same part in the sky. In the closely related dung beetle it is considerably larger, and in the fruit fly *Drosophila* it even runs all the way from the frontal to the caudal equator of the eye, but is only one or two rows of ommatidia wide. These differences often go along with differences in the optics. In crickets the array of corneal lenses is degraded to such an extent that the DRA photoreceptors have extremely wide and nearly completely overlapping visual fields. In contrast, in the cataglyphs acute vision is maintained even in the DRA.[70]

What do these differences in spectral sensitivity and spatial layout of the DRA mean in functional terms? They might mean that in con-

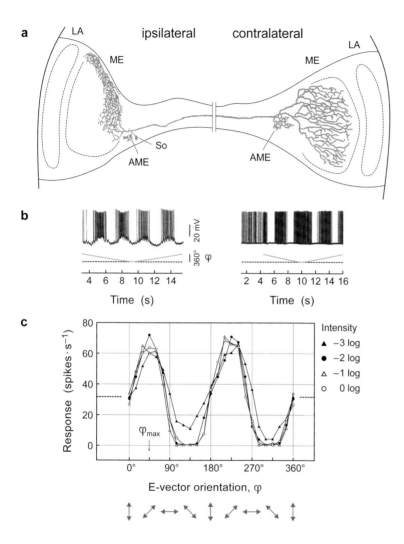

a

LA   ipsilateral   contralateral   LA

ME   ME

So

AME   AME

b

20 mV

360° φ

4   6   8   10 12 14   2   4   6   8   10 12 14 16

Time (s)   Time (s)

c

φmax

Intensity
▲ −3 log
● −2 log
△ −1 log
○   0 log

0°   90°   180°   270°   360°

E-vector orientation, φ

3.30 Polarization-sensitive neuron in the optic lobe (medulla). POL1 neuron of the cricket, *Gryllus campestris*

(a) Morphology. Dendritic arborizations in the dorsal rim medulla of the ipsilateral (input) side and axonal projection to the contralateral medulla. *AME,* accessory medulla; *LA,* lamina; *ME,* medulla; *So,* soma. (b) Physiology. Intracellular recordings from the ipsilateral *(left)* and contralateral *(right)* side of the neuron. As a polarizing filter rotates through 360° *(blue line),* excitation (bursts of spikes) and inhibition alternate every 90° of rotation. The contralateral recording starts in the dark and thus demonstrates the spontaneous spiking activity of the POL neuron. φ, e-vector orientation. (c) Responses to a continuously rotating e-vector at different light intensities. *Dotted line,* spontaneous activity in the dark.

veying different information to their downstream pathways different species make different use of their polarization channels and exploit polarized skylight in different ways and for different purposes. The primary preprocessing mechanisms leading to high polarization sensitivity, intensity invariance, and homochromacy are similar in all these cases, but how the preprocessed e-vector information is further handled and used in the nervous system certainly depends on the particular functional task to be accomplished. What the task is in a given

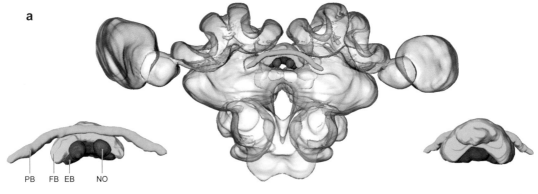

a

PB   FB EB   NO

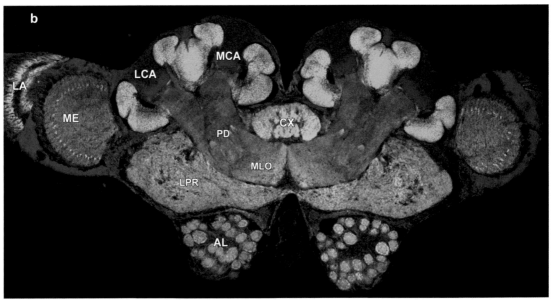

b

LA

ME

LCA

MCA

CX

PD

MLO

LPR

AL

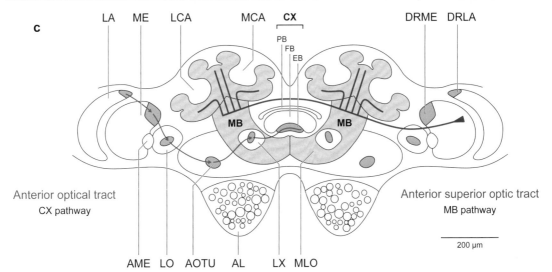

c

LA   ME   LCA   MCA   **CX**   DRME   DRLA

PB
FB
EB

MB   **MB**

Anterior optical tract
CX pathway

Anterior superior optic tract
MB pathway

200 µm

AME   LO   AOTU   AL   LX   MLO

3.31 The cataglyph's brain

(a) Three-dimensional reconstruction of the brain of *Cataglyphis* with the central complex shown in *color. Insets:* Central complex seen in caudal *(left)* and frontal *(right)* view. (b) Frontal section through the brain. Immunofluorescence staining: *red,* synapsin (anti-synapsin labeling); *green,* f-actin (phalloidin labeling); *blue,* cell nuclei. (c) Schematic drawing including the main relay stations *(orange)* of the sky-compass pathway to the central complex *(CX pathway, blue).* The *MB pathway (red)* starts in the medulla and leads to the calyces of the mushroom bodies in either brain hemisphere. *AL,* antennal lobe; *AME,* accessory medulla; *AOTU,* anterior optic tubercle; *CX,* central complex *(PB, FB, EB); DRLA,* dorsal rim lamina; *DRME,* dorsal rim medulla; *EB,* ellipsoid body (lower unit of central body); *FB,* fan-shaped body (upper unit of central body); *LA,* lamina; *LCA,* lateral calyx; *LO,* lobula; *LPR,* lateral protocerebrum; *LX,* lateral complex; *MB,* mushroom body; *MCA,* medial calyx; *ME,* medulla; *MLO,* medial lobe; *NO,* nodulus; *PB,* protocerebral bridge; *PD,* pedunculus. *(a) Cataglyphis nodus, (b) C. fortis.*

species can only be unraveled by a deeper understanding of the ecological requirements and lifestyle of the species in question, and then by developing proper experimental strategies to analyze the animal's e-vector-guided behavior in detail. For example, dung beetles, which roll their dung balls away from a dung pile along a straight path, use any celestial cue present at the moment to maintain their bearing. They just seem to acquire and use a celestial snapshot, even if in this snapshot the celestial cues—the sun and the polarization and spectral cues—are experimentally arranged in arbitrary ways.[71] It might well be that out of this flexible use of any skylight information currently available, a celestial compass system has evolved in which the animal is eventually equipped with some internal knowledge about the natural geometric relationship between these cues.

In proceeding further into the cataglyph's brain (Figure 3.31), we encounter two separate visual pathways leading from the optic lobes to the main higher-order integration centers in the ant's forebrain (protocerebrum). These pathways, which are highly conserved across various insect taxa, have been traced out in the cataglyphs by applying modern neurohistological techniques such as fluorescent dye injections, fiber tracings, confocal microscopy, and three-dimensional reconstructions.[72] One of them—shown in blue in Figure 3.31—transfers information from the photoreceptors of the compound eyes via at least four neural way stations to the central complex, CX (hence, CX pathway). This is the pathway that interests us here, because it conveys sky-compass information to the central brain. The other one—shown in red in Figure 3.31—leads from the optic lobes, particularly via projection neurons from the medulla, directly to the visual subregions of the mushroom bodies (hence, MB pathway). It will come into play only in Chapter 6, when we discuss how the cataglyphs use landmarks for guidance. An early hint that the cataglyph's brain receives sky-compass and landmark-based information most likely via different visual pathways came from behavioral experiments in which the ants exhibited interocular transfer with respect to the former but not the latter kind of information.[73]

Let us now follow the CX pathway in a bit more detail. In the cataglyphs this endeavor has been bedeviled by methodological problems, especially by the small size of the ant's brain. This renders intracellular recordings a difficult exercise even for experienced electrophysiologists, who have tried their hands on the cataglyphs. Hence, for analyzing the more central processing stages of sky information, larger insects such as locusts and crickets have become the organisms of choice. With a breakthrough paper published in 2007 in *Science,* Uwe Homberg and Stanley Heinze of the University of Marburg, Germany, have pioneered this field of e-vector coding in the insect brain.[74] Due to the phylogenetically conservative nature of the CX pathway we feel justified to build on the extensive analysis of this sky-compass pathway in the locust, all the more as more recent neurobiological investigations have largely confirmed the results obtained in the locust also for various other insect species studied so far, e.g., for honeybees, monarch butterflies, dung beetles, and fruit flies.[75] Whenever checked for the cataglyphs, the CX pathway conforms to that unraveled in the locusts.

We start in the optic lobes with the medullar POL neurons mentioned above. Some of them have dendritic arborizations in the dorsal rim medulla and on neurites running vertically through the main medulla. Hence, it could already be at this very peripheral site that polarized-light information from the DRA is combined with spectral / intensity information from specific azimuthal positions in the sky. This convergence of information is clearly represented further downstream in the first central brain neuropil, the anterior optic tubercle (AOTU). The next relay station is a specific bulb-like region within the lateral complex (LX). It connects the AOTU with the final processing stage of polarized-light information, the central complex (CX). This LX bulb houses a group of giant synaptic complexes, which have been discovered in locusts, but have also been found and characterized in the cataglyphs (Figure 3.32b, c). In these conspicuous glomerular structures—actually the largest synapses in the *Cataglyphis* brain—the AOTU projection neurons form large cup-shaped presynaptic terminals, which encase numerous fine postsynaptic processes of tangential (TL) neurons, the main input pathway of polarized-light information to the central complex. The function of the gigantic synaptic complexes, of which *Cataglyphis fortis* has about 100 in each lateral complex, has remained elusive in all species examined so far. Judged from comparable structures in the mammalian auditory pathway, one could hypothesize that they are responsible for the exact timing of polarized-light information arriving from the left and right side of the visual system. In *Cataglyphis* it has been shown that their number increases significantly at the start of the ants' foraging lives, or when interior workers, which have not yet reached their foraging age, are experimentally exposed to light.[76]

As mentioned above, the final common terminal of skylight information is the central complex. This "brain within the brain," as Nicholas Strausfeld has put it, is an intricate network of neurons in the center of the insect's forebrain.[77] It strikes by a remarkably stereotypic internal neuroarchitecture with an almost crystalline structure. Its three midline spanning components—the protocerebral bridge, the fan-shaped body (the upper division of the central body), and the ellipsoid body (the lower division of the central body)—consist of stacked horizontal layers and vertical columns symmetrically arranged from left to right around the midline (Figure 3.33). They are ventrally associated with a pair of noduli. Most importantly in the present context, the protocerebral bridge consists of a linear array of 16 slices, 8 slices in each hemisphere, with each slice housing neurons that respond best to a given e-vector orientation.

Let us embark on a little tour de force through the neural circuits involved in generating this e-vector mapping. As already mentioned, tangential cells (TL neurons) project from the giant synapses in the

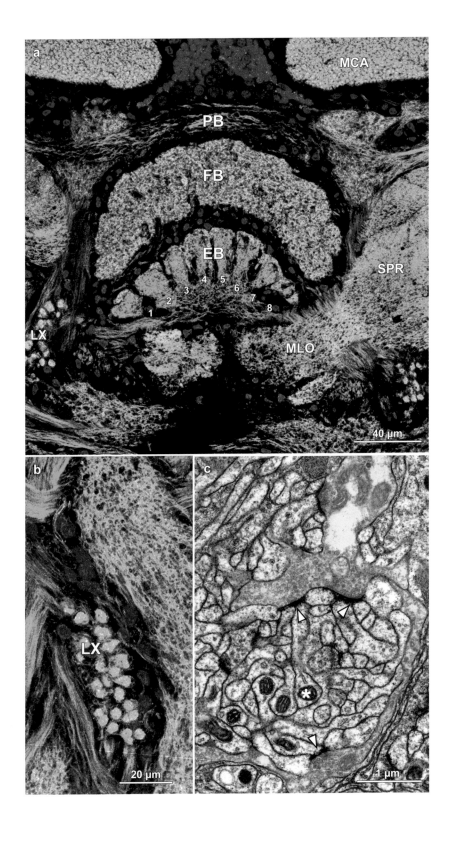

3.32 Close-ups of the central and lateral complex

(a) Partial view of central complex. Immunofluorescence staining as in Figure 3.31b. *EB*, ellipsoid body; *FB*, fan-shaped body; *LX*, lateral complex; *MCA*, medial calyx; *MLO*, medial lobe; *PB*, protocerebral bridge; *SPR*, superior medial protocerebrum. (b) Magnification of giant synapses in the lateral complex. (c) Ultrastructure of one giant synapse. A large cup-shaped presynaptic terminal (marked in *yellow*) encloses a core of postsynaptic profiles. *Arrowheads* point at active synaptic zones. *Asterisk:* mitochondrium. *Cataglyphis fortis.*

lateral complex (LX) to the ellipsoid body. Next, columnar cells (CL, see the CL1 type in Figure 3.33c) interconnect individual columns ('slices') of the ellipsoid body with individual slices of the protocerebral bridge, where they are presynaptic to tangential cells (TB neurons, see the TB1 type in Figure 3.33a, b). The CL1 neurons with their remarkably sharp e-vector tuning convey e-vector information to the protocerebral bridge. There, the TB1 neurons have their φ_{max}-values spatially arranged in such a way that the entire 180° e-vector scale is systematically mapped twice, once on either side of the midline (Figure 3.33b). When a linear polarizing filter is rotated above the head of the insect, an excitation wave with a marked activity bump of maximal spike rate sweeps across the left and right part of the protocerebral bridge, just as windshield wipers move in unison. As the TB1 neurons combine corresponding columns in the left and right hemisphere, the two halves could represent an integrated dual compass. Regardless of such integration processes, the twofold representation of the compass scale on the left and right side is a vivid demonstration that due to its evolutionary history the unpaired central complex consists of paired subunits, of which each is responsible for turning movements toward one side.[78]

The flow of information outlined above (TL→CL→TB) provides only a sketchy picture. In reality it is complemented by multiple recurrent loops and parallel pathways involving other types of neuron. In short, however, the ordered array of e-vector tuned TB1 neurons encode the animal's head direction relative to the e-vector in the zenith, but by the same token, to the azimuthal position of the sun (which is strictly correlated with that e-vector). Hence, these neurons can conveniently be called 'compass neurons.'[79]

3.33 E-vector representation in the central complex

(a) An individual TB1 neuron in the protocerebral bridge. Note the two prominent arborizations *(shaded in red),* which are located eight columns apart (in columns L5 and R4). (b) Arborization patterns of eight TB1 neurons in the protocerebral bridge. Each TB1 neuron invades two columns with prominent arborizations *(black).* The *red marking* refers to the neuron shown in *(a).* (c) Wiring scheme of 16 columnar neurons (type CL1), which receive their inputs from the lateral complex and project to the TB1 neurons. *EB,* ellipsoid body; *FB,* fan-shaped body; *LX,* lateral complex; *PB,* protocerebral bridge. The *blue double-headed arrows* depict neuronal e-vector tunings. Desert locust, *Schistocerca gregaria.*

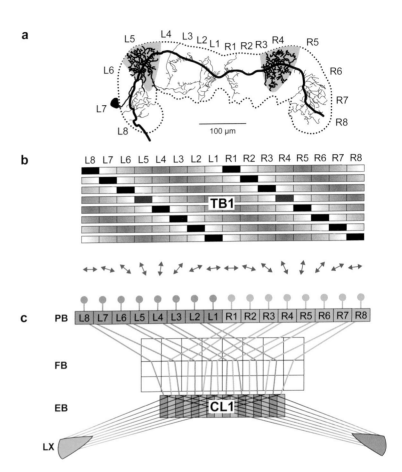

More recently, an elegant combination of behavioral, electrophysi-ological, and neurogenetic tools has shown that *Drosophila* flies use their CL1 neurons for encoding their headings with respect to the sun and the zenithal e-vector.[80] Tested in a flight simulator, the flies are able to maintain an arbitrary azimuthal position relative to these ce-lestial cues for periods of up to several hours, but lose this ability when their CL1 neurons have been silenced. In walking flies tested on a trackball within a virtual reality arena, the CL1 neurons can encode the fly's heading with respect to a visual landmark such as a vertical stripe. When the tethered fly turns, say, to the left, an activity bump rotates across a ring of such CL1 neurons by the same angular amount to the right. Again, the central complex provides the insect with an internal compass, in this case referring to a terrestrial cue, very similar to what the head direction cells do in mammals.[81] Moreover, as we have seen in this chapter, the cataglyphs derive compass information not only from various celestial cues but also from a number of global terrestrial cues such as the geomagnetic field and the direction of the prevailing wind as well as from idiothetic cues. The central complex—especially the ellipsoid body—seems to be the place where all this compass information is combined, and hence provides the animal with a common internal representation of azimuthal space. As this kind of circuitry has been found in almost all insect species studied so far, in locusts, flies, cockroaches, beetles, bees, and butterflies, it most certainly represents the internal compass in the cataglyphs as well.

Having dwelled for quite some time on neuroarchitectural aspects of the polarized-skylight compass, let us finally return to how the cataglyphs may use this compass. The preceding discussion has been concerned with individual e-vectors presented in the zenith. However, the higher-order polarization-sensitive neurons, and especially those of the central complex, have extended visual fields, which cover almost the entire celestial hemisphere. This means that they are not designed to specify individual e-vectors anywhere in the sky (and, for instance, might not determine the sun's position in the way outlined in Figure 3.12). Instead, they integrate polarized-light information across the entire celestial dome. In the locust's polarization pathway some wide-field neurons (of the TL type) have been described, in which the e-vector tunings within their visual fields are arranged in concentric circles and thus follow the global structure of specific e-vector patterns in the sky. Hence, it has been assumed that locusts possess a set of wide-field neurons each matched to the sky polarization pattern that pertains to a particular elevation of the sun.[82]

You may recall that in our behavioral experiments the desert ants and honeybees behaved as if they referred to an invariant e-vector distribution, actually to the most uniform one that is present in the sky when the sun is at the horizon (Figure 3.13b)—a strategy that is equivalent to using a zenithal e-vector, or to referring to the symmetry plane of the skylight pattern (on which the sun is located) and determining one's heading relative to this plane. In general, the use of large-field polarization-sensitive neurons is in agreement with the behavioral data, but the details of the neural processing need further clarification.

And what about locusts? Even though they have been the first to offer a window to the central neural pathways of polarization vision, the role that a polarized-light compass plays in their solitarious or gregarious stages of life has not been sufficiently investigated yet. Behavioral responses to polarized light have been recorded only in the laboratory. When an overhead e-vector is slowly rotated, the tethered walking or flying locusts show sinusoidal modulations of their headings.[83] In this case they use e-vector information to stabilize their flight course. On their long-range migratory mass movements the animals largely follow the downwind directions associated with synoptic-scale wind systems. Whether and how they use a celestial compass to adopt a preferred heading remains to be elucidated.[84]

What we face here could be dubbed 'reverse neuroethology.' I use this term in analogy to 'reverse genetics.' Since Gregor Mendel's times classical (forward) genetics starts with phenotypical traits and investigates the genetic basis of these traits, but since the advent of recombinant DNA techniques reverse genetics aims to understand what phenotypical effects result from a (usually manipulated) genetic sequence.[85] In the present context we may ask what guidance routines are mediated by the neural circuitries uncovered in the locust, and what parts of these circuitries are necessary and sufficient for generating these behaviors. Similarly, classical neuroethology starts with the analysis of particular behavioral traits—especially those traits that can also be studied in their natural contexts such as the cataglyphs' navigational repertoire—and then aims to understand the neurobiological underpinnings of this behavior. A prime example is Walter Heiligenberg's analysis of the jamming avoidance response in the weakly electric fish *Eigenmannia*. These fish use electric discharges in the millivolt range for electrolocating objects in the dark. Starting with the field observation that a fish shifts the frequency of its discharges away from the interfering one of a neighboring fish (upward or downward depending on the neighbor's frequency), Heiligenberg unraveled the entire sequence of neuronal events occurring along a series of relay stations from the sensory input to the motor output side.[86] Of course, in the best of all research worlds both lines of investigation should be combined, but with the rapid advances of neurophysiological and neurogenetic as well as modeling tools reverse neuroethology will most likely get in the lead—at least for some time.[87]

In the cataglyphs natural selection seems to have done everything to equip the insect navigator's peripheral visual system with sharply tuned, highly polarization sensitive e-vector detectors (in the DRA) as well as with polarization-blind detectors for spectral/intensity cues (in the non-DRA). The more we move centrally, the more is the information provided by the small-field e-vector detectors channeled into large-field units, and the more is it combined with information about the spectral/intensity gradients in the sky and other directional cues—as well as with information about the distances covered by the animal. The latter is the topic of the following chapter.

4 Estimating Distances

While marine navigators had used the magnetic compass since the twelfth century, technical devices for measuring speed or distance run through the water had become familiar only some centuries later. The chip log—a knotted rope, the *log* line, attached to a wooden board, the *chip*—was such a device. By letting the chip float in the water, where it roughly stayed in place while the ship sailed away from it, the navigator could measure the amount of rope paid out in a given period of time, and thus gauge the distance the ship had traveled during that time. The first vessel that in the beginning of the sixteenth century circumnavigated the globe, Magellan's *Victoria,* was also one of the first for which the use of a chip log has been reported.[1]

But how do the cataglyphs moving on terra firma keep track of the distances they have traveled? In principle, they could proceed along several lines and thus employ various kinds of distance estimators, so-called odometers. They could (i) measure the time spent moving (if travel speed were constant); (ii) monitor the speed of movement, e.g., by recording the image motion of the environment across the eyes (the self-induced 'optic flow') and integrate it over time; (iii) employ some kind of inertial navigation by double-integrating linear acceleration over time; (iv) measure energy consumption associated with forward movement; (v) use some proprioceptive cue that measures locomotor activity, e.g., monitor stride length and compute the product of stride length and the number of strides; or (vi) monitor the output of the central pattern generator that is responsible for the ant's locomotor

4.1 Ups and downs in ant navigation: the sawtooth paradigm

The cataglyphs perform their outbound journeys along a linear sequence of open-topped channels, which are alternately sloped upward and downward. For return to their home base, they are transferred to a straight channel aligned in parallel to the hilly racetrack (and partly visible at the *lower right*). For results, see Figure 4.11.

behavior. How different these potential ways of gauging travel distance might ever be, they have one aspect in common: they all depend on information derived from the animal's own movements.[2] There is no external surveyor's rod from which the ant could directly read the distance it has traveled.[3] Nevertheless, as we shall see in this chapter, the cataglyphs can estimate their travel distance quite exactly, and irrespective of whether they navigate in flat or hilly terrain (for the latter, see Figure 4.1).

All the strategies mentioned above have their particular advantages and limitations. A flying bee that moves within a medium that moves itself might like to monitor self-induced image flow, and hence might resort to strategy (ii). A walking ant, which is in physical contact with the ground, could preferably exploit cues correlated with the movements of its legs and hence adopt strategies (v). When we started our *Cataglyphis* work, energy expenditure was considered to be the decisive cue (strategy iv). Early studies in honeybees suggested that the distance traveled was measured in terms of the total energy expended during flight.[4] We hesitated to sympathize with this suggestion from early on, because the ant's odometer did not seem to be affected by the often heavy burdens the ants were carrying. Later experiments with artificially loaded ants running in test channels confirmed our early observations. Even when the ants drag extremely heavy load by walking slowly backward, their distance estimation is barely affected.[5]

Our way toward understanding what strategy the cataglyphs apply was a devious one meandering around in various directions and often ending in cul-de-sacs. The final solution that there are at least two odometers at work emerged only slowly. This might not have become immediately apparent in the final publications. In a witty essay the biologist and Nobel Prize laureate Sir Peter Medawar even pondered the question of whether the scientific paper is a fraud, not because it misrepresents facts, but because its rigid introduction-methods-results-discussion scheme "misrepresents the process of thought that accompanied or gave rise to the work that is described in the paper."[6] Even though in the following we shall not retrace all the tortuous sideways and wrong turnings taken in studying how the cataglyphs estimate their travel distance, the reader might get an impression of the "process of thought" that led to the final solution.

The Pedometer

Our investigations gathered momentum only after we had introduced the channel test paradigm (Figure 4.2). There the cataglyphs perform their foraging journeys in straight channels open at the top, which prevent them from seeing natural landmarks that could be used as odometric cues, but which provide them with a view of the sky (necessary for acquiring compass information). Within the channels the ants tend to run along the midline. This remarkable 'centering response' results from the animals' attempts to balance the angular heights of the channel walls on either side (Figure 4.2b).[7] For example, if one wall is increased in height relative to the other wall, the ants shift their trajectories away from that wall toward the lower one by exactly the amount necessary to balance the angular heights of the two walls. In studying distance estimation, we let the cataglyphs run within a channel from the nest to the feeder, from where they are displaced to a test channel. This test channel is aligned with the training channel, but considerably increased in length, so that upon release the ants are able to run as far as their odometric signal tells them. Upon release, they unhesitatingly hurry off in the home direction as if they had not been displaced. When they have covered about the homeward distance, they stop, make a rapid U-turn, and start to meander to and fro about the fictive position of the nest (Figure 4.2c).

A way to investigate whether the ants use some kind of 'stride counter' (pedometer) would be to change the stride length of the animals by manipulating the lengths of their legs. Of course, such manipulations are easier said than done. Nevertheless, Matthias Wittlinger, then a graduate student working at our North African field site, succeeded in shortening and lengthening the ants' legs by performing delicate microsurgical operations. He shortened the legs by severing them at mid-tibia level (stump-legged ants, 'stumps') and artificially lengthened them by gluing stilts—specifically cut bristles of pigs—to the tibia and the tarsomeres (stilt-legged ants, 'stilts') (Figure 4.3).[8] Surprisingly, the walking behavior of the ants remains almost unaffected by these manipulations, even though the shortened legs induce the ants to crawl with their bodies closer to the substrate, and the stilts substantially increase the mass of the ants' legs. It is

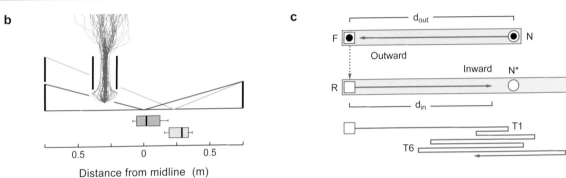

4.2 The channel paradigm

(a) Installing channels for studying ant odometry. Training and test channels are set up in parallel. Graduate student Stefan Sommer *(right)* and technical assistant Hansjörg Baumann *(left)*. (b) Centering response. The boxplots indicate the ants' positions in channels, in which the left wall is either as high *(red)* or twice as high *(green)* as the right wall. *Inset:* Walking trajectories (view from above) of ants starting at the feeder *(bottom)* and heading for home. (c) Experimental setup (not drawn to scale). Nest entrance *(N)* and feeder *(F)* in the training channel, point of release *(R)* and fictive position of nest *(N*)* in the test channel; d_{out} and d_{in}, outward and inward travel distance; *T1, T6,* 1st and 6th turning point in the test channel.

amazing and reassuring to see how successfully the manipulated ants stump and stilt through the test channels and subsequently through their foraging grounds, even several days after their legs have been manipulated—just as if nothing had happened.

In the critical tests, the ants walk in a training channel to a feeder at a distance of, say, 10 meters from the nest. There they are captured, subjected to the surgical operations, and transferred to a test channel.

4.3 Cataglyphs on stilts

The ants' legs are lengthened by gluing small bristles *(red)* onto the lower parts of their legs. In addition, in some experiments (see *inset*) the ventral halves of the ants' compound eyes are provided with light-tight eye covers *(yellow)*. *Cataglyphis fortis.*

While the untreated control ants accurately center their searches at the fictive nest position, i.e., at a distance of 10 meters from the start, the stilt-legged and stump-legged ants significantly overshoot and undershoot, respectively, the 10-meter finishing line (Figure 4.4a). However, after having been placed back to the nest and tested a second time, the ants now performing both their outbound and inbound runs with their shortened or elongated legs center their searches again at the correct 10-meter mark.

The overshooting and undershooting observed in the experimental animals on their first inbound run qualitatively correspond to what a 'stride integrator' hypothesis predicts. But do the data also meet the quantitative predictions? Detailed measurements of the stride lengths in the stump-legged and stilt-legged animals, and the ways of how these measures vary with body size and running speed,[9] indeed show that the predictions from the video analyses (see open boxes in Figure 4.4a) come very close to the experimental data obtained in the homing ants. Obviously, the ants employ a stride meter—a pedometer—in which the number and lengths of the strides are taken into account.

How could such a stride measure be obtained? More than 100 years ago the French physiologist Henri Piéron concluded from observations that he had made in some *Aphaenogaster* ants that these animals recorded and memorized all movements of the legs made during an outbound journey.[10] He was not specific about the *sens musculaire,*

4.4 Stepping on stilts and stumps

Homebound runs performed by ants with their legs either shortened ('stumps') or lengthened ('stilts') and **(a)** with their compound eyes untreated and **(b)** with the ventral halves of their compound eyes covered during both inbound and outbound runs. Nest-feeder distance: 10 m. *Upper figures:* Search density distributions of stump-legged *(blue)* and stilt-legged *(orange)* ants; *black:* untreated controls. *Lower figures:* Positions where the ants search for the goal (turning positions). *Filled boxplots:* experimental data; *open boxplots:* search positions predicted by the stride integrator. The *dotted red lines* indicate the degree of correspondence between experimental and predicted data. **(c)** Leg manipulations. *Cataglyphis fortis.*

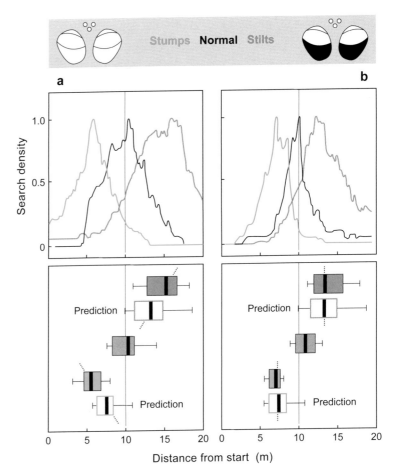

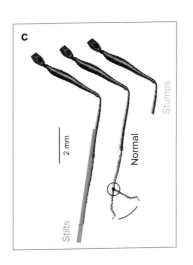

which he attributed to the ants, but at present it is a likely hypothesis that the tension receptors in the muscles and / or tendons of the ants' legs could provide the odometer with proprioceptive sensory feedback about stride width, and that force-sensitive mechanoreceptors such as the campaniform sensilla in the cuticle of the legs as well as mechanoreceptive hair plates at the bases of the legs could do so as well. Besides these various kinds of afferent information arising from steps actually taken, it is also efferent information copied from motor commands and hence related to steps still to be taken that could account for the necessary input to the odometer, either alone or in conjunction with afferent information.

One way to decide between these afferent and efferent mechanisms—between strategies (v) and (vi) listed in the beginning—is to disturb the ants' walking behavior to such an extent that the stride meter might finally fail. Such failures, however, are difficult to induce, even if one severely interferes with the animals' normal locomotion by manipulating either the walking substrate or the ants' locomotor apparatus. When the animals are trained to walk across heavily corrugated surfaces (see Figure 2.34b, c) or, even more surprisingly, when some of their six legs have been removed, their distance estimation and homing performance remain virtually unaffected.[11] This is even the case when the ants drag a heavy load by walking backward toward their home nest. While under these conditions the leg coordination patterns exhibit an enormous flexibility, with often more than three legs touching the ground simultaneously and with the swing phases considerably shortened, the rearward-walking ants are able to estimate their homing distances as accurately as the forward-moving ants.[12] In conclusion, the extreme resistance of the ant's stride integrator to severe changes in the ants' regular walking behavior is truly remarkable. It may support the hypothesis that the pedometer depends on sensory feedback from leg mechanoreceptors, which provide the animal with information about its stride length. Moreover, the observation that ants walking on corrugated surfaces or dragging heavy food items backward exhibit irregular stepping patterns, in which the usually strict correlation between stride length and stride frequency is largely decoupled, might even mean that the odometer receives separate input from each individual leg. In any way, the cataglyphs most likely exploit odometer mechanism (v). This seems to be the case also in fiddler crabs, *Uca pugilator,* the only other animal in which odometry has been shown to depend on a stride meter, and in which the stride meter is also able to flexibly account for considerable variation in stride length and frequency.[13]

The Optic-Flow Meter

Even though the pedometer enables the cataglyphs to gauge travel distances quite accurately, this does not exclude the possibility that a visually driven odometer works in parallel. The latter is a likely possibility all the more, as a walking ant keeps its eyes at an almost constant distance above ground, and thus can reliably infer its walking speed from the speed of the self-induced image motion of the ground across its eyes. Speed multiplied by time yields distance. To test this 'optic flow' hypothesis we let the ants run in channels of which the floor is lined with black-and-white stripes oriented perpendicularly to the channel axis (Figure 4.5a).[14] The stripes are stationary during the ants' outbound runs, but during the inbound runs they are moved with a particular speed either in the ants' direction of movement or opposite to it. The artificial forward and backward movements of the pattern should cause the ants to overestimate and underestimate their travel distances, respectively. This is exactly what happens—a first hint that an insect can take advantage of self-induced optic flow for gauging travel distance. As expected, optic flow is effective in the ventral but not the lateral field of view.[15] As in natural environments the distances of objects changes unpredictably along an ant's path, lateral optic flow does not provide reliable information about the ant's walking speed.

Even a cursory glance at Figure 4.5c shows that the effect of the experimental manipulation of the ant's ventral image flow is quite small. The stride integrator seems to dominate the ants' odometric decisions. That a visual odometer is operating anyway can also be inferred from a closer inspection of the lower part of Figure 4.4a. As this figure indicates (see the obliquely oriented dotted red lines), the stilt-legged ants overestimate and the stump-legged ants underestimate their training distance slightly but consistently more than would be predicted by the operation of the stride integrator alone. This excessive overshooting and undershooting could likely be explained by the contribution of a visually driven odometer. Note that when ants are provided with stilts, their bodies and hence their eyes are raised further above the ground, so that they experience a smaller ventral image flow per unit distance traveled than they did before the stilts were attached to their legs, and hence should overshoot the training mark merely on the basis of visual odometry. The opposite is the case for the stump-legged ants. When any optic-flow contribution to the odometer is shut off by occluding the ventral halves of the ants' eyes, the homing distances of the animals walking on stilts and stumps and being half-blinded coincide with the ones predicted by

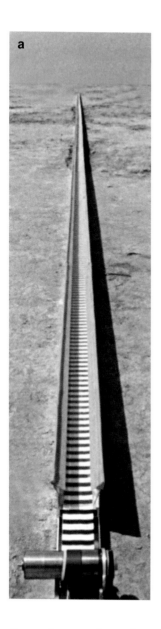

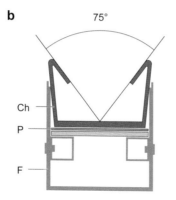

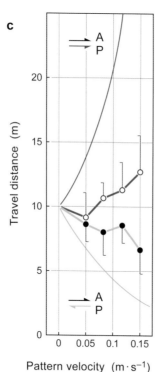

4.5 Manipulation of self-induced optic flow in walking ants

(a) Test channel in full view. (b) Test channel in cross section. *Ch*, channel (transparent Perspex); *F*, aluminum frame; *P*, moving belt with pattern. (c) Results. The ants overestimate (*open circles, red*) or underestimate (*filled circles, green*) the 10 m training distance, when the pattern (*P*) moves in the same or the opposite direction of the ant (*A*). The *curved thin lines* indicate the distances that the ants should travel, if they relied exclusively on a visual odometer. *Cataglyphis fortis.*

the pedometer hypothesis alone (see vertically oriented dotted red lines in Figure 4.4b, lower part).

Of course, for obvious experimental reasons it is well-nigh impossible to switch off the pedometer in a walking ant. In this seemingly hopeless situation we have the good fortune that the cataglyphs exhibit a special kind of social behavior, adult transport, which offers a solution. The foragers are able to transport adult nestmates along straight lines over tens of meters from one nest of a polydomous colony to another nest. When in the beginning of the twentieth century the British physician and naturalist Richard Hingston explored northern India, he was much impressed by "the remarkable mode of geographical dispersal by which one group of [*Cataglyphis*] workers established a formicary by transporting thither their comrades in their jaws" (Figure 4.6).[16]

When such a pair of workers is captured midway between the nests of departure and destination, and transferred to a distant test field, where it is released, the pair gets disengaged. First the two ants search around in narrow and then widening loops to get in contact again, but when tested one hour after their separation, the transporter hurries off in the direction of the goal nest, while the former transportee walks more slowly in the opposite direction, i.e., in the direction of the nest of departure, out of which it has been carried. In the early days of our *Cataglyphis* research, Peter Duelli did a decisive experiment. He let the pair of workers run on a manually operated conveyer belt within a little open-topped arena. The moving belt induced the transporter ant to run on the spot in a particular direction for a particular time. The arena walls occluded all natural landmarks and shielded off the wind, but nevertheless, later out in the field the transporter and the transportee selected their compass courses as accurately as the free-field controls.[17] Compass cues could have been derived only from the sky or the earth's magnetic field.

But what about the distance cues? Does the optic flow, perceived by the ant while it has been carried in its stereotypically fixed posture above the ground, provide the necessary information? The affirmative answer comes from a series of tricky experiments in which Matthias Wittlinger has invited the transporters to carry their nestmates through the very channels that we had previously used for unraveling the ant's pedometer, and which have now been spatially arranged so as to connect the nests of departure and destination.[18] When in this channel device a couple has left its nest for a given distance—e.g., 10 meters—the transportee is disconnected from the transporter and transferred to a parallel test channel. There it first searches about the

4.6 Adult transport in *Cataglyphis bicolor*

In a typical formicine manner, a cataglyph that is transported toward another nest of the colony (transportee, *left individual*) adopts a characteristic body posture. **(a)** Having been grabbed by the transporter *(right individual)* at the base of its mandibles, it quickly straightens up, **(b)** folds its legs and antennae tightly to the body, curls up its gaster under the transporter's head and alitrunk, and is finally carried away—in a 'pupal posture'—upside down but headfirst as a compact load hanging underneath the running transporter.

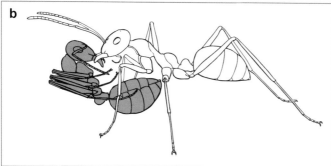

point of release and then sets out in the home direction, until it finally meanders back and forth around the fictive position of the nest at 10 meters' distance from the start (Figure 4.7a). When prior to being grasped by the transporter the ant had the ventral halves of its eyes occluded by light-tight paint, it does not move toward the fictive goal, but searches about the point of release (Figure 4.7b). This is clear proof that the optic-flow information acquired while the ant has been carried in the situation described in Figure 4.7a must have provided the proper distance cues. Quite unexpectedly, however, the same behavior as the one shown in Figure 4.7b is observed in carried ants in which the ventral eye covers are applied only before the animals start their

4.7 Ants carried by nestmates: optic flow suffices for gauging travel distance

Ants are released in a test channel after they have been carried by nestmates 10 m away from the nest entrance. Search density distributions of the transportees (a) with their compound eyes untreated *(black)* and (b) with the ventral halves of their compound eyes covered during both the outward trip in the carried mode and the inward trip in active locomotion (control). (c) Eye covers applied only during the inward run. Goal distance: 10 m. In *(a)* and *(c)* the search distribution of the controls is superimposed in *shaded blue*. *Cataglyphis bicolor.*

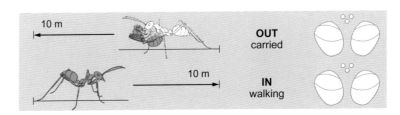

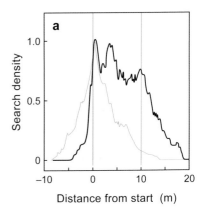

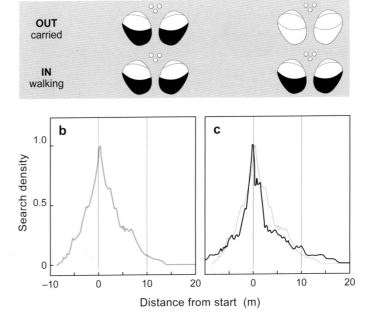

home run in the test channel. In this case the transportees are loaded with optic-flow information about the distance previously traveled in the carried mode, but their pedometer does not seem to have access to this information (Figure 4.7c). Hence, we hypothesize that the two ways of estimating distances represent separate mechanisms, which run independently and in parallel to each other, and are parts of separate integrators: a 'stride integrator' and a 'flow integrator.' Both can work on their own—the stride integrator when the ventral halves of the cataglyphs' eyes have been occluded or when the animals walk in the dark,[19] and the flow integrator when the ants have been carried by nestmates.

As an aside, it is worth mentioning that the visual odometer is used by the transported ants only in the actively acquired mode of transportation, i.e., when the ants have been carried by a nestmate. In this case they are not just grasped by the nestmate and passively transported to another nest of the colony. They actively invite themselves to be taken up and actively assume a characteristic body posture underneath the transporter (Figure 4.6). In contrast, when they are experimentally mounted on a sledge and passively moved at ant height across a patterned floor, with the same speed with which they would be carried by a nestmate, they obviously do not employ their flow integrator (Figure 4.8).[20]

If there are really two separate distance memories, one based on stride integration and the other on flow integration, should it then not be possible to let them discharge their information sequentially, one after the other? A further set of tricky experiments addresses this question.[21] Ants having arrived at the feeder, and thus with both their distance memories charged, are captured, provided with ventral eye caps, and released in the test channel. Consequently, in their home runs they can discharge only their stride integrator. Having ended their straight home run by performing the usual sharp U-turn, they are gently grasped again, freed from their eye caps, and released in the test channel for a second time. They now continue to travel in their home direction by discharging their optic-flow integrator, so that they finally end up, and commence their nest search behavior, far beyond the fictive position of the nest. This shows that the cataglyphs can discharge their two distance memories separately, but the quantitative details

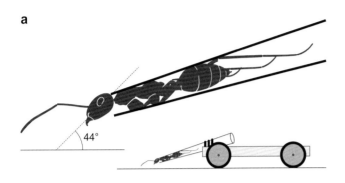

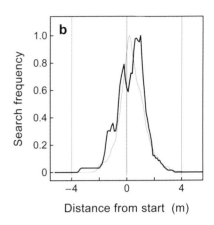

4.8 Passively transported ants

(a) An ant is fixed inside the tip of a pipette *(top)* with the head peering out and the rest of the body being immobilized by the pipette. The pipette is mounted on a small vehicle such that the pitch angle of the head and the distance of the head above ground correspond to normal running conditions. The vehicle is loaded with a zero-vector ant and moved with the ants' average running speed for 4 m across the floor of a patterned channel. (b) Search density distribution of the released ants in the test channel. Release at 0 m, fictive position of nest at 4 m. The ant attached to the vehicle has been passively moved *(black)* or has remained stationary (control, *blue*). *Cataglyphis fortis.*

of the distances traveled after the first and second releases also reveal that the two distance memories compete with each other in unloading their contents.

Let us quickly return to the 'carrier experiments,' which at the outset have provided clear-cut evidence that the cataglyphs can estimate travel distance by sensing optic flow. In these experiments it has been human interference that induced the transported cataglyphs to return to their nest of departure. But in the closely related *Rossomyrmex*—a slave-making semi-desert ant, which raids colonies of *Proformica* ants—this behavior occurs naturally.[22] Here the transport of adults is used for recruiting nestmates to the host colony. Inexperienced workers are carried by experienced workers to a newly discovered *Proformica* colony, where they participate in the raid and return with host brood to their home nest. Immediately thereafter they recruit new nestmates by themselves.[23] When returning home on their own for the first time, they must rely exclusively on information acquired while having been transported—as we have studied it in the cataglyphs.

Why do walking ants that are provided with an efficient stride integrator possess a flow integrator at all? The answer may lie in their evolutionary heritage. Ants are "modified wasps" or even "wingless bees."[24] According to recent phylogenomic data, they are most closely related to a group of hymenopterans, the apoid clade, that comprises bees and hunting wasps, so that their ancestors have certainly exploited optic flow as a means of monitoring travel distance. Flying honeybees

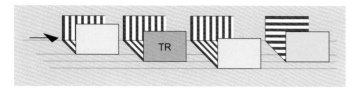

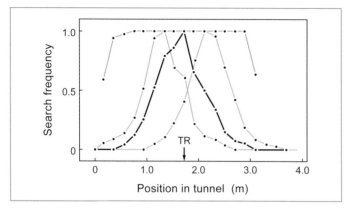

4.9 Flying bees employ a visually driven odometer

Honeybees enter (see *arrow*) a flight tunnel at position 0 m and are rewarded at position 1.7 m in the training tunnel *(TR)*. The inner walls and the floor of the tunnel are lined with black-and-white gratings oriented at right angles to the long axis of the tunnel. The curves depict the distributions of the bees' search frequencies in the test situations: control test *(black)* and critical tests in which the test tunnel is either narrower *(blue)* or wider *(green)* than the training tunnel, or in which the gratings run parallel to the long axis of the tunnel *(orange)*.

do so today. They cannot rely on counting numbers of wing beats as ants rely on counting numbers of strides, because in air—a volatile medium—the number of wing beats necessary for covering a certain distance strongly depends on the movement of the volume of air within which the bee flies. What actually counts is not the bee's air speed (the speed relative to the air) but the ground speed (the bee's speed relative to the ground), and this speed is equivalent to the speed by which the image of the environment moves across the insect's eyes.

In using two different experimental approaches, Mandyam Srinivasan at the Australian National University in Canberra and Harald Esch at the University of Notre Dame in Indiana have provided clear evidence that honeybees actually employ a visual odometer. In Srinivasan's paradigm the bees fly through narrow channels, in which the walls and floor are lined with black-and-white gratings (Figure 4.9).[25] As these gratings are oriented perpendicularly to the axis of the channel and hence to the bees' direction of flight, they provide the bees with powerful optic flow: the narrower the channel, the greater the amount of flow. When bees are trained in a channel of a given width to find food at a given distance from the channel entrance, and later tested in channels that are either narrower or wider than the training channel, they fly shorter or larger distances, respectively, before they start to search for the missing food source. Obviously, during training the bees have integrated and remembered the total amount of visual flow experienced on their way to the goal. If the test channels are narrower, the amount of visual flow per unit distance traveled is larger than in the training channel, so that the bees must fly a shorter distance to acquire the same total amount of flow that they have experienced in the training situation. The opposite holds for the wider test channels. When the channels are provided with axially oriented stripes, the bees while flying parallel to the stripes experience no or only very little image motion. In this case they fly without stopping directly from the entrance to the other end of the channel and hence seem to be unable to gauge any distance traveled.

In Esch's experimental paradigm the bees are trained in the open field.[26] One group forages high above the ground between the roofs of two tall buildings, while bees of the other group travel an equivalent distance close to the ground. Upon return to the hive the successful foragers perform their characteristic recruitment behavior—their 'waggle dances'—by which they inform other bees in what direction and distance to fly to the food source.[27] In these dances, they encode distance flown to a food source in waggle time, which the recruits conversely decode into distance. As predicted by the visual odometer hypothesis, the bees of the first group, which have flown at a high distance above the ground (and have perceived a smaller amount of image motion), signal a much shorter distance than the ones of the second group, which have traveled the same horizontal distance close to the ground.

However nice these demonstrations are that flying bees employ an optic-flow odometer, they also reveal a serious problem with this kind of distance estimation. The image motion generated by a unit distance of forward flight strongly depends on the distances of the objects relative to the bee. Indeed, dancing bees of the same colony exhibit considerably different distance calibration functions, i.e., functions that relate waggle time to distance, when they are induced to forage in differently cluttered environments.[28] This means that the system works accurately only when the bees fly along the same route during their outward and inward journey. For this reason, the scene-dependency of the visual odometer does not at all affect the reliability of the honeybees' recruitment dance, because the recruits take the same route through the same environment as the recruiters.

In the cataglyphs, which walk on the ground and look down to it, this problem of scene dependency does not exist. Hence, worker ants may have adopted the flow integrator from their flying ancestors as a reliable and supportive system. However, even when the function of the ant's pedometer is severely compromised, e.g., by the loss of one or two legs, it does not matter whether or not these handicapped ants are in addition ventrally blindfolded. The humbly walking animals are equally successful in determining their home distances, independently of whether or not they have access to optic-flow information.[29]

As to the neural mechanisms underlying the two ways of sensing distances, the pedometer receives its sensory input most likely from mechanoreceptors of the leg-alitrunk locomotor system, but how this information is handled on its way to the integrator is unknown. For the visually driven odometer, one would like to find neurons, which monitor translational optic flow and exhibit firing rates that are proportional to image velocity. In honeybees there are some hints that such neurons exist,[30] but the most elaborate analysis has been performed in the tropical bee *Megalopta*. There, a set of translational optic-flow detectors ('speed neurons') has been identified in the noduli of the central complex. In another compartment of the central complex, the fan-shaped body, the signals of these optic-flow detectors are integrated with the compass signals from the protocerebral bridge.[31] We shall return to this integration process in Chapter 5.

The Third Dimension

Until now we have considered the world to be flat, at least for the cataglyphs, but this is generally not the case. Even the surface of the *chotts,* the largely even terrain of the North African salt pans inhabited by *Cataglyphis fortis,* is rugged and ridged, and the *ergs,* the Saharan fields of sand dunes, the realm of the high-speed silver ants, are hilly to the extreme (Figure 4.10). How do the ants navigate within these types of habitat? In principle, they could follow some mathematically feasible procedures and perform path integration in three di-

4.10 Saharan sand dunes in the Grand Erg Oriental, the Great Eastern Sand Sea, Algeria

mensions. Alternatively, they could project all path segments taken in the three-dimensional (x, y, z) world down to a virtual (x, y) plane, so that path integration could again be performed within the familiar two-dimensional frame of reference. This mental 3-D to 2-D conversion might seem to be a fantastic leap of imagination, but believe it or not, this metaphor describes what the ants are actually doing.

Proof comes from ants riding on a roller-coaster. In this sawtooth-like device (Figure 4.1) the cataglyphs perform their outbound journeys along a series of alternating uphill and downhill channel segments, but return home in a straight channel aligned in parallel with the hilly outbound path. In this straight channel the ants perform sharp U-turns, and thus terminate their inbound journeys, after they have covered the ground (baseline) distance that corresponds to the distance actually traveled over the hills (Figure 4.11). They do so independently of the inclinations of the sloped ramps on which they are walking,[32] and even in 3-D channel configurations, which may give the impression of some bizarre pieces of landscape art. For example, in the channel array shown in Figure 4.12 the feeding station is placed high into the air, so that start and finish of the ants' foraging journeys are at different levels above ground. The cataglyphs must first climb a

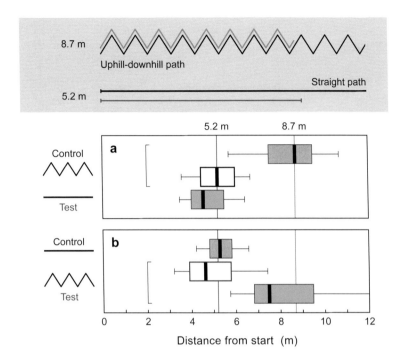

4.11 Uphill-downhill path

(**a**) Outbound runs along a hilly path (see Figure 4.1); subsequent inbound (test) runs in a straight channel. The *red boxplot* indicates the positions of the first U-turns. The *gray* and *white boxplots* depict the ants' travel distance in the training control along the hilly path and the corresponding ground distance, respectively. (**b**) Reverse situation. Outbound runs in the straight channel and inbound (test) runs along the uphill-downhill path. In the test the ants cover the uphill-downhill distance *(filled red boxplot)* of which the ground distance corresponds to the training distance in the straight channel *(open red boxplot)*. *Cataglyphis fortis.*

ramp (channel 1), then turn right, enter the horizontal channel 2, and after having performed another turn to the right reach the food source in channel 3. When captured at the feeder and released on level ground, they behave as if they had covered the ground distance of the ramped channel 1 rather than the distance actually traveled there.[33] Hence, they do not acquire a 3-D path integration vector. More recently, it has been shown that also the fiddler crabs mentioned above compensate for vertical detours by accurately traveling the horizontal distance, just as desert ants do, even though the crabs change their gate when walking on sloped and level surfaces.[34]

Whatever mechanisms the ants employ in performing the 3-D to 2-D conversion, first and foremost they must be able to measure the slope of the surface on which they walk.[35] Unlike vertebrates, ants do not have otolithic gravireceptors in their heads. Instead, they possess fields of mechanoreceptive hairs at the joints of all major body parts such as head, alitrunk, and gaster (see Figure 2.4b, c, and d).[36] Due to the influence of gravity these body parts alter their angular positions relative to

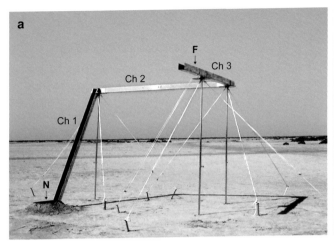

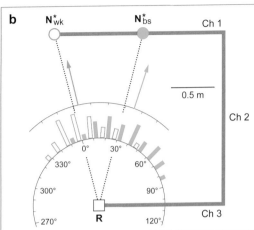

4.12 Three-dimensional maze

(**a**) 3-D training array. (**b**) Results. *Blue signatures:* Full-vector ants displaced from the feeder in the 3-D array to the flat test field. *Orange signatures:* 2-D control experiment with the training array of *(a)* laid down on the ground. *Ch 1, 2, 3,* open-topped channels 1, 2, and 3; *F,* feeder; *N,* nest; N^*_{bs} and N^*_{wk} fictive positions of the nest according to the baseline distance *(bs)* and the distance actually walked by the ants during training *(wk)*, respectively. The *arrows* indicate the ants' mean directional choices. *R,* release point. *Cataglyphis fortis.*

each other whenever the angle of inclination of the walking substrate changes, so that the fields of mechanosensors, the hair plates, which are associated with the body joints, could serve as some kind of distributed gravireceptor. If this were the case, changing or even abolishing the mechanoreceptive input by shaving the sensory hairs, or immobilizing the joints, or loading the animal with artificial weights, should severely impair the ant's ability to gauge the slopes of inclined paths and hence to compute the baseline distances correctly. But when the cataglyphs are subjected to these treatments, they still succeed in correctly projecting their sloped walking trajectories on the horizontal plane.[37] This is surprising indeed, as it has previously been suggested that especially the hair plates at the neck (head-alitrunk) joint and the alitrunk-petiole-gaster joints are involved in gravireception and slope detection. Due to technical difficulties, the hair plates associated with the legs, e.g., at the alitrunk-coxa joints, have not yet been manipulated.

If the hair-plate gravireceptors are not, or at least not predominantly, involved in measuring the slope of the ground on which the ants walk, could leg mechanoreceptors—campaniform sensilla or muscular strain sensors such as chordotonal organs—play a role? These sensors monitor the reactive forces experienced by the legs while the animal walks. Detailed measurements in cataglyphs walking on miniature force platforms show that the magnitudes and directions of the ground reaction forces of the legs change with the slope of the surface. These changes will certainly affect the strain on the leg segments and their joints.[38] Moreover, could vision provide the necessary cues? On inclines the angle between the head axis and the horizontal changes substantially. As can be deduced from high-speed video recordings, during 60° ascents or descents the ant's head is tilted so far that the mandibles are pointing either slightly upward or fully downward, respectively.[39] Consequently, the way of how the sky and its celestial compass cues are projected onto the ommatidial lattice of the eye is, at least to a certain degree, correlated with the inclination of the ant's walking substrate, so that theoretically it could be used as some means of monitoring the angle of inclination. However, such is not the case. When the ants are experimentally provided with polarization information that is identical in the even and uneven parts of a channel device, or when such information is absent at all, the animals are still fully able to distinguish between the two parts of the channel.[40] In conclusion, at the time of this writing the question of how the cataglyphs monitor the slopes of the terrain must remain open, but the proprioceptive inputs from leg mechanoreceptors are the most likely candidates.

As the foregoing has shown, the cataglyphs are able to measure the slope of the ground and use this information to compute a 2-D home vector, but they do not store this 3-D information in their path integration memory. However, the ant's memory system does not completely disregard the third dimension. For example, when the cataglyphs are trained within our 3-D channel systems to walk upward along a certain segment of their outbound path, they significantly more often climb downward on an unexpected ramp of their subsequent inbound path. As this downward climbing occurs at arbitrary points, the ants do not associate memories of inclined path segments with particular states of their path integrator.[41] Hence, the cataglyphs seem to acquire at least some information about the hilliness of a path along which they walk, but how this information is filed and linked to other behavioral routines remains to be explored. Yet with respect to path integration we can agree with Victor Cornetz, who long ago remarked that "for the ants the world is flat." Cornetz, however, did not have the ants' path integration system in mind.[42] In fact, in his days, path integration—the topic to which we shall turn in the next chapter—was not considered a routine that ants were able to employ.[43]

5 Integrating Paths

When Ulysses was on his long adventurous journey home from Troy, he finally reached the island of Kórkyra, modern Corfu. This island— as Homer tells us in his *Odyssey*—was inhabited by the Phaeacians, the finest seamen in the world of their day.[1] However, what the great epic poet of ancient Greece does not tell us are the techniques that these skillful sailors had used for navigation. The Phaeacians, he says, kept these techniques secret. The only remark King Alkinoos is told to have made was that "our ships have no steersmen, they know by instinct what their crews are thinking, and propose to do." What did they know "by instinct"?

We can only speculate about what the answer to this question might be. But in principle there was only one way by which the Phaeacians and other early navigators could have deduced their position after they had left their familiar coastal waters and ventured out into the novel, uncharted world of the ocean, out of sight of any land. They could have done so only by keeping track of their own locomotion, only by continuously integrating the directions they had steered (relative to some reference direction, say, the sun at daytime and particular stars at nighttime) and the distances they had covered in these directions. Whatever techniques they had applied to measure directions and distances, they must have been able to continuously integrate these two measures, i.e., to perform what in modern parlance is called path integration[2] or what was known to sailors at least since the sixteenth century as 'dead reckoning,' a term commonly used by the British

5.1 Filming path-integration studies

In August 1998 a BBC Green Umbrella team recorded studies on path integration in *Cataglyphis* at our North African study site near Mahrès, Tunisia.

Royal Navy.[3] As path integration results in the computation of a vector, the path integration vector, it has also been dubbed vector navigation.[4] We do not know what mental computations the Phaeacian navigators had performed to accomplish this path integration task, but they must have accomplished it in one way or another.

More than two thousand years later, Christopher Columbus had to rely on the same general strategy of path integration, of deducing his position "by account," from "the course the ship steered and her speed through the water, and from no other factors," when he was sailing on his first voyage from the Canaries to his final landfall in the Bahamas.[5] In contrast to the Greek navigators, he could determine his course by means of a magnetic compass, the only indispensable navigational instrument he had aboard, and he could measure speed by employing some precursor of the chip log,[6] but the way by which he finally had to deduce the position of his little fleet was the same dead reckoning strategy the Phaeacians had applied "by instinct." What "by instinct" means in *Cataglyphis* terms is the topic of this chapter.

We start by again consulting Figure 1.3, which I have redrawn in Figure 5.2a to emphasize that vector navigation is an ongoing, continuous process, in which the animal updates its path integrator by monitoring its change in position at every step, and thus keeps a running total of the distance and direction that it has traveled from the starting point. Rather than recording and memorizing the spatial details of its outbound path, rather than forming a route engram, so to speak, at any one moment the ant refers only to the resultant path integration vector and can thus select the direct route between its current location and home. This is illustrated in Figure 5.2a by the green lines that indicate what the path integrator tells the ant at three arbitrary points of the outbound path. A direct demonstration that the animals can behave according to what the green lines imply is provided by the learning walks that the animals perform around the nest entrance at the very beginning of their foraging lives, or by the relearning walks induced later in foraging life by changes in the landmark panorama of the nest environs. During these winding walks they often stop, turn, and look back in the direction of the starting point of their journey, the invisible nest entrance (see Figures 6.20 and 6.21). They can accomplish this turn-back-and-look task only by reading out their path integrator.

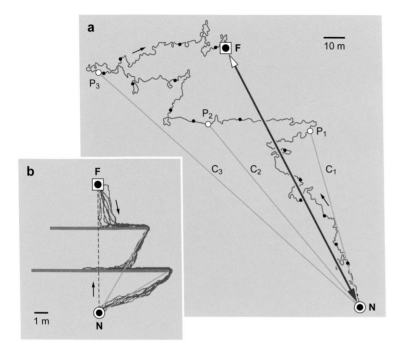

5.2 The cataglyphs continually update their path integration vector during their round-trip journeys

Figure (**a**) recapitulates the foraging round-trip shown in Figure 1.3. The states of the ant's continually running path integrator *(green lines)* are indicated for three arbitrary positions $P_{1, 2, 3}$, $C_{1, 2, 3}$, current vector states. The *red-filled and red-open arrows* mark the ant's final home and subsequent food vector, respectively. In (**b**) a forager has been trained to run straight from the nest to an artificial feeder. On its inbound journey *(blue lines)* it is forced to detour, nine times in a row, around two barriers *(gray bars)*. The *green lines* depict the paths to be taken if the ant had fully updated its home vector. *F*, food finding site (**a**) and artificial feeder (**b**); *N*, nest. *Cataglyphis fortis.*

Finally, when the ants have returned to the nest, their path integration vector is zeroed, but the vector leading to a profitable food site is memorized and can be retrieved and used on later forays.

In addition, Figure 5.2b presents a nice illustration of how successfully path-integrating ants redirect their courses after unexpected detours. In the present case the ants must bypass two barriers placed in the line of their direct inbound route. Once they have reached the end of a barrier and hence recovered from the enforced detour, they immediately reorient toward the position of the nest. As they also do so after they have been displaced (together with the barriers) to unfamiliar territory, their directional choices at the end of either barrier provide clear evidence that their path integrator is updated all the time while they are running their obstacle course.[7] In fact, path integration is the cataglyphs' basic navigational routine, which operates incessantly when the ants are outside the nest. It works not only in featureless or novel terrain, where it is the only means of navigation, but also—as we shall see in Chapter 7—when the ants' behavior is governed by other and more reliable routines relying, for instance, on

5.3 Coordinate systems for path integration: geocentric (earth-bound) and egocentric (animal-bound)

Within either system the ant's path integration vector *(PI vector, green)* can be defined in Cartesian coordinates (spatially defined, 'static' reference directions) or polar coordinates (variable, 'dynamic' reference directions). The geocentric angle η is measured with respect to an external reference direction provided here by the x-axis, e.g., the sun direction. The egocentric angle σ represents the angle between the ant's forward direction *(black arrow)* and the goal direction.

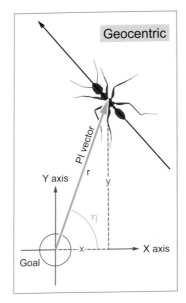

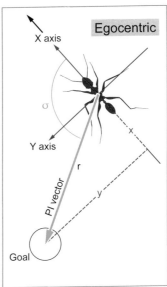

landmark-based information. Moreover, it works already when the ants start their outdoor lives, so that they can rely on it even under the most exceptional circumstances.

Some General Remarks

In the animal kingdom path integration is a rather ubiquitous navigational routine, which has recently become the focus of substantial theoretical work. Among the animals in which path integration has been investigated in considerable behavioral detail are ants, crabs, spiders, and rodents, but also bugs, bees, and birds, and even humans.[8] These studies, especially recent experiments performed in desert ants, have sparked the interest of theoreticians in modeling path integration. One of the first questions to be asked in this context refers to the frame of reference (Figure 5.3). Where is the path integration vector anchored—at the moving animal or at the point of departure? In other words, do we have to consider an egocentric (animal-bound) or a geocentric (earthbound) frame of reference? Moreover, the animal may integrate its path within a static system of coordinates, e.g., with two reference directions as in a Cartesian system, or the reference direc-

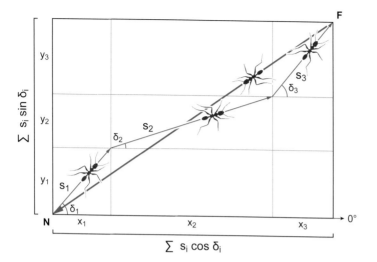

5.4 Updating the path integrator: the classic 'bicomponent model'

In this illustration of a geocentric system with static coordinates—two orthogonal (Cartesian) coordinates—the animal's outbound path is broken up into three straight path segments of lengths s_1, s_2, s_3, which deviate from a constant direction (x-axis, 0°) by the angles δ_1, δ_2, δ_3. The trigonometric components $\sin \delta_i$ and $\cos \delta_i$ are multiplied by their respective path segments s_i and added up. If s_i gets infinitesimally small, summation transforms into integration.

tions may be free to rotate (dynamic system of reference, in which polar coordinates are used for defining the animal's position).[9] When in updating its path integration vector the animal must record rotations, as in the dynamic system and in the egocentric static system, theory tells us that the accumulation of angular errors renders the animal's estimate about its position much less precise than in the case of the geocentric static system. In fact, in the classical 'bicomponent model' developed already in the 1960s at the Max Planck Institute for Behavioral Physiology at Seewiesen, Germany, Horst Mittelstaedt used just such a geocentric static system of reference (Figure 5.4). In this model the animal is considered to compute two Cartesian components, the sine and cosine component of its angular deviation from an external compass cue, and after modulating them by odometric data to integrate them independently.[10] In the wake of this approach several other path integration models have been devised. We owe it to the comprehensive and profound work of Allen Cheung at the Queensland Brain Institute in Brisbane, Australia, and Robert Vickerstaff at the Lincoln Research Centre in Christchurch, New Zealand, that the potentialities and constraints of the various types of models have been systematized.[11] For the moment, however, we are mainly interested in how the cataglyphs use their continually updated path integration vector rather than in the updating process itself.

Nevertheless, there is one important point to be considered here. Any path integration process is unavoidably prone to cumulative random errors, i.e., to the accumulation of noise from various internal and external sources. This problem becomes especially challenging when the path integrator acquires its information from internally generated, self-motion cues (idiothetic information) rather than from external references (allothetic information).[12] A telling example are the accelerometers of the inertial navigation system traditionally used in aeronautics, where the angular and linear accelerations experienced by the navigator during all the twists and turns of the journey are recorded and double-integrated over time.[13] Systems of that kind were once discussed for birds, but in a critical study performed in pigeons by the late William Keeton at Cornell University, Ithaca, New York, were later dismissed.[14] However, they might be used by some mammals. When foraging at night, rodents such as mice, hamsters, and gerbils have been described to depend on inertial navigation by exploiting signals from angular and linear acceleration sensors located in the vestibular system (in the semicircular canals and the otoliths, respectively).[15] Insects and spiders do not possess these systems, but in a few cases have been described to perform path integration based on idiothetic information alone. When nocturnal spiders, *Cupiennius salei,* wander about in the neotropical rain forest, they integrate their paths by drawing upon proprioceptive information from cuticular strain receptors, the so-called lyriform slit-sense organs, which are located near the joints of the spider's legs. If these organs are experimentally destroyed, the homing abilities of the spiders are abolished. Similarly, cockroach larvae perform path integration in complete darkness by apparently also relying only on idiothetic cues.[16]

Theoretical considerations and computer simulations show that idiothetic path integration is extremely sensitive to noise. Provided only with internally generated angular inputs, animals could not even move in a straight line far away from home and would soon end up moving in random directions.[17] In a study on humans Jan Souman and his collaborators at the Max Planck Institute for Biological Cybernetics in Tübingen, Germany, asked their subjects to walk for several hours in a straight line through a large flat forest area in central Europe or across desert terrain in southern Tunisia.[18] The walking trajectories were recorded by a global positioning system (GPS). Under overcast conditions, with no celestial compass cues available, the subjects repeatedly veered from the intended straight course, looped back, and walked in circles. As this extreme circling behavior was not observed when the sun was visible, it most likely resulted from the buildup of noise in a navigational system that had to rely mainly on idiothetic cues.

In conclusion, an external directional reference is indispensable for any organism and agent that navigates by path integration over long distances. Seen in this light, the intricate compass systems employed by the cataglyphs are some of the most precious sensory and neural gadgets in the ant's navigational toolbox, which is why we have dwelled on these mechanisms at quite some length in Chapter 3.

The Path Integration Vector

In returning to the cataglyphs, let us begin with outlining a set of experimental paradigms, which may help to understand how the ants might use their path integration system during their foraging journeys (Figure 5.5). We first consider a forager that has arrived at a food site *(1)* and is displaced sideways. After release, the displaced ant, still having its full home vector at its disposal (hence 'full-vector ant'), rotates until its body axis is aligned with the direction of its home vector and then immediately starts to run off this vector *(2)*. It is as if the animal, having filled a path integration accumulator during its outbound run, now tries to empty the accumulator on its way home. When the home vector has been zeroed, the ant (now being a 'zero-vector' ant) assumes to have arrived at the nest. As in fact the nest is not there, it starts a systematic search centered about the tip of the home vector. For an example, please have a look back at Figure 1.5. Another way to demonstrate that the path integrator has reached its zero state after the ant has arrived at the nest *(3)* is to capture the ant on its return journey just when it is about to enter the nest, and displace it to foreign terrain *(4)*. There it would again perform its systematic search behavior. In landmark-free environments one can displace the ant even back to the very feeding site from which it has returned a few minutes ago, and observe that it would now concen-

5.5 The ant's path integration vector: some basic experimental paradigms

(a) Standard displacement experiment with a full-vector ant, i.e., an ant that is displaced from a feeding site, *F*. (b) Displacement experiment with a zero-vector ant, i.e., an ant that is displaced from the nest, *N*, after it has returned from a foraging journey. The food vector *(6)* is the inverse of the home vector *(3)*. (c) Recalibration of the path integration vector after the ant has been subjected to an open-jaw experiment. For experimental procedures *(1–10)*, see text. *Long, heavy black arrows,* the ant's paths based on the path integration vector; *dotted arrows,* displacements; *blurred blue circles,* ever widening search patterns. *N*,* fictive position of nest; *R,* release point; α, angular deviation of R→N from R→N*; δ, angular deviations of the ant's inbound and outbound paths from the directions R→N* and N→F, respectively.

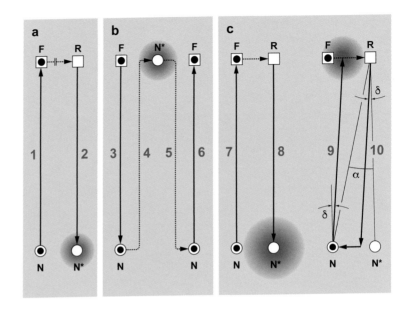

trate its search behavior in a radially symmetrical way around that site. However, when displaced back into the nest *(5)*, or usually when having actively returned to the nest, it later sets out for a new foraging journey to the former feeding site *(6)*. This behavior unambiguously shows that a 'food vector'—the inverse of the home vector—must have been stored in long-term memory.

Let us now ask a question that might seem peculiar at first glance, but will lead to a satisfying answer on second thought. Would one be able to train the cataglyphs to acquire and store separate inbound (home) and outbound (food) vectors, i.e., vectors that are not just 180° opposites of each other? One way to explore this question is to employ an open-jaw test paradigm (Figure 5.5c), in which the point of arrival at the feeding site does not coincide with the point of departure for the homeward trip. As already shown in Figure 5.5a, in this situation the displaced ants first run off their inbound (home) vector and then start their systematic search about the fictive position of the nest. If the displacement distance F→R is not too large, the search routine will finally enable the ants to arrive at the real position of the nest *(8)*. Now comes a surprising event. When the ants set out for the next foraging trip, they do not run straight toward the previously visited food site, but deviate from the direct N→F course by a small angular amount δ *(9)*. Obviously, the experimental open-jaw paradigm has caused the ants to recalibrate their path integration vector. If the recalibrated animals have subsequently arrived, after some search, at F and are then subjected to the same F→R displacement as before, they deviate from their previous homeward course R→N* by exactly the same angular amount by which they have deviated from N→F on their outbound run *(10;* $\delta_{in} = \delta_{out}$*)*.[19] They never choose the direct course to the nest, i.e., the angle α.

Figure 5.6 depicts the results from an open-jaw experiment. After the ant has reached the feeder (Figure 5.6a, *1*), it is displaced to R. Upon release it first walks directly back to the fictive position of the nest, N* *(2)*, and then employs its search routine, which finally brings it back to the real nest. There is nothing peculiar yet with this behavior.

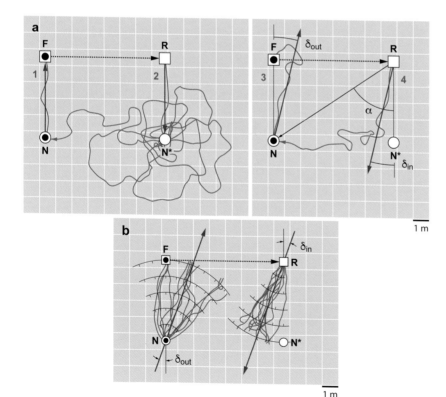

5.6 Vector recalibration: the open-jaw experiment

Experimental paradigm: Outbound runs from nest *(N)* to feeder *(F)* followed by sideways displacements *(dotted arrows)* to release point *(R)* and subsequent inbound runs. See Figure 5.5c. *N*,* fictive position of nest. The ant's paths *(blue)* are shown together with the mean directions of the outbound and inbound paths *(red arrows)*. **(a)** Example of the behavior of an individual ant. First *(left)* and second *(right)* open-jaw experiment. **(b)** Twelve inbound paths *(right)* and subsequent outbound paths *(left)* of one individual ant. For conventions, see Figure 5.5. *Cataglyphis fortis.*

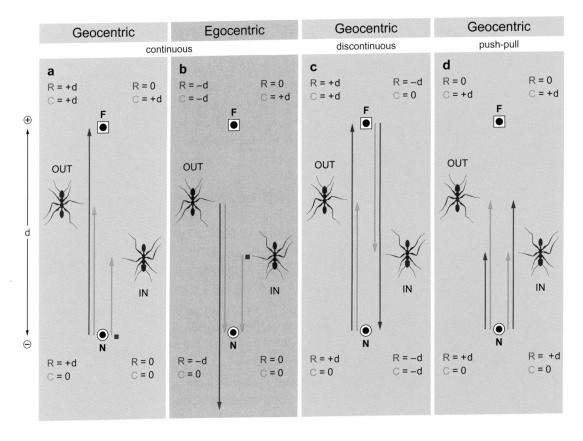

However, in the subsequent open-jaw procedure *(3, 4)* we encounter the very deviations $\delta_{in} = \delta_{out}$ that we have already seen sketched out in Figure 5.5c *(9, 10)*. This $\delta_{in} = \delta_{out}$ correspondence is further corroborated by the data shown in Figure 5.6b. There, an ant has been subjected to the open-jaw test paradigm 12 times in succession.

What happens during these open-jaw round-trips in the ant's perspective? Having returned to the nest after its first displacement from F to R, the ant experiences a mismatch between the current state of its path integrator and the expected zero state. It realizes, so to speak, that upon arrival at the nest a residual home vector remains (corresponding to the experimental displacement). Upon setting out for its next foraging journey, it uses a fraction of this residual vector to recalibrate the food vector in its long-term memory. The amount of recalibration may depend on a number of factors: on how far the animal

5.7 Potential modes of using path integration in navigation

(a, b) Continuous (geocentric and egocentric) mode, (c) discontinuous geocentric mode, (d) Push-pull mode of operation. *OUT,* outbound (foodward) run from the nest *(N)* to a familiar feeder *(F); IN,* inbound (nestward) run. The *red* and *green* arrows indicate the directions and lengths of the reference vectors *(R)* and current vectors *(C),* respectively, at the site of the *ant icons.* The resulting travel vector is defined by R-C. The values for *R* and *C* are given for the nest site *(bottom)* and the feeding site *(top). d,* distance; for sign conventions, see (+) and (−) on the left-hand side.

has been displaced (i.e., on the angle α in Figures 5.5c and 5.6a, right panel); on the familiarity with the previous food site; on how long the displaced ant had to search for either the nest or the feeder; or on the number of consecutive displacements. It may also depend on the species and the type of habitat within which the species has evolved. Note that all open-jaw experiments presented here have been performed with *Cataglyphs fortis* in featureless salt-pan terrain. The questions outlined above are topics for future research.[20] At present, we only know that recalibration is a quick process. It is completed after one to three forced detours. Even if the open-jaw procedure continues, the animals do not change their compass setting. In the described experimental paradigm we never observed that the displaced ants walked straightly back from the release site to the nest by choosing the direct R→N path. After more than 50 trials we, rather than the cataglyphs, lost patience and ended the experiment on this matter.[21]

At this juncture the question may arise how the cataglyphs use a memorized path integration vector, or several of them, in navigating to and from habitual food sites. Figure 5.7 sketches out some possibilities. At present, these possibilities are nothing but purely theoretical considerations. Based on some experimental evidence, we shall finally favor one of them, but for outlining the problem it might be helpful to go through a bit of exercise in discussing also some other ones.[22] First, as depicted earlier, path integration could work in an egocentric or geocentric frame of reference depending on whether the moving animal or the stationary starting point, respectively, is used as the origin. Please have a look back at Figure 5.3. In the following we assume that the ant has already discovered a lucrative food site, F, and has stored its coordinates, the reference vector,

R, in memory. Let us start with the geocentric case and have a look at Figure 5.7a. When the ant sets out for a new foraging journey to a habitual feeder F, the reference vector R anchored at the nest is loaded down in the working memory from some long-term memory store. While on the move, the animal builds up a current home vector C. By subtracting the latter vector from the former, the path integrator yields a travel vector $T = R - C$. This travel vector guides the ant to its goal, which is reached when T is zeroed. For the return run, the ant sets its reference vector $R = 0$, by inhibiting the downloading from long-term memory, and walks until its travel vector T has again reached its zero state. The use of an egocentric system of reference is shown in Figure 5.7b. Now all vectors are anchored at the animal, but otherwise, the same considerations apply. As at F the ant makes a U-turn, C should change sign.

Second, the path integrator could run continuously during the entire round-trip journey, as considered so far (continuous mode), or the current vector could be reset to zero not only at the nest site but also at the food site (discontinuous mode, Figure 5.7c). Each time resetting occurs, R changes its sign, meaning that the polarity of the compass is inverted. Hence, in discontinuous vector navigation the resetting of C and the concomitant rotation of the compass through 180° at the nest and the feeding site render path integration during the outbound and inbound journeys completely equivalent processes. Finally, one can think of a mode of operation in which not only the current vector C but also the reference vector R is dynamic (Figure 5.7d). In this 'push-pull' mode—what is pulled from R is pushed into C, and vice versa—the vectors R and C are no longer compared. Instead, R becomes the outbound travel vector, and C the inbound travel vector. During the outbound and inbound trips the animal tries to zero R and C, respectively.

After this theoretical exercise, let us go out to the fields and design experiments that may shed at least some light on the issues outlined above. First, we use a special array of open-topped channels to let the cataglyphs acquire the vectorial representations of two spatial goals, i.e., two reference vectors, R_A and R_B, pointing at two food sites, A and B (Figure 5.8). When later deprived of food at, say, A, the ants first search around A, but then move directly to site B along a route that

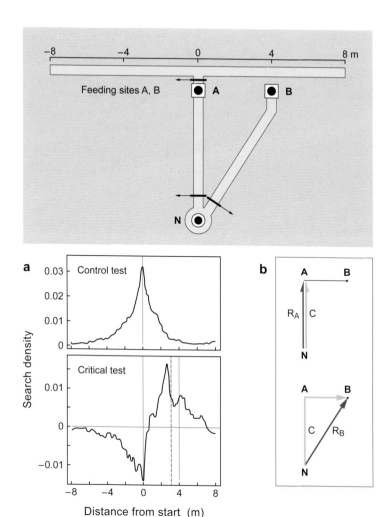

a

b

Distance from start (m)

5.8 Performing novel shortcuts by path integration

Top: Experimental paradigm. Open-topped channels are used to train ants from the nest *N* to two food sites, *A* and *B. Small black bars* associated with *arrows*, shutters. (**a**) Results. The ants are allowed to enter the test channel at site A. *Control test:* After training only to site A. *Critical test:* After training to sites A and B, in the test the ants are deprived of food at A. The graphs show the ants' search distributions in the test channel. The *lower graph* depicts the difference of the ants' search distribution and the control distribution. *Dashed vertical line:* Ant-subjective location of B within the channel array (see note 23). (**b**) Interpretation. Vector integration scheme for control *(top)* and critical test *(bottom). C,* current vector; R_A and R_B, reference vectors for sites *A* and *B,* respectively. *Cataglyphis fortis.*

they have never traveled before. This novel-route behavior can readily be explained on the basis of the ant's vector navigation system. Just assume that when the ant has reached A, its current vector C matches R_A, but when the ant does not find food at A, it would load down the reference vector for site B, R_B, and walk until $C = R_B$.[23] In this kind of reasoning we have tacitly assumed that the ant uses a geocentric system with R_A and R_B being anchored at the nest site. In an egocentric system, in which the vectors were rooted in the moving animal, these vectors would have to be updated continuously, step-by-step: not only the cur-

5.9 Entering the nest is necessary for resetting the path integrator

Top: Test arena. A cataglyph carrying a food item travels from the feeder to the nest hole in the center of the arena. Results obtained in the arena and in an open test field *(bottom)* do not differ significantly. **(a)** Experimental paradigm. Ants are displaced from the feeder *(F)* to the nest *(N)* in one of three ways: by releasing them close to the nest entrance *(test 1A),* by placing them in small wire-mesh cages into the nest opening and releasing them thereafter *(test 1B),* and by letting them vanish into the nest *(test 2).* **(b)** Results. The ants' subsequent walking trajectories in tests 1A *(black)* and 1B *(blue),* and after in test 2 *(red)* the ants have reappeared at the nest entrance—with their vector reset—and start for their subsequent foraging journeys. *Gray* trajectories underlying the *red* ones refer to control ants heading for the feeder without having been displaced in any way. *Cataglyphis fortis.*

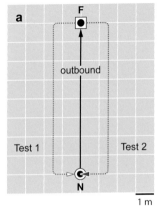

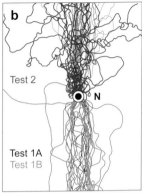

rent vectors in working memory, but also the reference vectors in long-term memory, even the ones that are not used at the moment. Moreover, any rotation of the animal even on the spot must be integrated to update the current home vector. All these effects would result in enormous noise accumulation and render an egocentric system extremely susceptible to cumulative errors and thus a rather unfavorable frame of reference for path integration.

In a geocentric system the origin and final destination of all round-trip journeys, the nest, is the place at which the ant can be absolutely sure about the coordinates of its location. It is thus the place where the path integrator is most likely reset. However, just being at the nest site does not suffice (Figure 5.9). When in largely featureless terrain ants are captured at the feeder (i.e., full-vector ants) and released close to the nest entrance, where they even might have some visual or olfactory nest-defining cues at their disposal, they do not enter the nest but run off their home vector by heading away from home. Even if they are allowed to have antennal contact with their indoor nestmates (by being confined to a small wire-mesh cage placed in the entrance shaft of the nest), they do not reset their path integrator. Moreover, conspicuous nest-defining landmarks, though demonstrably used by the ants for localizing the nest entrance, do not help either.[24] As it appears, it is only inside the colony that the cataglyphs can reset their path integrator from any nonzero state to zero.

Behavioral studies will rarely be able to finally decide which of the alternatives outlined in Figure 5.7 is implemented in the ant's neural cockpit. Nevertheless, as judged from the experiments described above, we favor the continuous geocentric mode of operation (Figure 5.7a). It is also theoretically that this mode would require the least amount of memory manipulation.[25]

Search Strategies

It was during three weeks of miraculously cloudless summer days in the early 1970s that my wife, Sibylle, and I were slowly wandering about, day after day, on a sandy plain next to the North African village of Mahrès. Our eyes were persistently fixed on the ground where a cataglyph was running at a steady pace. The ant continually looped around an imaginary point. There was nothing special by which this point could be discriminated from any other point on the desert floor, but it was the point where the cataglyph expected to find the entrance hole to its subterranean colony. Its colony, however, was more than half a mile away. We had captured the ant there after it had successfully finished a foraging journey, and we had displaced it from its nest site to that very point on the featureless plain about which it was now performing its search movements. How did these searches develop over time? This was the question we wanted to pursue in individuals that had already run off their home vector and thus expected to be at, or at least close to, their nest entrance hole. However, the nest was out of reach, so that we could let these zero-vector ants search unsuccessfully for unlimited periods of time. Actually we did so for one hour in each ant. As the searches went on, the loops remained strictly centered about the starting point, the origin, but expanded in size until they finally covered an area of almost 10,000 square meters (Figure 5.10a). Much impressed by this remarkable centering of the ants' search loops around the origin, we wondered whether terrestrial cues of whatever kind could have been used as a reference. Such was not the case. When we displaced the ants to a point some 10 meters away, they did not move back to the former origin. Irrespective of the displacement distance and direction, they continued their searches around the displaced fictive position of the nest as if nothing had happened. During the entire search time they relied exclusively on their path integration routine.

Note that for an ant homing by path integration, the tip of the home vector is not pointed but blurred. Path integration is an iterative process. The ant's estimated position at the end of each step is based on the ant's position estimated at the end of the previous step. All position errors made in acquiring and processing the necessary information add up continually. Hence, the longer the foragers are on their way, the less certain they are about the position of their starting point, the nest—and the more widely do the endpoints of their straight homeward runs spread out (Figure 5.11).[26] On the one hand, these data show that the ants' precision (meaning the random component of the ants' errors) deteriorates quite substantially as foraging runs get longer. On the other hand, the ants' accuracy (meaning the systematic component of the ants' errors) seems to be only slightly impaired. In the case of the larger foraging distance shown in Figure 5.11a (right part) the ants tend to break off their straight homeward runs a bit before the fictive position of the nest is reached. However, this tendency markedly increases as foraging distances get larger. For example, when the cataglyphs are trained and tested in the channel device described in Chapter 4 (see Figure 4.2), and when training distances increase up to 50 meters, the ants' underestimation of the homing distances gets quite substantial (see Figure 7.8b*). It can be described by a saturating exponential function, which is compatible with the assumption of a leaky integrator, i.e., with the idea that a certain fraction of the dis-

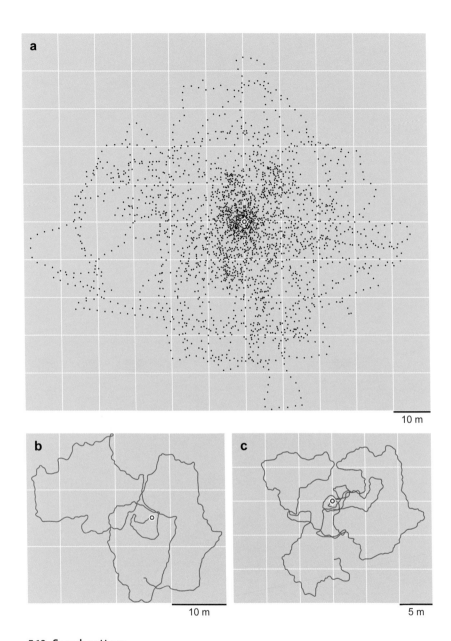

10 m

10 m   5 m

5.10 Search pattern

(a) The spatial distribution of the search behavior of eight zero-vector ants, each recorded for a 1 h search period. Fixes are taken every 2 m of path length. (b, c) Search paths recorded in two zero-vector ants; search time recorded: 18.75 min *(b)* and 17.50 min *(c)*. *Open red circle,* release point. (a) *Cataglyphis bicolor,* (b, c) *C. fortis.*

5.11 Positional uncertainties

(a) Spatial distribution of the end points of the ants' home runs *(filled circles)* and the centers of the subsequent 10 min searches *(open triangles)* in the test field. *Red circle,* fictive position of nest; d_{nf}, nest-feeder distance. As the ants frequent a familiar feeder, path length and beeline distance, d_{nf}, are almost identical. (b) Distances of the ants' search centers from the fictive position of the nest *(left)* and variances of these positions *(right)* as a function of d_{nf}. *Cataglyphis fortis.*

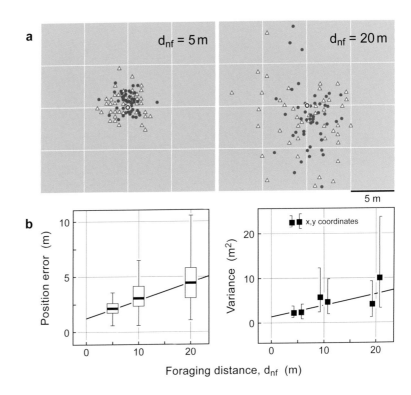

tance traveled is gradually leaking out from the integrator and thus decreasing its accuracy.[27]

Owing to these uncertainties and inaccuracies inherent in any path integration system, backup strategies are necessary for finally pinpointing the goal. Such strategies may exploit visual landmark cues,[28] tactile cues provided by surface structures of the desert floor,[29] and olfactory landmarks such as the carbon dioxide odor plume emanating from the nest entrance.[30] If these signposts are not available, the ant resorts to what could be considered its final emergency plan.

An efficient way to search that comes to mind would be to move along an ever-expanding arithmetic (Archimedean) spiral.[31] However, a spiral search is a reasonable strategy only for a perfect goal-detecting agent. For had a cataglyph missed the nest entrance while being next to it, there would be no way of getting as close to it again. Instead, by continuing to spiral outward, the ant would move farther and farther

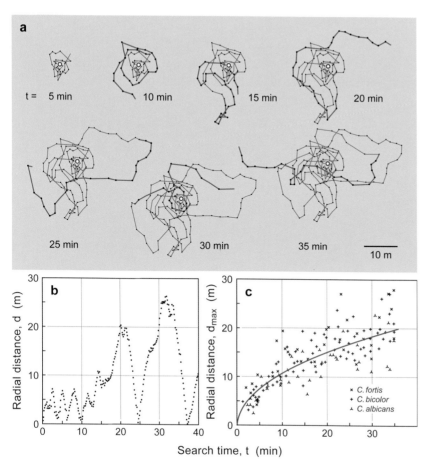

a

t = 5 min   10 min   15 min   20 min

25 min   30 min   35 min   10 m

b

Radial distance, d (m)

Search time, t (min)

c

Radial distance, d_{max} (m)

× C. fortis
+ C. bicolor
▲ C. albicans

Search time, t (min)

5.12 Characteristics of search patterns

(a) Search path of an individual zero-vector ant. *Open red circle, release point.* The sequence of search path segments shows how the ant's search pattern develops over time. The *heavy blue line segments* depict the routes taken in the last 5 min of each recording. (b) An ant's radial distance from the release point as a function of search time. (c) Maximal radial distances, i.e., the maxima in *(b)*, in three *Cataglyphis* species. *Red curve,* prediction from optimal search theory: search radius increases as the square root of search time. *(a, b)* time marks *(black dots)* are given every 10 s. (a, b) *Cataglyphis fortis*, (c) three *Cataglyphis* species.

away from the goal. In fact, the searching cataglyphs do not move along a spiral path. As Figure 5.12 shows, they perform a series of loops, which gradually expand over time, spread out in all directions, and are always centered about the point where the home vector has been fully paid out, the estimated home location (the zero-point of the search).[32] What results is a radially symmetric, roughly bell-shaped search density profile. As expected for an adaptive search strategy, the ants adjust the width of their 'search density distribution' to the uncertainty that has built up in their path integration system. This uncertainty can be described by the 'target probability distribution.' In Figure 5.11a it is represented by the distribution of the data points that mark the end points of the straight homeward runs. The larger the foraging distance has been, the wider do these points spread out, and the wider are the sweeps that the searching ant subsequently takes from the beginning (Figure 5.13).[33] In short: the search density distribution is adapted to the target probability distribution.

On a small scale, search movements were first observed around a fictive food site rather than a nest site. More than a century ago, Charles Turner, an eminent African American zoologist, recorded the meandering search movements of a small *Camponotus* ant, which he had displaced from its feeding place. Some contemporary myrmecologists referred to these search loops as "les tournoiements de Turner" (Turner loops) or "la recherche concentrique" (area-concentrated search). Many decades later, Vincent Dethier, then at Johns Hopkins University in Baltimore, described "fly dances," in which walking blowflies repeatedly looped around newly found food sources. This behavior has recently been studied in detail in fruit flies. When tested in darkness, the flies are even able to accomplish their search task by fully relying on idiothetic cues.[34] At daytime and at a much larger scale, the cataglyphs perform such focused local searches not only, as discussed above, around the nest, but also around frequently visited food sites. Depending on the kind of food offered at artificial feeding stations, the abundance and reliability of the food source, and the ant's familiarity with the source, *Cataglyphis* and *Melophorus* foragers are quite flexible in adjusting their searches.[35] For example, when they are presented with protein-rich and carbohydrate-rich sources, their searches are more concentrated for the latter. This corresponds to the natural situation, in which carbohydrate sources such as flower nectar and plant exudates are more likely to be renewable and spatially more concentrated than the usual protein sources, insect corpses.

5.13 Search density profiles recorded for zero-vector ants and different foraging (nest-feeder) distances, d_{nf}

The three-dimensional search profiles depict the relative amount of search time spent by 125 ants during 15 min searches. Size of unit squares: 2×2 m². *Cataglyphis fortis.*

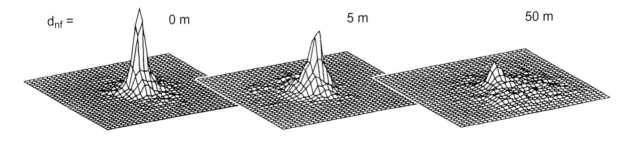

$d_{nf} =$ 0 m   5 m   50 m

5.14 Desert ant, *Cataglyphis bicolor*, meets desert woodlouse, *Hemilepistus reaumuri*

While on its way back from a nocturnal foraging trip, a woodlouse encounters a cataglyph, which had just left its nest in the morning hours. Being antennated by the ant, it exhibits a startle response. Both species engage in area concentrated searches when pinpointing the final goal of their journeys, the nest entrance.

Rather than contemplating about the details of such adjustments, let us finally switch to a desert dweller that lives along with the cataglyphs and searches in a similar way: the desert woodlouse, *Hemilepistus reaumuri*.[36] As these woodlice avoid the heat of a summer day and hence leave their subterranean burrows only after sunset and retreat to them already early in the morning, they usually do not meet the cataglyphs (for an encounter, see Figure 5.14). In terms of evolutionary relationship, they are quite apart from them. They are crustaceans. However, their marine isopod ancestors have invaded the land in such a successful way that irrespective of their original aquatic lifestyle they now inhabit vast arid areas of North Africa and Asia in high population densities. They form long-lived family groups, of which each is bound to a fiercely defended burrow.[37] When foraging for plant and dead organic matter they leave their burrows for a few decimeters or meters, but due to the error-proneness of their path integration system must engage in lengthy searches for the small entrances to their burrows, which they can finally localize only by olfactory means. As their parental care period lasts for 50–200 days, and as the adults perform dozens of foraging excursions per night, the number of outdoor round-trips per individual exceeds that of the cataglyphs by two orders of magnitude, but structurally their search patterns look like miniaturized versions of the cataglyphoid ones (scaled down in dimensions by one or two orders of magnitude). The same can be said for the searches of some seed-collecting beetle larvae returning to their burrows, and for young cockroach instars searching for their shelters.[38]

On a larger scale, even the behavior of lost people conforms to this pattern. The databases obtained from search-and-rescue operations clearly show that people lost in the wilderness may wander for great distances in convoluted paths, but then unwittingly return to where they started their search, so that they are often found relatively close to their last known position. In the same vein, the Swedish cultural historian Gunnar Hyltén-Cavallius describes how the nymph Skog-snuva guides lost people circuitously around in the woods and then always leads them back to the same place.[39]

In detail, what search strategy should people apply when they have gotten lost, or when they search for a lost object? Particularly during World War II, the theory of search became an urgent scientific endeavor. Then, Bernard Koopman headed a US Navy Operations Research Group, which worked on strategies for antisubmarine warfare, and in this context developed concepts and methods of optimal search. "When an object of search on the ocean such as a downed airplane life raft has had its appropriate position disclosed to a searcher, the searcher has the problem to maximize his chance of detecting the object. . . . The information regarding the object's approximate position may be derived, in the case of the life raft, from a radio communication from the aircraft about to crash. The point at which this information locates the target is called the *point of fix*." Correspondingly, in the case of the cataglyphs this point of fix is the point where the path integrator has reached its zero state. "If the fix," as Koopman continues, "were a perfectly accurate one, the searcher's task would be simple, but such accuracy of fix is seldom if ever obtained: only a probability distribution of target positions at the time of fix is given. This probability density function will have its greatest density at the point of fix and will fall continuously to zero at a distance. In an important group of cases, this distribution can be regarded as symmetrical about the point of fix and can indeed be taken with satisfactory accuracy as a circular normal one."[40]

A loop search strategy based on these considerations and on the subsequently developed optimal search theory was applied when in January 1966, during the Cold War, the US Navy had to search for a hydrogen bomb lost by a US Air Force plane in the Mediterranean Sea near the fishing village of Palomares (Spain). A Boeing B-52 Strato-fortress plane carrying the bomb had collided with a tanker plane during a midair refueling attempt, and subsequently crashed. A local fisherman, who had observed the bomb entering the water, was able to provide the US Navy with the point of fix, where the loop search started. After several weeks of search, the bomb was finally located and rescued. It is also in the cataglyphs that the searches might take quite some time, but in the end they are almost always successful.

Finally, we may ask whether the ants' search behavior—this 'emergency plan,' as we had dubbed it before—is really a separate navigational routine. Could it not directly result from the ant's continually running path integration system? A homebound ant slows down when its path integrator approaches its zero state, and maintains this slower speed during the subsequent search,[41] but the integrator keeps running while the ant searches. This may be one of the reasons why in the test field the 'turning point,' the *Abknickpunkt*, where the straight home run ends and the search begins, is usually difficult to detect, and why researchers have tried hard to develop proper criteria for defining it. Note that when the ant has passed beyond the fictive position of the nest, its path integration vector increases and thus would eventually 'draw' the animal back toward the zero-point. Then the path integration vector shrinks again, so that the ant's directional uncertainty increases and the next loop might lead the ant in another direction. If we further assume that with increasing search time the path integrator somehow 'loses its strength' by tolerating increasingly larger states of the travel vector to develop, the search loops would grow in size, again as observed. Increased uncertainty as it accumulates during longer and longer foraging journeys may have a similar effect. Indeed, modeling approaches are able to describe the main features of the entire homing process by a uniform mechanism, which results in straight home trajectories when the animals are far from the goal and looping search patterns when they are close to it.[42]

Forced Detours

Quite surprisingly, when a century ago some myrmecologists became interested in topics of navigation, they did not take path integration into account. Felix Santschi and Victor Cornetz did not mention it at all, and Rudolf Brun deliberately dismissed it. Even though Brun had observed that some wood ants were able "to close the polygon," he attributed this behavior to the use of landmark information and considered it beyond an insect's ken to solve the path integration problem—"a task which even a mathematically versed human could accomplish only by tedious computations and constructions performed with pen and paper, but the ants would have to do it off the cuff."[43]

Now that we know that the cataglyphs continually integrate their paths when they are on their round-trip journeys, let us apply the test channel device to demonstrate that they are indeed able "to close the polygon" by exclusively relying on path integration. In these experiments the ants perform their outbound runs within an L-shaped channel array, but are unconstrained during their subsequent inbound runs.[44] There, out in the fields, something unexpected happens (Figure 5.15). The animals do not head directly for home, but deviate consistently from their true homeward direction by exhibiting systematic inward errors. They turn through larger angles than necessary. The amount of this over-turning depends on the geometrical layout of the two-leg outbound path, i.e., the lengths of the two straight sections and the angle between them. Generally speaking, the ants' straight inbound trajectories deviate from the homeward direction in a clockwise (or counterclockwise) way, whenever the outbound journeys have been biased by clockwise (or counterclockwise) turns. These deviations can be simulated surprisingly well by again adopting a leaky path integration model, as we had inferred it in the preceding section for the underestimation of travel distances. An animal following an L-shaped, two-leg outbound path would underestimate the length of the first leg more than that of the second leg, and thus produce the inward errors observed in the experiments. What in Figure 5.15b looks like a curve fitted to the experimental data is in fact the theoretical prediction from this model.[45]

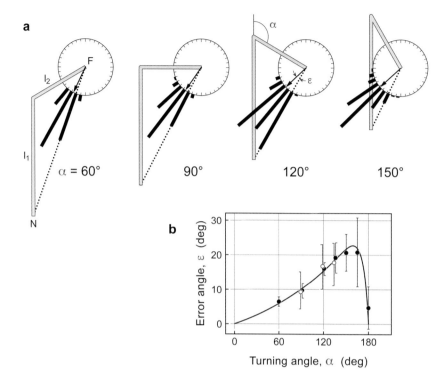

a

$\alpha = 60°$ 90°   120°   150°

b

Error angle, ε (deg)

Turning angle, α (deg)

5.15 The ants' behavior after one-sided turns (L-shaped routes): inward errors

(a) Forced-detour training paradigm. The ants perform their outbound runs in a two-leg channel array (N, nest; F, feeder; $l_1 = 10$ m, $l_2 = 5$ m; turning angle α) and their inbound runs in the open test field, where they deviate from the home direction (F→N) by the error angle ε. (b) Inbound error angles as a function of the outbound turning angles. During their outbound runs the ants can use either the sun (open circles) or the polarized skylight (closed circles) as compass cues. *Cataglyphis fortis.*

These systematic errors are not an artifact of the experimental manipulation within the channel array, but occur also under natural conditions whenever outbound paths are biased toward one side, as it often occurs especially in cluttered environments. In these cases, the home vectors do exhibit the inward deviations shown in the detour experiments (Figure 5.16). But what appears to be a deficiency, if the path integration system is considered in isolation, could turn out to be a functionally adaptive trait, if the system is seen in the context of the ant's overall navigational behavior. We shall return to this point in Chapter 7, when we discuss error compensation strategies. At this point, let me mention that during a subsequent literature search we came across some data showing that the directional errors unraveled in the cataglyphs had occasionally been observed also in other arthropods such as honeybees and spiders. Similar studies performed in mammals including humans arrived at the same conclusion.[46]

5.16 Two examples of one-sided outbound paths and the resulting inbound errors

Top: Outbound paths in open terrain. *Bottom:* Displaced to the test field, the ants deviate by the error angle ε from the true homeward course. *F,* food finding site; *N,* nest; *N*,* direction toward fictive position of nest; *R,* point of release.

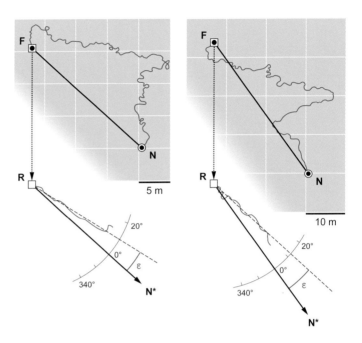

Direction and Distance Information Combined

The cataglyphs have several visual and nonvisual sensors for directions and distances at their disposal. How do these navigational gadgets interact? In principle, one possibility would be that irrespective of the sensory systems involved in acquiring information about directions and distances, this information is fed into a common compass unit and a common odometer unit, respectively, and only the outputs of these two units are continually combined. Alternatively, and as suggested in Chapter 4, there could be separate integrators, in which compass information is gated by information from sensors that monitor either leg motion or optic flow.

Beyond suggestions, let us perform an experiment in which the ants are deprived of celestial compass information during particular parts of their foraging journeys, while their stride meter is continuously running. This is achieved by constraining the ants to perform their outbound runs in a straight channel that consists of alternating open-topped and covered sections. When the animals then run home in a channel that is open-topped over its entire length, they cover only the distance traveled under the sky (Figure 5.17a). A similar

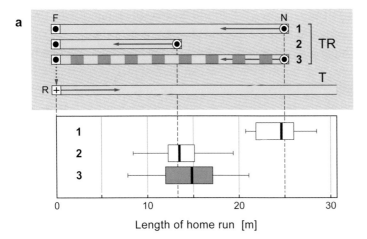

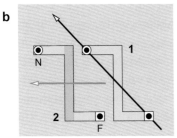

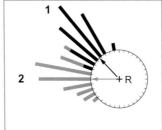

5.17 Combining sky compass and distance information

(a) One-dimensional paradigm. The ants perform their outbound runs (*TR*, training) in a straight channel, which either is open-topped over its entire length (*1, 2*) or consists of alternating open and fully covered sections (critical test, *3*), and their inbound runs (*T*, test) in an 'endless' test channel. *F*, feeder; *N*, nest; *R*, point of release. (b) Two-dimensional paradigm. Outbound runs in a Z-shaped maze, in which the central leg is either open-topped like the other two legs (*1*) or covered (critical test, *2*). Inbound runs are performed in the open test field. *Cataglyphis fortis.*

conclusion can be drawn from a two-dimensional analogue of this experiment (Figure 5.17b).[47] Hence, for computing the path integration vector paid out under open sky conditions, only the odometric information is used that has previously been associated with concomitant information from the sky. Of course, this does not rule out the possibility that the information acquired within the top-closed sections of the training channel could later be used in some other situational context.[48] When honeybees are subjected to the straight-channel paradigm described above for the cataglyphs, they behave as the ants have done. They ignore the distance traveled in the absence of celestial compass information. However, when they signal the travel distance to their nestmates, their waggle dances encode the total distance flown, irrespective of concurrent input from their celestial compass.[49]

This finally leads to the question of where and how path integration occurs in the insect brain. Soon after a couple of neural network models for path integration in ants had been developed, and a ring-like array of compass neurons had been proposed,[50] Uwe Homberg and his collaborators pointed out that owing to its neuroarchitectural features the central complex is a likely place at which such models could be implemented, and thus favored this midline set of neuropils (see Figure 3.31) as the seat of the path integrator.[51] Just recently, this presumption has brilliantly been confirmed, and an elaborate circuit model for path integration based mainly on structural and functional characteristics of the central complex has been developed by an international team around Stanley Heinze from Lund University, Sweden (for a sketch, see Figure 5.18).[52] The model combines the compass network (of the protocerebral bridge) outlined in Chapter 3 with the newly discovered 'speed neurons' (of the noduli), which monitor translational optic flow. Simply put, both kinds of information converge in columnar cells of the fan-shaped body, the CPU4 cells, which receive inputs from both the TB1 compass neurons in the protocerebral bridge and the TN speed neurons in the noduli. Once the home vector has been computed in the CPU4 cells, it must be compared with the animal's current heading to generate a steering command—a task that is accomplished by another set of columnar cells in the fan-shaped body, the CPU1 cells, which represent the major output pathway to the lateral complex and thus convey the steering signal to premotor command centers.[53] It is satisfying to see that this circuit model of path integration, which is based on neuroscientific evidence, is able to simulate many aspects of the animal's navigational behavior equally well as the former theoretical models have done.[54]

The model has been derived from studies in tropical bees *(Megalopta genalis)* and is in accord with evidence from *Drosophila*. Given the fact that the neuroarchitectural features on which it is based are so highly conserved across almost all insect taxa, it certainly applies to the cataglyphs as well. The latter offer the additional challenge of implementing the behaviorally unraveled stride integrator and wind compass into the central complex-circuitry. Finally note that the model deals with the updating of the home vector and its use in guiding the animal back to the nest, but does not yet address the question of how reference vectors stored in long-term memory (see *R* in Figure 5.7) come into play.

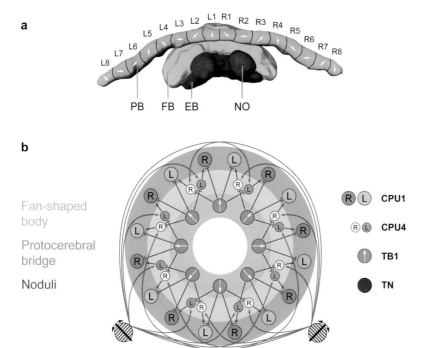

a

L8 L7 L6 L5 L4 L3 L2 L1 R1 R2 R3 R4 R5 R6 R7 R8

PB   FB EB   NO

b

Fan-shaped
body

Protocerebral
bridge

Noduli

Ⓡ Ⓛ CPU1

ⓡ Ⓛ CPU4

↑ TB1

● TN

5.18 Ringlike topology of the main types of central-complex neurons involved in path integration

(a) 3-D reconstruction of the central complex of *Cataglyphis nodus* (caudal view). *EB,* ellipsoid body (lower division of the central body); *FB,* fan-shaped body (upper division of the central body); *NO,* nodulus; *PB,* protocerebral bridge. Note the eight columns in either hemisphere of the protocerebral bridge. The head-direction tunings of their TB1 neurons (with respect to celestial compass cues) are indicated by *white arrows.*
(b) Hypothesis of how certain types of neuron (columnar cells CPU1, CPU4; tangential cells TB1, TN) in particular subcompartments of the central complex may interact. For the white arrows in the TB1 symbols, see *(a).* In the noduli *(red)* the *black arrows* indicate the direction of translational optic flow recorded by the TN cells. For explanation, see text. Based on studies in *Megalopta* bees.

We may end this chapter on path integration by making a first step toward a more unified understanding of the ant's navigational tool set. Of course, until now we have dealt only with path integration, but this routine in itself may already be able to account for a number of behavioral traits that have originally been considered as relying on their own, distinct steering systems. As we have seen, this might be most prominently the case with the ants' systematic search behavior. We might even dismiss the idea of a common odometer unit. At present, we can best speak of stride and flow integrators, in which the odometric inputs—proprioceptive and visual inputs, respectively—might be inherently combined with the directional one. The overarching question, then, is what role path integration—this constantly running and apparently omnipotent routine—plays in the cataglyph's larger system of navigation, in which guidance by landmarks, just due to their geostable nature, might be of utmost importance. This is the topic of the next chapter.

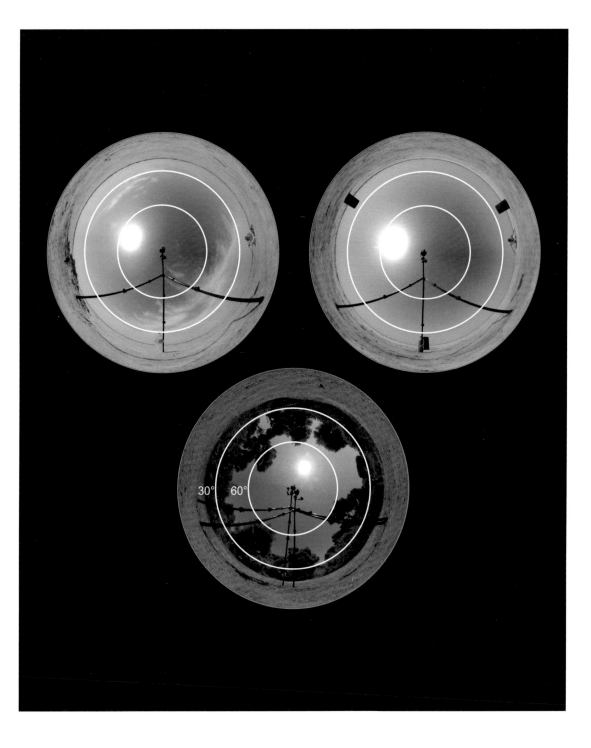

6 Using Landmarks

In the sand flats of the Wahiba desert in Oman we came across a small orange-colored cataglyph, *Cataglyphis aurata,* which was difficult to see because it nearly blended into the color of the desert floor. It ran from where we had met it, straight for at least 60 meters, until it vanished into its nearly invisible nest opening after only a short search in the immediate neighborhood of the hole. Had it exclusively relied on its 'omnipotent' path integrator? Most certainly not. To account for the ant's high accuracy and precision, nearby signposts—a larger stone here and one or another piece of dry vegetation there—may have become important. Indeed, many experiments should later show that if the goal area is completely devoid of any visual beacon, the path-integrating ants having missed the goal by only a few centimeters may spend several minutes of extensive search movements, which again may lead them several meters away from the goal, until they finally hit it. A single small, nearby landmark, natural or artificial, suffices to decrease the search time highly significantly (Figure 6.2). As in marine navigation, in which piloting by landmarks ('pilotage') is used during the approach to land in the final stage of the voyage,[1] the cataglyphs use mechanisms of local landmark guidance once they have arrived by path integration close to, but not exactly at, their final destination.

If landmarks are experimentally introduced and then displaced a bit to the side, the cataglyphs start to search at the position defined by the relocated landmarks, even though the real goal as indicated by the

6.1 Full-sky fish-eye pictures of two extreme *Cataglyphis* habitats

Each picture is centered about a nest opening. *Top row:* Flat-horizon environment: Tunisian salt pan inhabited by *C. fortis.* In the *right-hand picture* the nest opening is surrounded by an array of three cylindrical landmarks. *Bottom:* Heavily cluttered environment: open Greek pine forest inhabited by *C. nodus.* In all cases a camera used for recording the ants' learning walks is mounted on a tripod.

6.2 Search times needed for finally pinpointing the goal are largely reduced by even a single landmark

Left: Setup. Nest search times are recorded in ants that on return from their foraging trips (mean distance: 31 m) have crossed a recording circle (radius 0.4 m) around the nest entrance *(N)* for the first time and then engage in systematic search movements until they finally enter the nest. *Right:* Result. Search times **(a)** in a featureless plain and **(b)** with a small familiar landmark *(LM)* within the recording circle. As the ants moved with a mean speed of 0.36 m · s⁻¹, they could have reached the nest directly within about a second. *Ocymyrmex robustior.*

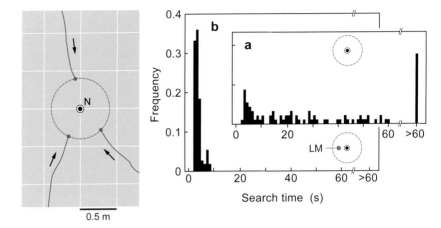

path integrator is just within reach. In such experiments the behavior of the cataglyphs is reminiscent of what Eugène Bouvier and Niko Tinbergen observed when they performed their classic studies on digger wasps in the sand dunes of Normandy and the heathland pine forests of Gelderland.[2] In order to provide the larvae with food, these solitary wasps—*Bembix rostrata* and *Philanthus triangulum* in this case—routinely return to the burrow they have dug in the sand. When a ring of pinecones, which had previously been placed around the burrow entrance, was shifted to a new position nearby, the returning wasp searched in the center of the relocated pinecone ring and completely disregarded the real entrance, which had now come to lie outside the ring. Sphecid digger wasps such as *Ammophila campestris* can even learn multiple locations and relocate these places, hours or even days later, on the basis of surrounding landmark cues. Parasitoid ichneumonid wasps tell an even more amazing story. They may monitor the locations of multiple potential hosts—egg clusters of a particular species of nymphalid butterflies—for weeks, until the host larvae have become ready for oviposition by the wasps. Visual landmark experiments performed within mesh enclosures show that the wasps relocate the host egg clusters by learning and memorizing the position of the clusters relative to visual landmarks.[3]

What exactly is it that the wasps and the cataglyphs learn about the spatial layout of their landmark environment? First, they could take advantage of landmarks to define a place merely on the basis of the view of the landmarks seen at that place. However, as the landmark scenes are usually complex, and the visual resolution of insect compound eyes is rather poor, one might wonder whether the cataglyphs recognize and remember individual local landmarks, whether they instead rely on rather unprocessed panoramic views, or whether they refer only to the skyline, i.e., the contour at which terrestrial objects meet the sky.

Second, irrespective of the answer to the latter question, the cataglyphs could combine visual memories with motor commands (actions), e.g., move in a certain direction when a familiar place has been reached. This strategy could be further expanded and used in charting a route and thus enabling the ants to employ landmark information for moving from one known place to another. Finally, if the cataglyphs were able to learn more than one place and more than one route, could they then use and combine these multiple landmark memories to freely navigate between places and routes independently of how they had acquired this landmark information in the first place? As path integration could act as a scaffold for landmark learning, what linkages and interdependencies between the two systems are to be expected? These questions lead us directly to the main design features of the cataglyph's navigational tool set. In the following we shall consider a set of experimental paradigms, which have been influential in research on these topics. We use them to retrace the conceptual steps that during the past 40 years have been taken to understand how the cataglyphs use landmarks for navigation, and that have finally led to the concept of view-based landmark guidance.

However, we cannot start this endeavor without first widening our scope. Even though the cataglyphs are primarily visually guided navigators, they make effective use of other sensory cues as well. Various kinds of chemical cues are involved in finding and locating nest sites, food sites, and even sites along a route—performances to which we shall return in a bit more detail below. Tactile landmarks are used for pinpointing the goal, and in particularly designed experiments the cataglyphs have even been able to associate artificially presented magnetic and vibrational cues with the location of the nest entrance.[4] All these findings show that the ants' overall navigational performances result from flexibly integrating information from many sensory modalities.

Visual Place Learning

Imagine a simple experiment, in which the entrance to a *Cataglyphis* colony has been marked by a set of artificial landmarks (two or three small, equally sized black cylinders) placed around the entrance hole (Figure 6.3). Further imagine that after the ants have become familiar with the new situation around their nest, the constellation of landmarks is altered in various ways, e.g., the sizes, number, and spatial arrangement of the landmarks are systematically varied. The way the ants respond to these alterations should then help us in inferring what information the animals have originally derived from the familiar landmark array, and how they later use this information to guide their final approach to the goal. For example, does a two-landmark array provide the cataglyphs with a geographic reference such as "the goal is located in the middle of two cylinders of particular metric heights and widths and placed at a particular distance to the east and west of the goal"? In other words, do the ants reconstruct the three-

6.3 'Landscape art': artificial landmark array used in studying local visual homing

A triangular array of black cylinders installed in a test field near Mahrès, Tunisia, and used in the experiments described in Figure 6.4b. Grid width: 1 m.

dimensional array of the landmarks and thus acquire and use information that we would read off a topographical representation of this two-cylinder landscape?

Experiments of the kind just described were performed in the early days of our *Cataglyphis* research, and the answer to the question phrased above is a clear 'no.'[5] Just have a look at Figure 6.4a. When the ants as well as the two-cylinder array surrounding their nest entrance are displaced to a new area, the ants search at the fictive position of the goal, i.e., midway between the two cylinders, as they have routinely done in their habitual nest-site area (test condition 1). However, this goal-centered search breaks immediately down, when the cylinders are separated by twice the training distance (test condition 2), but it reappears, when the cylinders separated by twice the training distance are presented in twice the training size (test condition 3). Various versions of this experiment have been performed, and the results have been very similar indeed. For example, when in a modification of the experiment described above the sizes of the two cylinders are equal during training but different in the test, the peak of the ants' search distribution shifts away from the center toward the smaller of the two cylinders, i.e., to the point where the small and large cylinder again appear under the same angular size.[6]

Figure 6.4b depicts the ants' behavior in a triangular rather than linear array of landmarks. Again, the ants search at the fictive position of the nest whenever the visual surroundings viewed from that position match the ones viewed during training (test conditions 1 and 3). It is as if the ants had acquired and stored something akin to a photographic image (a two-dimensional retinotopic image conveniently called a 'snapshot') of the skyline panorama seen from the place to which they intend to return. Upon return they retrieve the stored reference image, the template, and move so as to reduce the discrepancy between the current retinal image and the template, until finally both images match. Note that in test situation 3 the search peak is broader, meaning that the ants search in a wider area about the goal than they do in situation 1. This is indeed to be expected, because in the expanded landmark array in situation 3 the ants must cover larger walking distances to experience a given change in the angular sizes of the landmarks, and hence can zoom in on the goal only less precisely.

6.4 Local visual homing within landmark arrays consisting of (a) two or (b) three black cylinders

(1) Landmark array as during training. The fictive position of the nest *(red circle)* is in the center of each of the two arrays of landmarks *(black dots)*. (2) Landmark array at twice the training distance. (3) Landmark array at twice the training distance and in twice the training size. In (a) the same data obtained by Wehner and Räber in 1979 (see note 5) and replotted by Möller in 2001 (see note 16) are shown in the form of the ants' actual walking paths *(left)* and as search density plots *(right)*. In (b) data are presented as 3-D search density profiles. (a) *Cataglyphis bicolor* and (b) *C. fortis*.

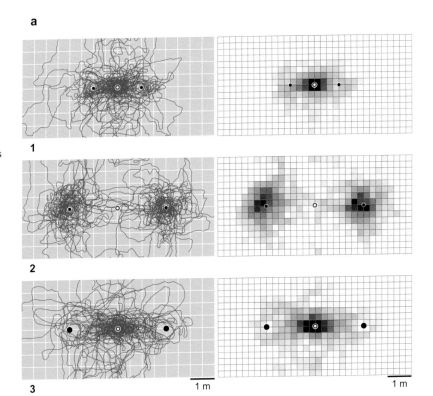

a

1

2

3   1 m   1 m

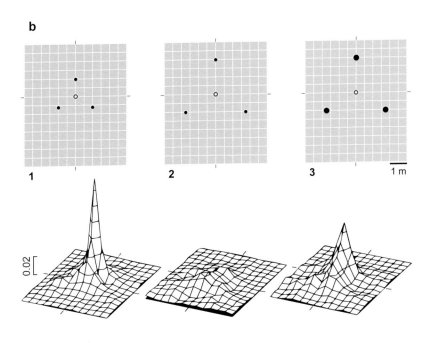

b

1   2   3   1 m

0.02

That this effect of 'motion parallax' is so nicely reflected in the ants' behavior lends further support to the view-based nature of the ants' visual homing strategy. A closer look at the paths actually taken by the ants within the linear and triangular arrays of landmarks shows that in test situation 1 the ants walk rather straight to the goal, but in situation 2, in which the focused search pattern breaks apart, they usually move in the direction of one or the other cylinder whenever it appears at a smaller angular size than expected, and thus alternately switch in their search from the vicinity of one cylinder to that of another along relatively straight paths. When in the laboratory wood ants, *Formica rufa*, are provided with similar sets of landmarks, their behavior nearly coincides with that of the cataglyphs in the field. In addition, video recordings show that the wood ants scan the route ahead over about 60° horizontally, and during these scanning cycles fixate one or another cylinder in their frontal visual field.[7]

Support for view-based visual homing comes already from a number of classic studies on local visual homing in bees[8] and wasps.[9] Moreover, visual place learning has been investigated beyond the hymenopterans' realm. The males of syrphid flies hover stably in midair, and after an unsuccessful chase of a female regularly return to the very same spot. By moving visual targets and filming hovering flies, Thomas Collett and Mike Land from the University of Sussex at Brighton, England, provided a first detailed analysis of this kind of behavior.[10] More recently, Michael Reiser and his collaborators at the Janelia Research Campus of the Howard Hughes Medical Institute have used a novel experimental setup to demonstrate that *Drosophila* flies are able to form and retain visual place memories.[11] Within an arena the fruit flies walk on a multi-tile platform of which all tiles except one are heated to an aversive 36°C. In this heat-maze arena the flies, which according to Tennessee Williams's phrase are virtually walking 'on a hot roof,' quickly learn to locate the preferable cool (25°C) spot by means of visual cues displayed by an array of light-emitting diodes on the walls of the arena. When later the temperature of all tiles is raised to the aversive level, and the visual surround is rotated, the flies consistently search where they expect the cool tile to be relative to the rotated visual panorama. It remains to be elucidated how much the learning strategies involved in these various kinds of visual place learning actually have in common, and to what extent one of these behaviors may be a modified form of the other. For example, while in the rapid place learning of hover flies the necessary visual information seems to be acquired almost instantly, in the cataglyphs—as we shall see toward the end of this chapter—this information is derived from elaborate and time-consuming learning maneuvers.

Before we return to the desert ants in the fields, let us illustrate the power of behavioral experiments performed in the laboratory. There Thomas Collett and his collaborators have investigated how wood ants determine their direction of travel when they are confronted with simple black shapes within an otherwise uniform arena. Based on such studies of short-range navigation the authors try to unravel perceptual mechanisms that are most likely also used by ants navigating on a larger scale in their natural environment. As an example of this short-range approach, let us train the ants to an inconspicuous feeder located at the base of a prominent landmark feature, e.g., the right vertical edge of a large black rectangular shape displayed on an LCD screen. When the ants head for this goal, they progress along sinusoidal, zigzag-like paths. Every now and then, usually just after a zig or a zag, they perform saccadic turns in the direction of the goal. These rapid turns end in short, about 200-millisecond goal-fixation events. Apparently it is through these behavioral episodes that image matching occurs.[12] When the endpoint of a saccade is reached, the ant's current view of the visual scene matches the memorized one (black to the left and white to the right of the goal). Thereafter the ant while approaching the goal proceeds along its sinusoidal path, performs another saccade, and so on, and thus displays a discontinuous control behavior.[13]

Next we consider a situation in which the goal is inset from the vertical edge (Figure 6.5, left part). In this case, a single snapshot memory does not suffice. While the animal moves forward, the image of the edge shifts to an increasingly peripheral position within its field of view. As the right part of Figure 6.5 shows, the ants indeed adjust their saccadic turns leading to the goal angle $\gamma = 0°$ in such a way that the edge angles α correspond to the ones appropriate for the particular goal distance, d. During their previous runs to the feeder they must have

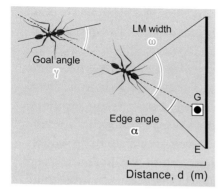

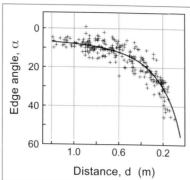

6.5 Short-range navigation in wood ants

Left: Experimental paradigm. Within a laboratory arena the ants approach an inconspicuous feeder (goal, *G*) located inset from the vertical edge *(E)* of a landmark *(LM, black bar)*. During the approach they repeatedly perform short saccadic body turns (by the goal angle γ) to fixate the feeder. *Right*: The edge angle α after the ants have completed their saccadic turns plotted against the ants' distance from the landmark. *Formica rufa*.

learned this correlation by either memorizing a sequence of discrete snapshots or encoding the α / d function in a continuous way. Earlier experiments revealed that when wood ants depart from a goal-defining landmark, they frequently turn back and fixate on the landmark at discrete locations, and that upon return to the goal they perform the saccadic turns just described.[14] Hence, one could surmise that during these fixation episodes performed at various distances from the goal the ants retrieve and match multiple discrete views in which the landmark is held at several discrete positions on the retina. But how are the animals informed about the distance to the goal? The apparent angular width of the landmark (the increasing angle ω as the ant approaches the goal) could provide the decisive cue, for when the test landmark is made either wider or narrower than the training landmark, the ants walk leftward or rightward, respectively, from their training route. Nevertheless, does the animal's brain really relate alpha to omega (α / ω), i.e., put a local feature, the edge, in relation to a global feature, the angular size of the landmark? There may be other cues and combinations of cues by which the ants encode the visual scenes used for their navigation. For example, while facing the goal, the ants may learn the retinal position of the center of mass of the shape, or the fractional position of mass, i.e., the proportion of the shape that lies to the left and right of the goal. Moreover, local features like edges may dominate these global features in controlling the ants' directional choices. Recently, the researchers performing these elaborate experiments in wood ants concluded that "directional control by a panorama is interestingly complex, even when the panorama consists of just a single shape on an otherwise flat skyline."[15]

Whatever cues the ants finally extract from their visual world, the preceding considerations already indicate that the ants do not seem to process and evaluate passively acquired images of their surroundings, but actively shape their visual input by performing characteristic movement patterns. The role that such movement patterns—saccadic body rotations and other motor routines—may play in the ant's spatial behavior will continue to occupy us throughout this chapter. In any way, the experiments performed in the sparse visual world of the laboratory arena may gradually unravel the visual properties of scenes that are actively acquired and used as well by the cataglyphs in their natural environment. With this in mind, we return to the desert navigators and first try to model the results of some key behavioral experiments.

Modeling Approaches

Since the early 1980s several neural architectures and memory networks have been proposed to account for local visual homing in ants and bees. Of course, any match between a model and the experimental data does not tell us whether the animal actually employs the very strategy proposed by the modeler. Different models can conform to a set of experimental data equally well. By the same token, the models do not yet provide us with a glimpse into the ant's brain and thus do not claim to describe underlying neural mechanisms. Algorithmic modeling attempts have mainly been driven by the desire to generate hypotheses that are as parsimonious and robust as possible and guide us to new experimental paradigms. In the following we elaborate a bit on the diversity of such models, and concomitantly try to retrace the historical path along which these models have been devised. However, as neuroethologists, we should always keep in mind that the final goal must be to link the ants' navigational performances, and the model architectures derived from them, to the types of neural circuitries actually found in the insect brain. This goal will be approached toward the end of this chapter.

a   Template Matching model

Average Landmark Vector model

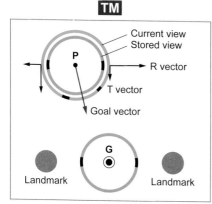

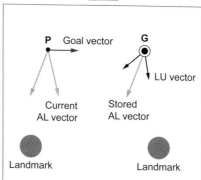

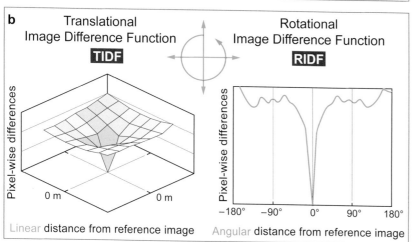

b

6.6 Models of local visual homing (a) based on visually isolating discrete objects and (b) relying on global panoramic scenes

(a) In the example shown here the visual scene consists of two identical landmarks. In the TM model the circles represent the stored view (template; *orange*) acquired at the goal location (G) and the current view *(blue)* at an arbitrary point (P) in the neighborhood of the goal. The goal vector *(red)* is the vector sum of all radial (R) and tangential (T) component vectors. In the ALV model the difference between the current *(blue)* and stored *(orange)* average landmark *(AL)* vectors yields the goal vector *(red)*. The AL vector is the vector sum of the two landmark unit *(LU)* vectors pointing at the landmarks. (b) The TIDF represents the pixel-wise image differences between views taken at the goal (0/0) and at different x, y distances from the goal. The RIDF represents the differences between views taken in different orientations relative to the orientation of the original view.

We start with the classic snapshot model. In this Template Matching (TM) model the animal—or an autonomous agent (robot)—looks for features in a visual scene and then pairs these features in the current view with corresponding features in the snapshot acquired at the goal.[16] In particular, the ant is supposed to determine the difference in bearing and retinal size between any landmark image in the stored view and its closest landmark image in the current view (Figure 6.6a, TM). The tangential and radial component vectors computed from these individual image differences are summed up and yield the goal vector. Using this TM strategy, the animal will be able to return to the goal from any location within the catchment area of the snapshot, i.e., from any location at which the animal's current view overlaps sufficiently well with the stored view, the 'attractor.'[17]

In template matching we have tacitly assumed that the cataglyphs take advantage of their wide, omnidirectional visual field of view (see Figure 1.9). Some indications that this may actually be the case can be gained from experiments in which parts of the compound eyes are blackened out.[18] Among four groups of ants, which are trained and tested with the dorsal, ventral, frontal, or caudal halves of their compound eyes occluded, only ants of the first group fail to localize the goal in the center of the landmark array (Figure 6.7a). Even with their frontal vision abolished, the ants are still able to accomplish the task. In conclusion, the dorsal halves of the eyes are necessary and sufficient not only—as we had seen in Chapter 3—for compass orientation by skylight cues but also—as we see now—for navigation by panoramic terrestrial cues. Whether particular parts of the 360° range of the visual field are especially important remains to be explored.[19]

Finally, there is one important requirement that must be met for the TM image matching strategy to work. As the animal continually moves while comparing reference and current views, both views must be aligned with respect to an external compass reference. Otherwise, the probability of false matches will easily result in erroneous home vectors. One way of achieving this goal would be to mentally rotate either the current image or the snapshot, but such mental rotations might be difficult to achieve within the insect's brain. That landmark views are fixed relative to retinal coordinates is supported by exchanging eye caps between the eyes or parts of the eyes, before the critical tests start. The result is quite surprising (Figure 6.7b). When the ants have acquired a view with one eye (or the frontal or caudal halves of both eyes), they prove unable to retrieve it with the other eye (or the other halves of the two eyes). There is no sign of either interocular or intraocular transfer. Moreover, binocular ants, i.e., ants with both eyes open, are not able to locate the landmark-defined goal when during training one eye has been occluded.

When the cataglyphs use retinotopically fixed snapshots, the matching process is largely facilitated by a common feature of the ants' foraging and homing behavior. The animals usually approach a goal,

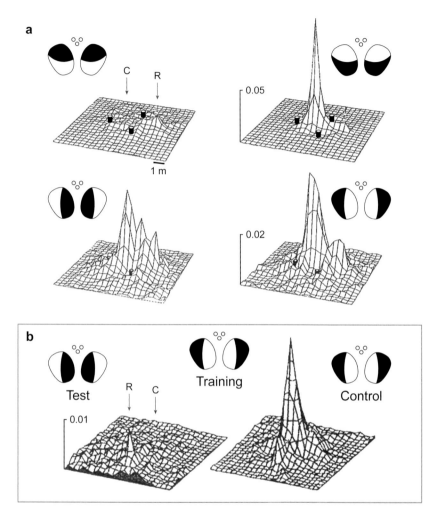

6.7 Half-blinded cataglyphs

(**a**) Ants tested for local visual homing after the dorsal, ventral, frontal, or lateral halves of their eyes have been occluded during both learning and memory retrieval. The ants are trained and tested to locate the goal (nest) in the center *(C)* of a triangular array of cylindrical landmarks (*shown in black,* see Figure 6.3). *R,* point of release of zero-vector ants. (**b**) Test for intraocular transfer performed with the same triangular array of landmarks as in *(a)*. During training the lateral halves of the compound eyes have been occluded. In the critical tests the eye caps are moved from the lateral to the frontal parts of the eyes, while in the subsequent control tests the training situation is reestablished. *Cataglyphis fortis.*

be it the nest or a frequently visited feeding site, from a familiar direction, and thus will routinely face in this direction when taking and subsequently employing their snapshots. Just take the example of an ambiguous localization task—ambiguous in the sense that without an additional reference the ants would face a symmetry problem. In a square array of four cylindrical landmarks the goal (nest) is located in one corner of the square, namely, in the corner that lies in the direction from which the ants habitually return from the feeder and from which they are thus used to face the landmarks (Figure 6.8).[20]

6.8 Local and global frames of reference

Left: Experimental paradigm. During training the ants approach a goal (nest, *open red circle*) located asymmetrically within a square of four cylindrical landmarks *(black dots)* from the southeast *(black arrow)*. In the test field zero-vector ants are released in four different compass directions *(open circles)*. *Right:* Results. The *false color images* depict the search density distributions for *(a)* the NE and *(b)* the SE releases. The *large open circle* marks the fictive position of the goal. *Cataglyphis fortis.*

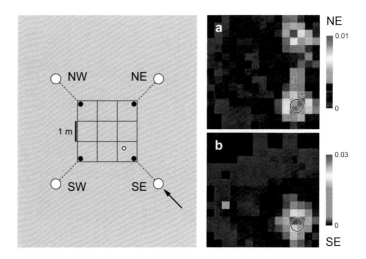

Zero-vector ants are later tested in unfamiliar terrain and released from four different directions outside the landmark array. When they approach the array from the familiar direction, their search density peaks exclusively in the correct corner. In the other three cases they initiate their searches at the corner that is closest to the current direction of approach, but then proceed to the habitual corner, where snapshot, current view, and whatever external reference are in register. The necessary reference need not be derived from the ant's celestial compass system, but could consist of skyline cues.

In parallel with performing these experiments in the field we have implemented the TM-plus-compass model into an autonomous agent, a robot dubbed Sahabot, short for *Saha*ra Ro*bot*.[21] For technical convenience the robot runs on wheels rather than walks on legs, but this difference in locomotor behavior notwithstanding, it operates in the very same habitat in which the cataglyphs roam the desert floor. A first version, Sahabot 1 (Figure 6.9a), performs path integration by using a celestial (polarized-light) compass. The same compass is built into Sahabot 2 (Figure 6.9b), which returns home by TM landmark guidance. In the tests, it navigates within the very artificial landmark arrays that we have used in the cataglyphs (of course, somewhat scaled up for robot dimensions). The landmarks are projected by a conically shaped mirror on a 360° CCD camera. The full panoramic view obtained this way is later processed on board by a number of computa-

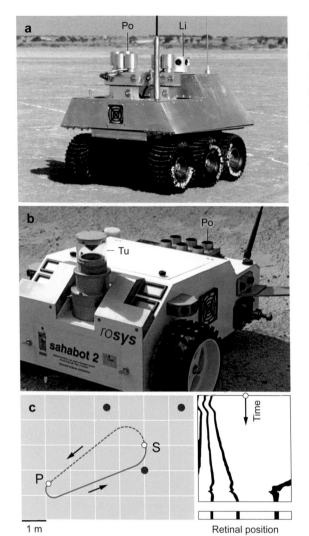

a
Po   Li

b
Po
Tu
rosys
sahabot 2

c
S
P
1 m

Time
Retinal position

6.9 Autonomous agents: Sahabots

(a) Sahabot 1 is equipped with a polarized-skylight compass and navigates by path integration. The compass consists of three pairs of zenith-centered polarized-light sensors *(Po)* each followed by a log-ratio amplifier (for principle, see Figure 3.29c). These three polarized-light detectors, which are tuned to e-vector orientations differing by 60°, provide the robot with a sky compass. For mode of operation, see Lambrinos et al. (2000), note 21. The polarized-light sensors are mounted in vertical tubes to exclude direct sunlight *(Po)*. *Li,* ring of eight light-intensity sensors used to resolve the solar / antisolar ambiguity in the sky. **(b)** Sahabot 2 navigates by landmark guidance (template-matching-cum-compass strategy). For mode of operation, see Lambrinos et al. (2000), note 21; Möller et al. (2001), note 16. *Tu,* Perspex turret with conical mirror inside. **(c)** Sahabot 2 on the move. *Left:* The robot's outbound path from start *(S)* to point *P (dashed blue line)* by remote control. Inbound path of the robot operating autonomously *(solid blue line).* Black dots, artificial landmarks (black cylinders). *Right:* Images of the three landmarks as seen by the robot during the inbound journey. The stored image is shown at the *bottom.*

tional steps such as brightness adjustments and thresholding. In the field the robot is first told to acquire and store a snapshot of the scene around the starting position. Then it is guided by remote control on an outward journey (Figure 6.9c). At the end of this journey, control is handed over to the robot, which now returns to the starting point by autonomously employing its inbuilt image matching routine. When the discrepancy between the stored view and the current view becomes smaller than a given threshold, the robot stops. Sahabot 2 has reached its goal. It does so with considerable accuracy from a rather wide area around the landmark array.

What one would certainly like to accomplish in future research is to 'improve' the wheeled Sahabots by adapting their visual input and computational processing properties more and more to those unraveled in the cataglyphs' eyes and brains and discussed later in this chapter. Such attempts may finally raise the question of what we finally gain by this biorobotics approach. Certainly it provides us with a proof of concept, a rigid test of biological hypotheses. The Sahabots must accomplish the same tasks under the real (and not just simulated) environmental conditions that are encountered by the cataglyphs. However, the objective is not to merely mimic the ants' solutions, but to use the real-world artifacts to improve our understanding of how the animals' navigational system may work, to test various insect-inspired solutions and inquire about their potentialities and constraints, and thus to generate new hypotheses and to develop new experimental paradigms for investigating the cataglyphs' behavior in the field.[22] After all, what we have personally enjoyed the most in our Sahabot endeavors is the intellectual ping-pong between researchers working on animals and robots, and the creative way of thinking that results from this two-way communication between biologists and engineers.

Let us now return to modeling and consider a method of local visual homing that does not require any image matching at all. It operates with vectors pointing from the goal location to individual landmarks or selected image features. Having determined such unit landmark vectors, the animal is supposed to compute the vector sum ('average') of all unit vectors, the Average Landmark Vector (ALV, Figure 6.6a). This means that only a single vector rather than a snapshot image must be stored. Later, on return to the goal the stored ALV is subtracted from the ALV pertaining to the current position. This vector subtraction process yields the home vector.[23] As there is no need for any image matching, the ALV method is computationally much cheaper than the TM method, but it still shares with it two essential preconditions: the visual scene must be dissected into individual landmarks or landmark features such as edges, and an external compass reference is needed for aligning either images or vectors.

In contrast to the TM and ALV models, which operate with local features, global models consider panoramic views as a whole and thus might come closer to what the ants actually do. As in natural scenes the global image differences between template and current views rise gradually and monotonically as the animal moves away from the goal, an Image Difference Function (IDF) can be computed, which has its global minimum at the goal (translational Image Difference Function, TIDF, Figure 6.6b).[24] Jochen Zeil and his collaborators at the Australian National University, Canberra, who have developed this elegant approach, have used raw unprocessed arrays of grayscale pixels to determine pixel differences between reference image and current image. The animal is assumed to move-and-compare, i.e., to compute the global image difference between a memorized reference view taken at the goal location and a set of views taken at different distances from the goal. This way it samples different directions of movement and finally selects the direction that reduces the global image difference the most. In other words, the animal is assumed to descend an image difference gradient.

As outlined in Figure 6.6b, a translational Image Difference Function (TIDF) can be devised alongside a rotational one (RIDF). This RIDF is computed by visually scanning the environment, i.e., by rotating the current image relative to a reference image taken at a particular place. It has a marked minimum when the animal is oriented in the direction in which the view has been encountered during learning. The direction in which the animal then faces is the direction of travel. This renders the RIDF a useful and quite robust means for traveling along a familiar route. It is robust insofar as in the ant's world panoramic scenes usually change rather smoothly with distance, so that the best match between current and reference image can be detected even at some distances from where the reference image was taken. Then the minimum of the RIDF may be shallower, but still detectable. In conclusion, a stored image can be used as a local visual compass, which enables the animal to recover a course to steer. How finely tuned this panoramic compass is, and how well it can be employed for finding directions, will depend on the visual structure of the environment and on the type of information that the ant deduces from a panoramic view, e.g., the whole unprocessed image or just the skyline.

An alternative to the scanning procedures mentioned above would be to determine the best image match not by physically moving but by internally simulating movements and predicting how views would change if one moved. This can be achieved by systematically distorting ('warping') one of the images—the stored or the current image—and comparing it to the other one until the best match is achieved.[25] However, as with the mental rotation of memorized images, it seems doubtful that such a demanding image warping strategy would fit into an insect's head.

In summary, a match achieved by the TM, ALV, or TIDF method tells the animal that it has arrived at the goal, i.e., where it is, while the RIDF method informs the animal about the direction leading to the goal, i.e., where to go. The models also differ in the kind of visual information used in image matching: discrete landmarks (TM and ALV) or full panoramic scenes (IDFs). However, what is a 'landmark' out there in the desert landscape? In the experimental situations portrayed in Figures 6.3 and 6.4, in which two or three cylinders were presented against an otherwise rather flat horizon, the question might be answered straightforwardly, but what would be the answer in a more structured visual world, in which bushes and loosely scattered trees prevail.

To investigate this question in the Australian *Melophorus bagoti*, Antoine Wystrach, then a PhD student at Macquarie University, Sydney, designed a series of revealing experiments.[26] He placed a prominent artificial landmark, a huge black screen, right next to the nest entrance. After a certain training period he recorded the homing trajectories of zero-vector ants when the landmark either was in place or shifted within its natural surroundings by various angular amounts to the side (Figure 6.10a). From a human perspective, the results are extremely surprising. Already after some small sideways shifts of the landmark (e.g., by 32°), the ants do not 'recognize' it any longer. They cover only about half their usual homing distance, perform a U-turn, and start searching. Finally, none of them reaches the fictive position of the nest in front of the prominent landmark. This behavior can be

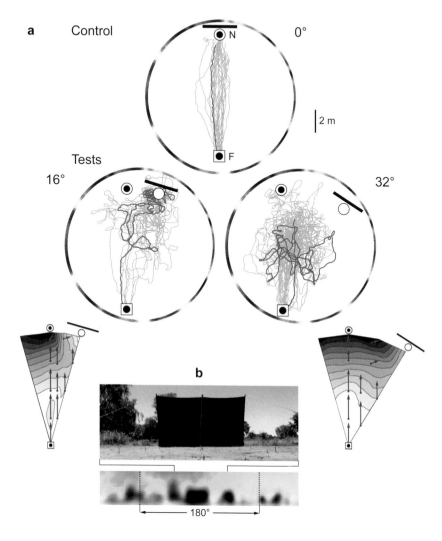

a Control 0°

2 m

F

N

Tests

16° 32°

b

180°

6.10 Ants use panoramic views rather than filter out discrete landmarks

(a) Zero-vector ants released at the feeder *(F)* either in the training situation *(top)* with a large landmark *(black bar)* just behind the nest *(N)* or in two test situations *(bottom)* with the landmark shifted 16° or 32° to the right from the feeder-nest direction. Trajectories of all ants tested and of an individual one are depicted in *light blue* and *dark blue,* respectively. The *textured outer circles* indicate the landmark panorama. *Insets:* Maps of panoramic mismatches. The darker the shade, the higher the mismatch between reference and current views. The *red arrows* indicate the steering directions resulting from RIDF matching (length proportional to matching values). (b) Two images taken of the setup from the same location in human *(top)* and ant *(bottom)* perspective, with the latter adjusted to the ant's visual acuity. *Melophorus bagoti.*

explained on the basis of how much the panoramic scenery recorded by miniaturized 360° fish-eye optics is altered by displacing the landmark from its familiar position. With increasing differences between memorized and current views, the approaching ants get more and more disoriented even when they are already rather close to the landmark. Obviously, they do not separate the landmark from its visual surroundings. While we humans endowed with acute foveal vision and particular neural pathways for object detection focus on the landmark and perceptually isolate it from the background, the ants do not seem to treat a 'landmark' as a particular object, but instead rely on the coarse-grain panoramic visual scenes that are perceived across large parts of their wide-angle compound eyes.

This leads us to the question: What do the foraging ants actually see in their natural environment? As the visual information content of panoramic scenes can be extremely high, we wonder what kind of information the cataglyphs actually extract from these scenes. In this respect, the low spatial resolution of compound eyes might be a first step in reducing potentially superfluous information. Computational analyses show that for view-based navigation with large panoramic visual fields, robust orientation may even profit from rather low resolution.[27] We can get an impression of the images seen by a desert ant in cluttered environments by low-pass filtering the animal's visual surroundings. As demonstrated in Figure 6.11a, the panoramic images of natural scenes can be spatially processed by Gaussian filters of increasing field size (corresponding to photoreceptors of increasingly larger visual fields), so that the images become more and more blurred and similar to what the ant's low-resolution visual system might receive from its visual world. Such low-resolution wide-field systems are not at all suited to recognize individual objects, but provide robust information for navigation. For the RIDFs based on low-pass filtered scenes exhibit marked minima and thus are able to inform the insect about its desired direction of travel rather reliably (Figure 6.11b).

Note that the cataglyphs have much smaller eyes and lower visual resolution than the utmost visual predators among the ants—the Australian bull ants *Myrmecia,* the Asian jumpers *Harpegnathos,* and the neotropical giant ants *Gigantiops*—which strike by the highest-resolution compound eyes of all ants. However, in *Cataglyphis* the high-level visual integration centers in the mushroom bodies, the collars, are considerably larger than suggested by the size of their eyes and optic lobes.[28] Obviously, the cataglyphs' visual system has put a premium on higher-level processing and storing of coarse-grain, wide-field spatial information.

For detecting the skyline, the ultraviolet receptors may play a major role. As skylight provides a global source of ultraviolet radiation, while

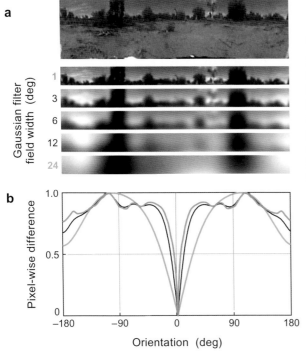

a

Gaussian filter
field width (deg)

1
3
6
12
24

b

Pixel-wise difference

1.0

0.5

0

−180   −90   0   90   180

Orientation (deg)

6.11 Panoramic scenes and rotational image differences

(a) Panoramic image in the habitat of *Melophorus bagoti* and five versions of the same scene processed with Gaussian filters. Filter field width at half maximum increases from 1° to 24°. Gaussian filtering roughly mimics the effect of the size of the visual fields of photoreceptors. The panorama sampled by the ants' eyes corresponds to the 3°–6° filtered images. (b) Rotational Image Difference Functions (RIDFs) for three of the scenes shown in *(a)*.

most terrestrial objects absorb ultraviolet to a substantial degree, it is a likely hypothesis that during evolutionary times ultraviolet receptors have primarily been used as some kind of skylight detector.[29] Recent measurements and computations based on thousands of image points in celestial and terrestrial scenes under a variety of illumination conditions clearly show that ultraviolet channels—or ultraviolet / green contrast systems—suffice for reliably differentiating between sky and ground.[30] When in experiments with artificial skylines ultraviolet light is blocked, the orientation of desert ants based on skylines is severely impaired.[31] The earthly part of the ant's panoramic view may contain coarse grayscale textures and distinct boundaries, but it might well be achromatic. At this juncture it is worthwhile to note that blue photoreceptors have not been found yet in either *Cataglyphis* or *Formica* ants.[32] If they have not been overlooked, these advanced formicine ants could have lost the blue type of photoreceptor, which—together with the UV and green type—is highly conserved in hymenopterans and all pterygote insects, and is retained as a possibly ancestral trait in the Australian bull ants, *Myrmecia*.[33] In desert ants some experiments show that the skyline alone can provide sufficient information for view-based homing. For example, in their studies on homing behavior of *Melophorus bagoti,* Paul Graham and Ken Cheng mimicked the natural panorama by crude black plastic sheeting.[34] In a distant test field the ants took this homogeneously black facsimile of the natural panoramic scene for the scene itself. When the artificial scene was rotated through 90°, their homeward courses rotated accordingly.

At present, the most likely strategy by which the cataglyphs use landmarks for navigation is panoramic view matching. The RIDF operation provides the animal with information about the best possible direction in which to move rather than about the best possible place at which to stay, and leads us directly to the second experimental paradigm to be discussed in the context of view-based landmark guidance. Before that, however, let us digress for a while from the visual world and turn to the cataglyphs' chemosensory means of navigation.

Olfactory Landmarks

Somewhat hidden in the cataglyphs' overt navigational repertoire is the use that the ants make of olfactory landmark information, for detecting and localizing nest and feeding sites and for navigating along habitual routes. After all, the observation that the cataglyphs exploit chemical cues for navigation should not come as too much of a surprise, as in many aspects of behavior the wingless, ground-dwelling workers of almost all ants strongly depend on chemosensory information, much more so than their winged cousins, the spheciform wasps and bees. Until recently, the use of such information has been investigated primarily in two contexts. One is pheromone-guided route traveling. "The principal instruments of the ants are chemical trails," as Bert Hölldobler and Edward O. Wilson succinctly remark,[35] but this instrument is not used by the cataglyphs. The other context, in which ants continually employ their chemical sense, is nestmate recognition based on blends of cuticular hydrocarbons—a task just beginning to be studied in the cataglyphs.[36] In either case the chemical compounds involved in these behaviors are produced by the animals themselves. Environmental odors and their use in fine-scale orientation in the vicinity of nest and feeding sites have attracted attention only within the past decade, and this time especially in the cataglyphs. We owe it to the pioneering multifaceted work of Markus Knaden, and his collaborators at the Max Planck Institute for Chemical Ecology in Jena, Germany, odor-based place recognition has become a new focus of research.[37] At first sight—or better smell—the salt-pan habitat of *Cataglyphis fortis* may appear rather uniform in chemical composition and texture, but with its small-scale clefts and ridges in the salty surface, its scattered pieces of dried vegetation, encrusted debris, and small halophilic plants, it provides a complex array of place-specific odor blends. Many dozens of such compounds have been detected and analyzed by mass spectroscopic means, and whenever tested in behavioral experiments, the cataglyphs were able to distinguish them from one another.

Note that due to the airborne nature of olfactory cues, olfactory landmarks can be detected only from one—the downwind—direction. This constraint renders the channel paradigm a useful means of exploring the ants' responses to olfactory signposts (Figure 6.12). In this experimental device the cataglyphs learn to associate a goal location with odors such as nonanal, decanal, and other alkyl aldehydes present in their habitat as well as with non-habitat plant volatiles. The ants even learn to locate the goal in the center of a small quadrangular array of different odors, and differentiate between variations in the positions of the odors within the array, meaning that simultaneous input from both antennae is desirable, if not necessary.[38] Moreover, olfactory landmarks are learned as fast as visual landmarks, but when combined both cues are learned much faster. After extended training with the combined cues, the effect of this consistent bimodal sensory input becomes so strong that the ants are no longer able to pinpoint the goal when presented with the individual cues alone. Out in the fields such multimodal sensory integration may help to disambiguate the similarly sized and shaped visual landmarks that usually occur in the ants' low-scrub foraging terrain.

Finally, the cataglyphs' main food sources, dead insects, are always detected and pinpointed by the odor plumes emanating from them

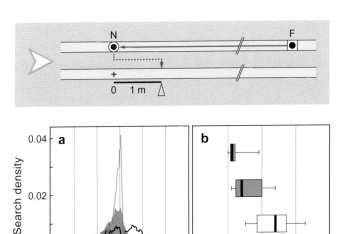

6.12 Recognition of a learned nest-associated odor

Top: Experimental paradigm. Ants trained in a channel to return from a feeder *(F)* to the nest entrance *(N),* which is marked with the odor indole. Wind direction is indicated by the *green arrowhead.* Upon arrival at the nest entrance the zero-vector ants are displaced *(dotted blue arrow)* to a test channel and released 1 m downwind *(open arrowhead)* from the odor source *(red cross). Bottom:* Results. **(a)** Search density profiles when the odor is indole *(blue),* or a blend of indole, nonanal, decanal, and methyl salicylate *(orange),* or when only the solvent is presented (control, *black).* **(b)** The ants' search locations as measured by the turning-point criterion. The *open arrowhead* indicates the release point. *Cataglyphis fortis.*

(see Figure 7.6). Naturally, these odors are necromones, mainly linoleic acid, derived from decaying arthropod corpses, but in carefully designed experiments more than a dozen common plant volatile odors consecutively offered in association with rewarding cookie crumbs can be learned by the cataglyphs. These odors are learned quickly—one experience suffices—and stored in long-term memory for the rest of the ants' lives. In contrast, memories for nest odors are acquired more slowly and restricted to the last odor learned. These distinct differences in the organization of olfactory memories for food and nest odors may be functionally related to the style of the ants' foraging life, in which the cataglyphs will encounter different food items, and may establish olfactory search images for such items, but always return to the same nest site, and thus refer only to the last odor learned. With this excursion in the desert navigator's olfactory world, let us return to the visual tasks that the navigator must accomplish.

Local Steering Commands

When traveling back and forth between the nest and a familiar food site, the cataglyphs acquire directions of travel that are associated with landmark views. Such 'local vectors' can later be retrieved when the ants encounter the familiar visual scene again. In this case, stored views are used to set directions.

The concept of local (site-based) vectors, which can work independently of global (path integration) vectors, was developed, and the term was coined, in the late 1990s.[39] Three of the key experimental setups, which gave rise to this concept, shall be discussed in a bit of detail, because they demonstrate how subtle differences in experimental design can have decisive influences on the animal's directional choices. Incidentally, the results of these experiments are graphically depicted in the three different ways that are generally used in such studies: as search density plots, traces of ant trajectories, and search frequency distributions (Figure 6.13a, b, and c, respectively). In the first type of experiment the ants must detour around a barrier, which has been placed right in their way, from a familiar food site back to the nest (barrier experiment: Figure 6.13a).[40] After they have bypassed the obstacle to either the right or the left, they follow their updated path integration vector and head directly toward the goal. In addition to the information provided by the path integration vector (PI-V) they could have memorized a local vector (LC-V) and then move from the left edge in the direction α relative to some distant cue (LC-V_{dt})—the distant panorama or celestial cues—and/or use nearby cues (LC-V_{nb}) and deviate by the angle β from the direction of the barrier. When in the critical test zero-vector ants are presented with the barrier rotated through 45°, and displaced to novel territory, they largely follow the rotation of the barrier (LC-V_{nb}).

In the second type of experiment the ants are forced to routinely return from a feeder along an L-shaped route, of which the first segment consists of an open-topped channel hidden in a trench. The second segment leads the ants from the channel exit across open desert ground back to the nest (trench experiment: Figure 6.13b).[41] Along this two-legged homeward route the ants associate the distal end of the channel with a local vector oriented at right angles to the channel. When in the test situation the channel is shortened to half its training length, the directions of the path integration (global) vector and the local vector do no longer coincide. Upon departure from the channel, some full-vector ants immediately express their global vector, but most of them first retrieve their local vector memories and disregard infor-

6.13 Local vectors: classic experimental paradigms

(a) Barrier experiment. During training the ants are forced to detour around a V-shaped barrier *(heavy black bars)*. In the tests the barrier is rotated through 45°. Zero-vector ants are released at *R*. The *orange arrows* deviate by the angle α from the north, either to the right or to the left. The *green arrows* deviate by the angle β from the barrier and the ants' paths along the barrier. For further conventions, see (c). *Cataglyphis fortis.*

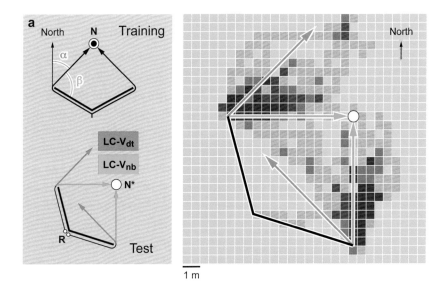

mation from the global vector, until after a while the latter—continually updated in the background—takes over. Learning effects might account for this difference. When the shortened channel is rotated through 45°, zero-vector ants choose either the LC-V$_{nb}$ or the LC-V$_{dt}$ direction (Figure 6.13b, upper right), but prefer more the former, the longer the preceding channel segment has been.

A quite striking example of a local vector outcompeting the global path integration vector is shown in a third type of experiment (U-turn experiment: Figure 6.13c).[42] In this case, the ants face a more difficult task. They are trained to run in a channel from nest to feeder along a U-turn detour path, and are then transferred from the feeder to a straight test channel. When the detour path is designed such that the starting direction of the homebound ants points in the opposite direction to that of the global path integration vector (condition 1), the local vector takes precedence over the global one. In the test the ants invariably express the vector memory of the first segment of their return route.

As it appears from these three types of experiments, the directional component of a local vector can be derived from a variety of cues delimited here as nearby and distant cues. However obvious at first glance, without sufficient information about the visual cues involved, this distinction might appear artificial. Moreover, guidance can be supported by learned motor patterns and control systems that rely on proprioceptive input.[43] Finally, one important point must not be overlooked. Once a zero-vector ant has been released, its path integrator starts running again and provides the ant with a global vector that points back to the point of release and thus may interfere with whatever local vector the ant has retrieved. Take the trench experiment in which zero-vector ants are released in a rotated test channel (Figure 6.13b, upper right), so that the continually updated global vector and the local vector associated with the end of the channel point in different directions. With increasing length of the rotated channel segment, the length of the backward pointing global vector—and with it its reliability and thus its weight (see Chapter 7)—increases. When combined with the constantly southward-pointing local ($LC\text{-}V_{dt}$) vector, the resulting travel vector increasingly rotates westward, until it may finally align with what we have called the $LC\text{-}V_{nb}$ direction.

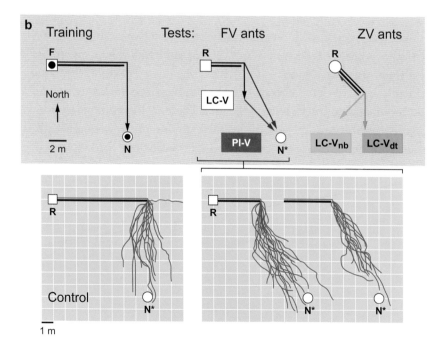

6.13 Local vectors: classic experimental paradigms (cont.)

(b) Trench experiment. *Top:* Training *(left)* along a two-legged inward path with the first leg in a channel (indicated by the *heavy black bar*). In the critical tests the channel is shortened *(middle)* and in addition rotated through 45° *(right)*. *Arrows* indicate the directions taken by the ants in various tests. In the *right figure (ZV ants)* the *red arrow* depicts the ant's PI vector. *Bottom:* Ant trajectories *(blue)* are shown for full-vector ants. *Left:* Control test. *Right:* Critical tests with the channel shortened, so that the global and local vectors point in different directions. The ants behave in one of two different ways; see text. For conventions, see *(c)*. *Cataglyphis fortis.*

6.13 Local vectors: classic experimental paradigms (cont.)

(c) U-turn experiment. Paradigm *(top):* During training homebound ants are forced to follow a U-turn path with the first leg pointing either *(1)* opposite to or *(2)* directly in the direction of the nest *(blue arrows).* Control training without U-turn *(red arrow).* In the tests full-vector ants perform their home runs in a straight channel. Results *(bottom):* The graph depicts the ants' first turning points after release *(boxplots)* and the subsequent search distributions *(blue curves* for U-turn training and *red-filled curve* for direct training). *F,* feeder; *FV ants,* full-vector ants; *LC-V$_{dt}$* and *LC-V$_{nb}$,* local vectors based on distant and nearby cues, respectively; *N,* nest; *PI-V,* path integration vector; *R,* release point; *ZV ants,* zero-vector ants. *Cataglyphis fortis.*

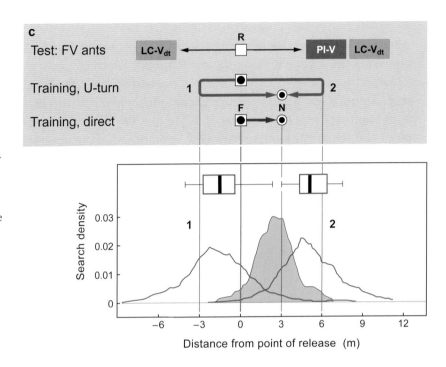

We now may also understand why zero-vector ants almost never run off the entire length of a local vector, and why this is not to be expected from the outset. While a local vector is reeled off, the continually updated global vector increasingly points in the opposite direction and counterbalances the local vector.[44] This interpretation already indicates that the local-vector experiments may finally lead to a more unified view of how the ants compute their travel vectors, namely, by continually integrating global information from their path integrator with information from local vectors. In this view, the use of 'local vectors' may finally merge in the ant's general view matching routine. How, then, do the path integration and view matching routines interact? This question will be addressed in the final chapter of this book (Chapter 7).

Route Traveling

Foraging along well-defined routes is one of the main traits of space use patterns in ants. To mention only a few, albeit spectacular, examples, many species of harvester ants (*Messor, Pogonomyrmex, Pheidole*), leaf-cutter ants *(Atta),* and wood ants *(Formica)* generate dendritic systems of trunk trails, which split into branches and terminal twigs and thus provide the colonies with easy-access roads to valuable resources in their foraging grounds. At our North African and Namibian study sites such ant highways, which special 'road workers' clear of small pebbles, debris, and other obstacles, cover wide areas of the desert floor (Figure 6.14, left).[45] In the context of pheromone-aided mass recruitment, they lead large numbers of foragers to profitable and persistent food sources. Simultaneously they assist newly recruited ants in acquiring route memories of the surrounding visual sceneries. After the animals have traveled the route for some time they rely even more on these visual memories than on the previously dominant chemical signals.[46] For example, when Jon Fewell Harrison and his collaborators used a Y-maze decision device to test foragers of the giant neotropical ant *Paraponera clavata* for the animals' preference of visual or pheromonal cues, naive (newly recruited) ants followed the pheromone trail, while experienced ants were visually guided by the landmarks of the surrounding tropical rain forest. Only when the landmarks had been screened off, did the experienced ants again follow the odor trail.[47] More recently, almost identical results have been obtained in the trail-forming meat ants, *Iridomyrmex purpureus.* As we have seen at the end of Chapter 2, these aggressive ants ecologically 'dominate' *Melophorus* species across the entire Australian continent, to which both are endemic, to such an extent that the latter—the thermophiles—start to forage only at temperatures that the meat ants can no longer stand.

One major advantage of preferring visual cues for guidance is the increased foraging speed observed under these conditions. In following chemical markers, the ants must engage in the somewhat time-consuming process of continually antennating the surface of the ground for olfactory probing. Another advantage of visual route memories lies in the fact that visual landmarks are likely to persist longer than chemical signposts. Wood ants have been shown to be able to recall their visual route memories even many months after they have acquired them, when due to adverse seasonal weather conditions they have been prevented from foraging for quite some time.[48]

6.14 Route traveling

Left: Mass recruitment trunk trail of the harvester ant *Pheidole tenuinodis*, Namibia. *Right:* Individually traveled route (schematized) of *Cataglyphis fortis*, Tunisia.

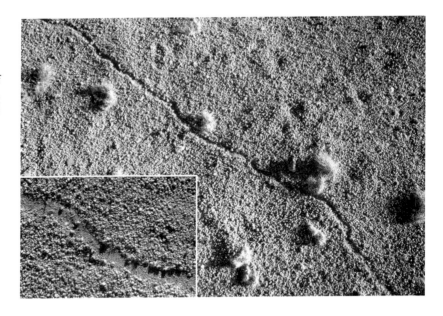

The cataglyphs are different, at least as far as their space use strategies are concerned. In contrast to the ant species mentioned above, which are mass recruiters employing multicomponent chemical trail markers, the cataglyphs forage solitarily, with each individual going its own ways. They do not deposit chemical signposts for route guidance, but nevertheless learn and maintain private, idiosyncratic outbound and inbound routes when running to-and-fro between their central place, the nest, and their foraging grounds (Figure 6.14, right). They accomplish this route fidelity task by acquiring visual landmark memories and recalling these memories whenever needed from long-term memory stores, even without any aid from guidance by path integration (Figure 6.15a).[49]

These are amazing navigational abilities. Foragers of *Melophorus bagoti* travel within natural mazes of grass tussocks interspersed with some sparsely distributed trees. These tussock landmarks all belong to the same plant species, are similarly sized and shaped, and are rather uniformly distributed across the otherwise bare sandy floor. In weaving around these obstacles through a labyrinth of sandy walkways, the ants acquire and maintain idiosyncratic routes, which usually differ be-

tween outward and inward trips, and between individuals of the
same colony visiting the same feeder (Figure 6.15b).

To facilitate further analyses let us experimentally increase the spatial difference between the outward and inward journeys by placing
long narrow barriers close to the nest and the feeder. This induces the
ants to take a looped route from the nest to the feeder and back to the
nest (Figure 6.15c).[50] When along this one-way circuit homebound
ants are captured midway on their inward route and released on their
outward route, they unhesitatingly follow their predisplacement path
integration vector, but when the displaced ants are captured after they
have already completed their homebound trip (zero-vector ants), they
do not choose any of the following three options but behave as though
lost. Neither do they (i) follow the outward route in the reverse direction or (ii) in the direction in which they have traveled it just before
to the feeder, nor do they (iii) directly head for the nest. Instead, they
start their usual search routine around the point of release. However,
when during these searches they accidentally encounter their homeward route, they immediately rejoin it and follow it back to the nest
as if nothing had happened.

6.15 One-way routes

(a) First demonstration that ants can travel visually guided idiosyncratic routes independently of the state of their path integrator. Homeward paths of two ants in the full-vector state *(dotted lines)* and zero-vector state *(solid lines)* of their path integrator. Low shrubs are depicted in *green (black contour lines* in 0.15 m height intervals). (b) Route density plots of five successive outward paths *(1)* and the corresponding inward paths *(2)* of the same ant. In the latter figure the area covered by the outward paths is shown in gray. (c) Forced detours in a one-way circuit. *(1):* Experimental setup. Two low barriers *(black bars)* force the ants to acquire different outward and inward routes from the nest *(N)* to the feeder *(F)* and back to the nest *(orange line:* looped round-trip). Homebound ants are captured either after arrival at the nest (zero-vector ants, *ZV ants)* or on their homeward route (vector ants, *V ants)* and released at *R* on their outward route. *(2):* Path density plot of three round-trip paths of one ant. *(3, 4):* Examples of home runs of displaced V ants *(3)* and ZV ants *(4)*. (a) *Cataglyphis fortis,* (b, c) *Melophorus bagoti.*

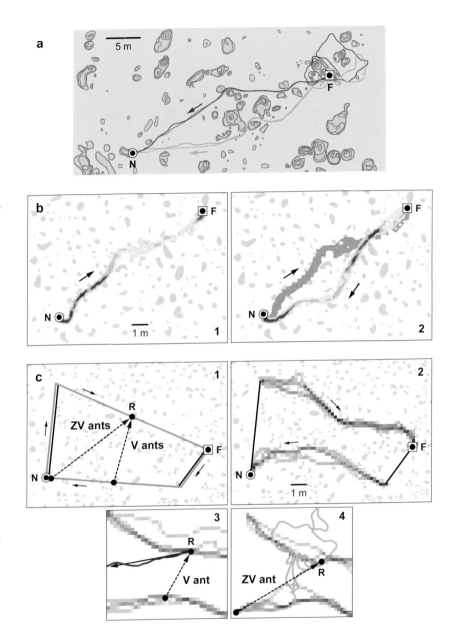

The outcome of this experiment suggests that routes are recognized only in the direction in which they have previously been traveled and along which the foragers had the chance to actively learn them. Route memories are not incidentally acquired in reverse: see experimental result (i) mentioned above.[51] Furthermore, route memories are activated according to the animal's motivational state. Inbound ants do not retrieve their outbound route memories: see experimental result (ii).[52] Finally, and most importantly in general, the ants do not acquire positional memories of the layout of their habitual routes: see experimental result (iii). Rather, they follow procedural instructions, which are bound to both the external stimulus situation present during learning and the internal motivational state.

Some additional impressive examples of the ants' idiosyncratic route memories are given in Figure 6.16a. Here *Melophorus bagoti* travels in a large desert plain highly cluttered with low grass tussocks but almost devoid of large, distinct landmarks. When zero-vector ants are displaced sideways of their habitual routes, they first perform their search routines about the point of release, but when later hitting the route at arbitrary places yet roughly in the right direction (Figure 6.16b), they immediately channel in and follow it 'unerringly' back to the goal, the nest site. This shows that for recognizing the nth part of a familiar route it is not necessary to have previously traveled the *(n-1)*th part, and that routes are recognized only in the direction in which they have been traveled before. Never did an ant choose the reverse direction.

After many remarkable route memories observed in *Cataglyphis* and *Melophorus* desert ants, to our surprise we have not yet found a situation in which the ants failed to recognize a familiar route in the right direction. Just have a look at a particular learning paradigm in which each experimental animal is trained to return to the nest along different paths from a set of different release points (Figure 6.16c, d).[53] Along each route the ants are experimentally induced to perform a fixed number of training runs followed by control runs on the same route and test runs on the previously traveled routes. In the three-route paradigm depicted in Figure 6.16d this procedure results in the recording of 32 successive return paths performed by each ant during a period of three days. In all control and test runs the experimental animals retrieve their previously acquired route memories extremely well, can switch from one route to another, and do so even after they

6.16 Learning and recalling idiosyncratic routes

(a) Recalling familiar routes on first encounter. Each of the five panels depicts the route-specific homeward path *(blue line)* of a zero-vector ant released at *R (black dot).* The ants' preceding 7 homeward paths from the feeder *(F)* to the nest *(N)* are depicted in gray. Note that in all five cases the approach angle (α) is less than 90°. Grass tussocks not shown. (b) Route recall depends on the approach angle α. (c) Experimental field site near Alice Springs, Australia: uniformly distributed tussocks of Buffel grass with a few scattered trees at a distance. The grid of strings (mesh width: 1 m) is used for mapping the tussocks and removed before the experiments start. (d) Learning multiple routes. *I, II,* and *III* depict the sequence in which one individual ant has learned three different inbound routes. *Left:* Inbound runs along an enforced detour from *F* to *N.* The ant's 3rd to 7th inbound paths are shown for each route. *Middle:* Control runs *(orange, subscript C)* performed by the zero-vector ant after 7 training runs. *Right:* Test runs *(red, subscript T)* performed by the zero-vector ant after all three routes have been learned and the ant has been prevented from foraging for five more days. (a, c, d) *Melophorus bagoti,* (b) *Cataglyphis velox.*

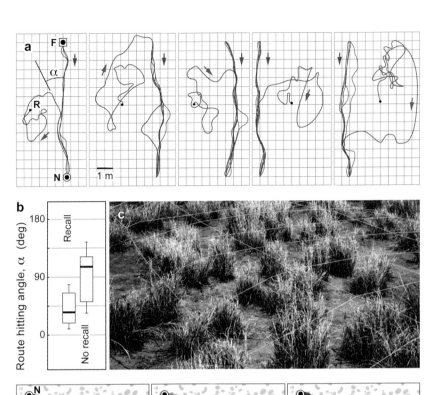

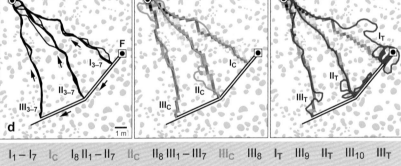

have finally been kept in captivity for five more days. Certainly, the ants' spatial memories have not reached their limits yet. Even though in this particular training paradigm we had let the ants perform seven training runs per route, later observations showed that one to three such runs would have sufficed. In conclusion, the routes are learned rapidly, are stored in long-term memories, and can be retrieved at any time independently of the sequence in which they have been obtained.

Until now we have taken the obvious for granted by assuming that the ants are exclusively guided by visual landmark memories. Direct proof, however, can come only from manipulations of the landmark environment in the field and from appropriate tests in laboratory arenas.[54] Guy Beugnon and his collaborators at the Research Center for Animal Cognition of the University of Toulouse, France, have performed one of the most impressive laboratory studies on visual route learning—not in desert ants, though, but in tropical ants of the large predatory species *Gigantiops destructor*.[55] Unlike the cataglyphs, these huge-eyed formicine ants (Figure 6.17a), which inhabit the Amazonian rain forest, prey on live insects, but just like the cataglyphs they forage solitarily, do not lay scent trails, and develop idiosyncratic landmark-guided routes both in the field and in the laboratory. The gigantic size of their eyes is certainly related to their predatory behavior rather than to navigational needs. When tested in a circular arena, which contains a set of different three-dimensional landmarks on its floor, they return to habitual food sites along distinct routes (Figure 6.17b). These routes are learned quickly and traveled uninterruptedly. When the entire landmark array is rotated through 90° or 180°, the ants shift their courses by the same angular amount. Removing all landmarks causes them to completely abandon their route behavior and to move around in arbitrary ways.

Experiments of this kind fully support the hypothesis that individual route guidance is based on visual route memories. The ants could acquire these memories incidentally while using their path integration vector as a scaffold for assembling these landmark memories, just as in chemically recruiting ants a scent trail could serve this purpose, but such a dependency of route acquisition on path integration need not necessarily be the case. In any way, for retrieving route memories the ants do not require their path integrator. For example, when foragers captured at the feeder are released on their familiar route halfway between nest and feeder, the ants still loaded with their full home vector experience a part of their route that they would usually pass with their home vector already run off by 50%, but the ease, speed, and precision with which the full-vector ants travel this 'unex-

6.17 The 'slalom racer' strategy of *Gigantiops destructor*

(a) The ant's huge compound eyes, each containing 4,100 ommatidia, cover much of the sides of the head. **(b)** *Top:* Visually guided route behavior *(blue trajectories)* of one individual ant in a circular arena provided with various artificial 3-D landmarks *(gray circles)*. The ant enters the arena in the center *(red circle)*. Food sites (access holes depicted by *small black squares*) are evenly distributed along the edge of the arena. *Bottom:* Any route fidelity is gone when the landmarks are removed.

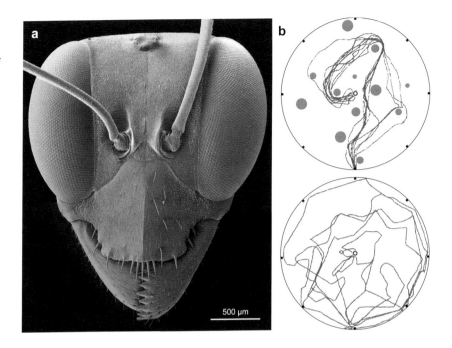

pected' part of their route do not differ at all from those of zero-vector ants.[56]

When it is not the path integrator, what else is it that binds individual landmark memories together? Given what we have already discussed about place learning (stored views) and associated steering commands (local vectors), it has been a likely hypothesis that in route guidance the ants rely on a series of discrete views taken and stored at points along the route and linked together like pearls on a string. In this sequential matching of a series of views, one signpost would prime the memory of the next (stepping-stone hypothesis),[57] so that a visually guided route would be composed of a sequence of habitual heading directions, or local vectors, which the ant would follow until it reaches the catchment area of the next stored view. However, a number of field and laboratory experiments show that ants do not seem to apply this kind of visual sequence learning (sequential priming). As we have already seen, *Melophorus* can join a familiar route at any arbitrary place to which it has been shifted, and can freely switch from any point of a familiar route to any other point of a second or third familiar route, which it then unhesitatingly continues to travel. Hence, these desert ants are clearly able to access route memories out of sequence. Moreover, laboratory experiments in wood ants have shown that it is really difficult to train the animals to sequentially link one set of stimuli to another.[58] In reality, such linkages might not even be essential, because the binding problem can largely be externalized. The sequence in which waypoints appear is already given by the environment ('external memory') and hence need not be specifically represented in the animal's visual memory.

At this juncture let us emphasize an important behavioral trait. Habitual routes are traveled in one direction only (Figures 6.15 and 6.16). Along these one-way routes the ants move in a sinusoidal way at a quite steady pace, but time and again perform short and rapid rotatory movements. During such scanning episodes they most likely employ the rotational image matching strategy to align themselves with the previously traveled route. Moreover, traveling by 'alignment image matching' could take place in a rather continuous way.[59] A simple sensory-motor routine—descending the rotational image gradient by merely increasing the size of alternating turns in proportion to the current image difference ('klinokinesis')—would suffice to keep the animal on course.[60] In fact, conspicuous scanning movements are observed most frequently when the ants have veered away from the route, but rarely when they are on route, and hence seem to occur especially in situations of increased spatial uncertainty. We shall consider them in more detail in the next section.

As an alternative to any multiple-view hypothesis—to ways of matching current inputs with sequences of distinct memories—Bart Baddeley and his colleagues have suggested that routes are learned in a holistic way.[61] In their model the simulated ants while following their path integration vector acquire multiple landmark memories, but they neither associate discrete views with particular places nor extract and learn unique landmark features. Rather, the learning algorithm provides them with a 'familiarity' measure of views encoded in a holistic route memory. When the ants later attempt to recapitulate the route, they employ scanning movements to compare the views currently experienced with all the memorized views, and move in the direction that yields the best match, i.e., the smallest pixel-wise difference, across

6.18 Route-following behavior of a simulated ant

(a) Training run *(red path)* from feeder *(F)* to nest *(N)* and several test runs *(blue paths)* within a cluttered environment. During learning 520 views of the landmark panorama were taken to train an artificial neural network by using an 'Infomax' learning rule. During the tests the simulated ant endowed with the trained network visually scanned the environment and moved in the direction that was most similar to the views encountered during learning. *Lw,* learning walks performed prior to the training run; *Ts,* tussocks and bushes in the simulated world.
(b) Four example views taken along the training route.

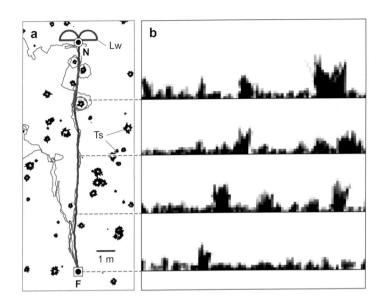

all stored views. This is the direction that is determined by the trained model network as the 'most familiar' one. The landmarks, which appear as silhouettes against the sky, have been simulated in a way that the resulting panoramic views correspond to those that are usually experienced by the ants in their visually cluttered natural environment. Under these conditions the habitual routes of real and simulated animals look very similar indeed. Just compare the natural routes portrayed in Figures 6.15 and 6.16 with those of a simulated ant in Figure 6.18.

In particular, the route behavior generated by this holistic model of scene familiarity shares many characteristics with that of real ants: The routes exhibit a distinct polarity and are channeled into a narrow corridor that can be accessed at any point. They are not separated into a series of waypoints that must be encountered in a strict sequential order. The scanning movements, which are periodically performed by real ants and are introduced in the simulated animals, do not occur at particular points along the route, and not in order to select discrete views. They occur whenever a certain degree of uncertainty has been reached, so that there is an increased need to consult the neural or artificial network for reestablishing familiarity. Moreover, the simulated ants can learn and maintain multiple routes without forgetting the earlier ones and, most remarkably, without holding them in separate memories. In addition, they can use the same mechanism for following a route and searching for the final destination. Finally, the model is even able to reproduce the ants' peaked search distributions occurring within the landmark arrays of Figure 6.4, and it does so even without the need of storing a view from the goal position itself.[62] Given these striking correspondences between the cataglyphs and their model companions, the question inevitably arises whether there is any way to implement such a 'familiarity' algorithm—or a similarly powerful one—into the insect's brain. We shall treat this question toward the end of this chapter.

At present, there is yet a tacit assumption to challenge. We have assumed that in view-based navigation the ants stabilize the orientation of their heads against pitch and roll movements, and thus keep their visual system in level orientation. This may almost be the case when the cataglyphs are running on flat ground (see Figure 3.11),[63] although even there the ant's stride cycle causes rhythmic changes in body orientation. In uneven terrain head pitch may rapidly vary in the range of 20°–30°. Such variations would significantly degrade image matching and impede the use of a visual panorama compass. The minima of the RIDF would be shallower, and additional minima indicating incorrect heading directions would appear.[64] How do the cataglyphs cope with this problem? Do they use information about the pitch and roll of their heads to mentally rotate panoramic images in their brains—a quite demanding task that the ants might not be able to accomplish? It is more likely that they recall snapshots only intermittently when their heads adopt a defined orientation in roll and pitch. Indications that such head stabilization events indeed occur come from Australian bull ants, which position their heads horizontally even when they scan the surrounding panorama while walking on undulating terrain or descending vertically from foraging trees.[65] The question that now remains to be answered is how the cataglyphs acquire their snapshots in the first place.

Learning Walks and Visual Scans

So far we have used the term 'snapshot' as a metaphor of how the cataglyphs obtain view-based landmark information later used for defining the location of the goal. In particular, we have assumed that the ants take such a snapshot directly at the goal. However, as we shall see now, acquiring the necessary landmark information is a process that is spread over a series of exploratory orientation walks, which the ants routinely perform when they start their outdoor lives (pre-foraging learning walks) and thereafter whenever the landmark panorama around the nest entrance has changed (aptly called relearning walks). In these particularly structured walks, during which the cataglyphs anticipate what to learn next, they most likely form view-based memories of heading directions toward the goal. As highlighted in *The American Practical Navigator,* "more than in other phases of navigation," in piloting, the final approach to the goal, "proper preparation and attention to detail are important."[66]

In the cataglyphs this "proper preparation" consists in the particular way in which the ants structure their learning walks and thus the visual information they foresee to acquire through these 'directed learning' maneuvers. In the first of these exploration walks,[67] they leave the nest for only some tens of millimeters, make one or a few loops around the nest entrance, and return to it within a few seconds. These novices are so timid and easily scared that if captured for getting individually marked, they usually reappear outside the nest only on the next day. In three to seven further learning walks, which may last up to one minute and are spread over two to three days, the ants gradually increase their walking speed, meander further and further away from the nest, and cover all points of the compass (Figure 6.19). Finally, they set out for their first foraging runs. In marked contrast to their behavior in the preceding learning walks, they now move much faster and leave the nest in a straight line.[68] As the *Cataglyphis* foragers are active outside the nest for only about a week (see Figure 1.10), they invest quite a substantial part of their outdoor activities in this early learning phase, and thus bear witness to how important this phase is for later accomplishing their navigation tasks. Almost the same sequence of behaviors is observed in the relearning walks, with the obvious difference that the relearning walks end not with a return to the nest but with the start of a foraging run.[69]

What navigational information the novices acquire can be studied best by looking at the structural details of these learning maneuvers.

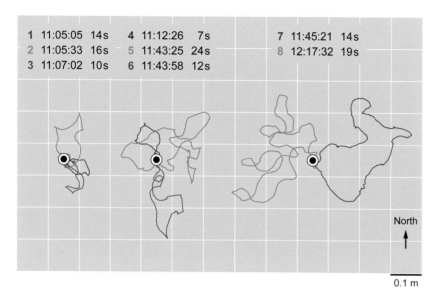

1 11:05:05 14s   4 11:12:26   7s   7 11:45:21 14s
2 11:05:33 16s   5 11:43:25 24s   8 12:17:32 19s
3 11:07:02 10s   6 11:43:58 12s

North

0.1 m

6.19 Successive learning walks of an individual ant

During eight learning walks (*color coded* and depicted in three groups placed side by side) the ant explored different sectors around the nest entrance. The 8th learning walk was followed by the first foraging run. Between the 3rd and 4th and the 6th and 7th learning walks the ant appeared above ground, but due to disturbances did not perform learning walks. These disturbances might have caused the ant to perform an unusually large number of learning walks before starting foraging. *Top:* Time of day and duration of learning walks. *Cataglyphis nodus.*

Embedded in the loops and sweeps of the general walking pattern are particular turning movements, of which the most conspicuous ones are graceful 'pirouettes' (Figure 6.20).[70] The ants stop walking forward, rotate rather slowly about their vertical body axis until they are aligned with the direction to the nest entrance, stop for a brief period of about 150 milliseconds, rotate rapidly back with the turn-out speed being about twice the turn-in speed, and continue their forward walks. For small turning angles the turn-out rotations occur in the counter direction of the initial turns, but for larger ones, the sense of rotation is maintained, so that full 360° rotations result. Obviously, the slow turn-in movements lead to an adjustment of the ant's longitudinal body axis with the ant-goal direction, whereas the fast turn-out movements are just to reestablish the ant's former walking direction. Most importantly, this alignment with the goal direction—this turn-back-and-look behavior—does not depend on any visual signpost defining the goal. The ants orient themselves toward the invisible goal—a small nest entrance level with the ground—by reading out their path integrator. Even if a prominent artificial landmark is placed sideways of the nest entrance, the ants face the invisible entrance rather than the landmark. During a learning walk they may perform more than 10 such pirouettes, and thus 'look back' at the goal from several directions around

6.20 'Relearning walk' induced by an unfamiliar landmark in the neighborhood of the goal (nest, N)

(a) First relearning walk *(blue trajectory)* of a forager, which on its preceding return to the nest had experienced a newly established landmark (dark gray cylinder, *LM*). The locations of pirouettes are depicted by *open red circles*. The *black lines* mark the ant's gaze directions during the stopping phases. In two instances *(enlarged red circles)* details of these rotatory movements are shown: slow turn-in movements (v_{in}, *bold red arrow*) and fast turn-out movements (v_{out}, *thin red arrow*) in either the counter *(1)* or the same *(2)* direction. (b) Spatial and temporal characteristics as well as temperature dependencies of the pirouettes. *d,* ant-goal distance; *l,* path length. *Ocymyrmex robustior.*

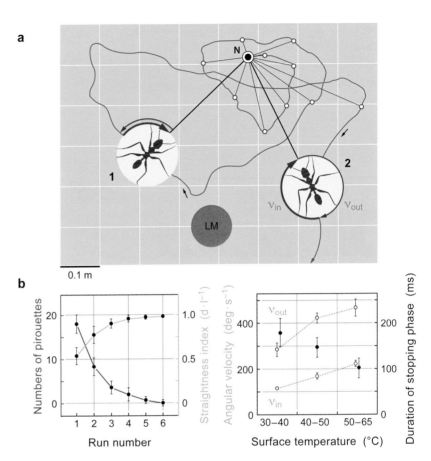

the nest. With this well-choreographed learning behavior, *Ocymyrmex* gave us a first hint that landmark information may be acquired by taking a set of goal-centered snapshots around the goal rather than, as previously surmised, a single snapshot at the goal itself.

Structural details of these learning walks may differ among species. Pauline Fleischmann and Robin Grob in our Würzburg research group have used high-speed videography to analyze such peculiarities (Figure 6.21). *Cataglyphis nodus,* which inhabits the scrubland *(garrigue)* areas of the Peloponnese, from where it was first described,[71] as well as the open pine forests at our Greek study site, includes several stopping phases in its pirouettes. Only the longest stops are oriented toward the nest entrance. In contrast, the salt-pan species *C. fortis* does not pirouette at all, even when the nest

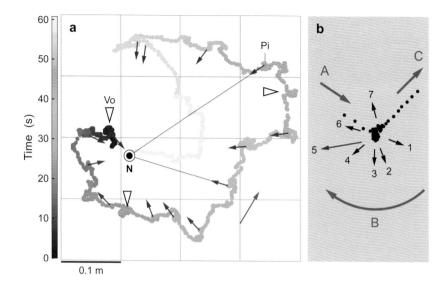

6.21 Learning walk recorded by high-speed videography

(a) Learning walk of a novice starting its outdoor phase of life. The walk contains several pirou-ettes *(Pi)*, in which the ant's gaze directions during the longest stopping phases are indicated by *red arrows*. Occasional voltes *(Vo)* are marked by *open arrowheads*. The band-like (time-coded) trajectory is derived from record-ings of the ant's head-alitrunk axis. (b) A pirouette in detail. *Blue arrows: A* and *C,* translatory in and out movements; *B,* rotatory movement. Gaze directions during the stopping phases *(1–7)* are denoted by *black arrows;* the longest and goal-directed one *(5)* is highlighted in *red.* The ant's positions are given in 10 ms intervals *(black dots). Cataglyphis nodus.*

entrance is surrounded by a set of artificial landmarks. It exhibits another type of turning movement ('voltes'), in which it walks in tight circles that are rarely interrupted by stopping phases. Such voltes are observed in *C. nodus* as well, but to a much lesser extent. Do these species-specific differences depend on the information content of the visual environment? In terms of habitat structure, the two species adopt the extreme ends of a spectrum covered by the cataglyphs: virtually featureless barren land in *C. fortis* and richly cluttered environments in *C. nodus.* Does this mean that pirouetting occurs the more readily, and contain the more stopping phases, the more visually structured the environment is that the ants encounter? Moreover, where does *C. nodus* look when it stops while not being oriented in the ant-goal direction, and what information do the ants acquire when they perform their voltes? As recently shown by Jo-chen Zeil and his colleagues, the learning walks of the Australian jack jumper ants include rapid turning movements, in which the ants alternately face toward the nest and away from it. To what ex-tent are such particular spatial and temporal characteristics ob-served in the learning walks of different species' genetically fixed traits or adaptions to the landmark information available in various environments? Are their certain motor motifs that commonly occur in all the ants' learning behaviors?

At present, we have more questions than answers. Nevertheless, let us assume that while turning on the spot—while pirouetting and gazing toward the goal—the ant acquires a goal-directed view of the landmark panorama at that site, so that a multi-pirouette learning walk would provide the animal with a set of reference snapshots taken from different directions around the goal. Paul Graham and his colleagues have theoretically shown that already a few panoramic memories taken during such nest-directed turns would suffice for later enabling the animals to determine goal directions from a rather large area.[72] This is because panoramic views taken at a particular location will only smoothly change with distance from that location. Hence, a cataglyph may later be able to determine the direction to its goal, the nest entrance, even at places at which it has not been before. The spatial range over which this kind of 'view extrapolation' may work, i.e., over which panoramic reference views can provide proper information for guidance, depends on the number and spatial distribution of the turn-back episodes performed during the learning walks, and on the visual structure of the landmark panoramas surrounding the goal. Even though the ant must read out its path integration coordinates for acquiring the goal-facing views, there is no direct need to associate the landmark views taken at particular sites with these coordinates. The path integration system may serve only a temporary and auxiliary function restricted to the view acquisition phase.

Moreover, are there other kinds of visual information than just panoramic views that the ants may acquire during their learning walks? For example, do they tag these views with the compass direction in which the views are taken? Or do they, in addition, exploit the depth structure of their environment? Owing to their small body size, ants—and insects in general—cannot use binocular stereopsis. Hence, their most likely means of acquiring distance information is through motion parallax, i.e., by exploiting cues derived from image motion.[73] We do not yet know enough about the geometry of the cataglyphs' learning walks to address such questions, just as research on how the cataglyphs employ 'active vision' strategies by moving and gazing in particular ways has just begun.

For comparative reasons, let us take a quick look at the functional equivalent of the ants' learning walks: the stereotyped learning flights of wasps and bees. Since the middle of the nineteenth century these conspicuous flight maneuvers have attracted almost worldwide attention. The German gentleman scientists Baron August von Berlepsch and Hugo von Buttel-Reepen, the English naturalists Henry Bates in the Amazon rain forest and Thomas Belt in Nicaragua, the American entomologists George Peckham and Elizabeth Peckham in Wisconsin, and the Russian zoologist Wladimir Wagner in St. Petersburg were so impressed by the lengthy "orientation flights" and "locality studies" of honeybees and bumblebees, of sphecid wasps and vespid wasps that they carefully observed and described many aspects of the bees' and wasps' unique movement patterns.[74] From these early accounts we can already learn that upon leaving the nest, a wasp or bee turns back, faces the nest entrance, and gets engaged in a series of arc-like flight maneuvers by swaying to and fro in alternating clockwise and counterclockwise directions. The widening arcs grow in size and increase in height as the learning flight proceeds, until finally the insect, having moved farther and farther away from the nest, turns around and starts its foraging journey. When upon return from this journey the wasp or bee approaches the nest, it recapitulates certain movement patterns of the learning flight so that it can encounter the same sequence of views that it acquired during the preceding learning flight.

Recently, these learning and return flights have been video-recorded in ground-living solitary wasps and bumblebees at high temporal and spatial resolution, and analyzed in unprecedented three-dimensional detail.[75] In spite of some species-specific and location-dependent differences, a general pattern emerges. One component of this movement pattern is especially intriguing. While the insect pivots about the nest entrance within a cone-like flight corridor and has reached the end of an arc, it performs saccadic head movements resulting in short nest-fixation phases. Hence, the behavioral routine to quickly turn toward the goal and then most likely take and store a goal-centered view of the visual surroundings seems to be shared by flying wasps and walking ants, i.e., could have been adopted by the ants from their flying ancestors. However, it still needs to be investigated in ants as well as in wasps and bees what and when images are actually taken and later retrieved,[76] and whether besides static views dynamic ('short movie') scenes are exploited as well, but there is every reason for hope that the incessant refinement of recording and evaluation techniques may soon bear further fruit in tackling these questions. In any way, the study of learning walks and learning flights has become one of the most promising fields of research in insect navigation. Here is a spatially restricted and structurally stereotyped behavioral routine of which the functional significance is immediately apparent.

Functionally, the entire suite of the cataglyphs' orientation walks is a necessary requirement for successful visual homing. Just have a look back at Figure 6.4, where the ants show sharply peaked search density distributions when they are presented with a familiar landmark panorama surrounding the goal. Now we can add that the accuracy and precision of pinpointing the goal gradually develops as the ants proceed in their learning walk sequence (Figure 6.22). The final learning plateau is reached only after the ants have completed this sequence and start to venture out for their first foraging journeys. Virtually the same sequence of events occurs in learning and retrieval, when experienced foragers are suddenly confronted with new landmark conditions around the nest, e.g., with the very same three-cylinder array that the novices had to learn.[77]

In conclusion, it is not a single snapshot taken at the goal, but a suite of goal-directed snapshots taken at several locations around the goal that provides the animal with the view-based information subsequently used in homing behavior. If this conclusion is correct, the ants would need "plenty of room at the bottom" to accomplish their learning walks around the nest entrance. Indeed, when novices are experimentally deprived of enough space around the goal and thus prevented from properly performing their learning walks, they do not acquire the necessary goal-centered landmark views. This can be experimentally achieved by installing a system of water moats around the nest entrance. These water-filled channels sunk into the ground restrict the ants' range of movement, but not their view of the surrounding landmark panorama. Having reached the moat and stumbled into the water, the ants immediately return to the nest. When novices, which have been spatially constrained this way, are later subjected to particular displacement experiments, in which their path integrator leads them in the opposite direction to the nest (Figure 6.23), they get lost. In contrast, when they have been provided with enough

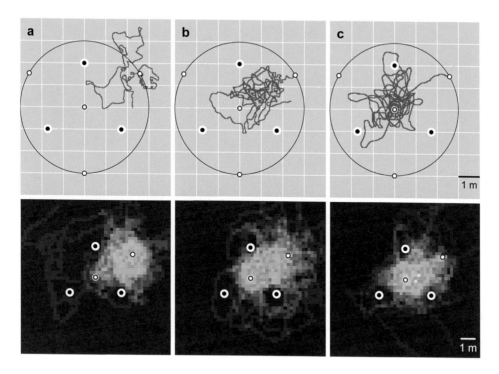

6.22 With increasing numbers of learning walks the cataglyphs gradually improve their accuracy of pinpointing a goal

Ants are trained to locate a goal (nest entrance) in the center of a triangular array of black cylinders *(black dots)*. After displacement together with the landmarks to a distant test field, they are released outside the array at three locations *(small white dots)*. Their subsequent search trajectories are shown in *blue*. The *red circle* marks the fictive position of the nest. The tests are performed after the ants have completed **(a)** their 1st and **(b)** their 2nd to 4th learning walks, and **(c)** after they have already become experienced foragers. The panels depict individual examples *(top)* and search density profiles of 15–20 individuals each *(bottom)*. *Cataglyphis fortis.*

space to perform their early landmark-learning routines, they finally return home successfully.[78]

Turn-back behaviors similar to the ones around the nest entrance are also observed, though hitherto less well studied, when the ants depart from the nest in new foraging directions, and when they start their homebound runs from newly discovered profitable food sites. Just like view-based site memories, view-based route memories are most likely acquired through active learning processes rather than merely en passant. When the ants are prevented from performing their

6.23 The water-moat experiment: restricting learning walks

(a) Novices perform their learning walks within a 2×2 m² area around the nest *(N)* or (b) are prevented from doing so. The different setups are indicated by *red arrows*. The areas available to the ants *(shaded in gray)* are bounded by water moats. Novices having reached the feeder for the first time are displaced *(dotted blue arrow)* as full-vector ants in the direction opposite to *N* and released at *R. N*,* fictive position of nest. Having originally followed their home vector, the ants are finally able to home successfully only in *(a)*. In either case, paths are shown for 15 ants *(light blue trajectories);* an individual path is indicated in *dark blue*. The outer gray borderline indicates that the nest is located within a heavily cluttered environment, in this case an open pine forest, rather than an empty plain. *Cataglyphis nodus.*

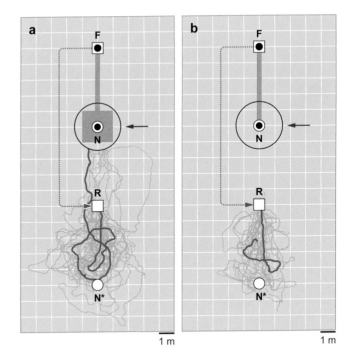

turn-back maneuvers along their outbound paths (by confining their paths to narrow channels, which also obstruct the natural landmark panorama), their homing behavior is severely impaired.[79] Route learning and site learning are sequences of events that blend into each other, so what the ants might finally learn and retain are independent landmark-based outbound (foodward) and inbound (nestward) paths.

The look-back routines in the learning phases have their counterparts in scanning movements later performed in view retrieval. When in the early 1970s Andreas Burkhalter, now a professor of neurobiology at Washington University, St. Louis, joined me as one of my first graduate students in desert ant research, we were fascinated by a remarkable behavioral phenomenon. Every so often, especially when displaced sideways of its habitual inbound route, *Cataglyphis bicolor* would climb stones or twigs and perform rotations on the spot or slight swaying movements to the left and right. What spatial information did they acquire when scanning and swaying?

Almost half a century later, an answer has come within reach. In two Australian ants, the thermophilic scavenger *Melophorus bagoti* and the predatory jack jumper *Myrmecia croslandi,* which both are

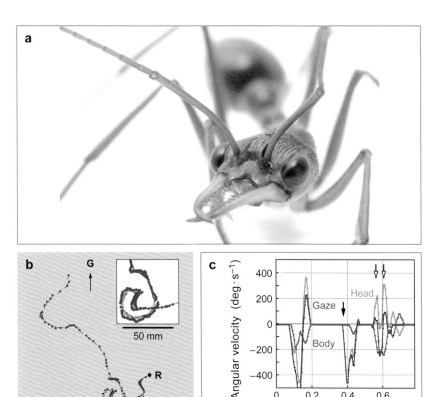

6.24 Scanning routine

(a) The Australian bull ant, *Myrmecia nigriceps*, turning its head. (b) Path of a zero-vector ant, *Myrmecia croslandi*, after release at R. The ant's positions are given every 100 ms (40 ms in the *inset*). The *red line segments* pointing in the direction of the front of the head *(dot)* indicate gaze directions. The *arrow G* marks the direction to the goal (nest). (c) Angular velocities of head (relative to body, *green*); longitudinal body axis *(red)*; and gaze direction *(blue)*. *Filled arrow*: head orientation is followed by body orientation; *open arrows*: head movement compensates for body orientation.

solitary foragers inhabiting open countryside, such systematic scanning routines are now studied in detail and recorded by high-speed videography.[80] During their scans the ants rotate at a fairly constant rate of 90° per second in one direction, and—usually before a full turn is completed—in the reverse direction. Superimposed on these turns are fast, finely attuned head and body saccades performed at rates of up to 500° and 200° per second, respectively. These saccades result in short periods of constant gaze directions (Figure 6.24). Similarly to what has been mentioned above for the pirouettes in the learning walks, these gaze directions are not correlated with any feature in the surrounding landmark scene. Most likely, the ants, while periodically scanning the visual world around them, employ their rotational image matching strategy, i.e., compare the remembered nest-directed views with the currently experienced panoramic view.

The mere fact that for comparing stored and current views the ants engage in locomotor scanning maneuvers may be another hint that they are not able to mentally rotate their stored views. Given the functional relationship between the learning-walk pirouettes (the acquisition of panoramic view information) and the foraging-walk scans (the retrieval of panoramic view information), the question arises as to how these two correlated modes of behavior compare in the details of their structure and dynamics. An answer to this question, obtained at best in a variety of species and environmental conditions, will be one of the most decisive steps in understanding the navigational stratagems used by the insect in finding familiar places and traveling habitual routes. At this juncture, however, let us again journey into the cataglyph's brain and ask where the snapshots may be stored.

Mushroom Body Dynamics

The most likely sites at which the acquisition, long-term storage, handling, and use of landmark information take place are the mushroom bodies, the 'corpora pedunculata.' Ever since in the middle of the nineteenth century Félix Dujardin had first described these prominent lobate neuropils housed in the insect's forebrain (protocerebrum), they have been considered to be higher-order processing centers associated with sophisticated behavioral repertoires and involved in multimodal sensory integration, learning, and memory. Together with the central complex considered in Chapter 3, they are the most prominent sites of multisensory convergence. Moreover, and in line with Dujardin's bold claim that the mushroom bodies provide the insects with "free will" and thus elevate them "above the instinctive," they have been hypothesized to play a major role in decision making and mediating behavioral plasticity.[81] As David Vowles remarked more than half a century ago, "All the fashionable mysteries of behaviour have been successively attributed to them."[82] The way of how the ants acquire the necessary navigational information from the unpredictable, diverse, and varied landmark environment, within which they perform their foraging activities, and how the underlying neural circuitries accomplish these learning tasks, is certainly such a "mystery."

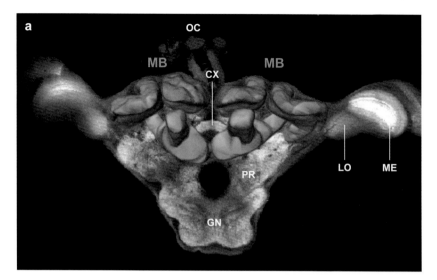

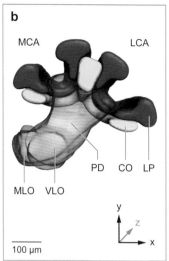

Among the hymenopterans the mushroom bodies are most conspicuous, voluminous, and elaborate in social insects such as wasps, bees, and ants. This is especially the case in the worker caste of ants.[83] Figure 6.25a provides a vivid impression of how prominently the mushroom bodies feature in the cataglyph's brain. Their intrinsic cells, the so-called Kenyon cells, account for almost half of the total number of neurons in the brain of *Cataglyphis*. For comparison, in *Drosophila* flies the relatively much smaller mushroom bodies contain less than 4% of all brain neurons.[84]

It is likely due to the small body size of ants that their prominent mushroom bodies have become a focus of research only rather recently. When Charles Darwin asked his son Francis to dissect for him the "cerebral ganglia" of a wood ant, he was surprised to find that they were "not so large as the quarter of a small pin's head." In fin de siècle times, the early insect neuroanatomists focused on honeybees, orthopterans, and cockroaches.[85] Ants were included only to emphasize the remarkable caste-specific size variations of their forebrains, especially the much smaller relative brain sizes in queens and males than in workers.[86] In this view, we may consider it a pity that in the beginning of the twentieth century Santiago Ramón y Cajal, who along with his student Domingo Sánchez y Sánchez performed the most beau-

6.25 The mushroom bodies in the cataglyph's forebrain

(**a**) 3-D reconstruction of the mushroom bodies *(orange)* superimposed on a frontal cross section through the brain. (**b**) A mushroom body isolated. *CO (yellow)*, collar; *GN*, gnathal ganglia; *LCA*, lateral calyx; *LO*, lobula; *LP (blue)*, lip; *MB*, mushroom body; *MCA*, medial calyx; *ME*, medulla; *MLO*, base of medial lobe; *OC*, ocelli; *PD*, peduncle; *PR*, protocerebral neuropiles; *VLO*, base of vertical lobe. Coordinates: *x*, lateral; *y*, dorsal; *z*, caudal. (**a**) *Cataglyphis bicolor*, (**b**) *C. nodus*.

6.26 Microglomeruli in the calyx of the mushroom bodies

(a) Overview of immunilabeled microglomeruli in the lip *(LP)* and collar *(CO)* region; inner branch of medial calyx. Pre- and postsynaptic structures are shown in *red* and *green,* respectively. (b) Detail in higher magnification. (c) Scheme depicting the organization of a microglomerulus. *Bo,* presynaptic bouton; *Kd,* Kenyon cell dendrite; *Ks,* Kenyon cell spine (postsynaptic site); *Pn,* projection neuron; *Sv,* agglomeration of synaptic vesicles. *Cataglyphis fortis.*

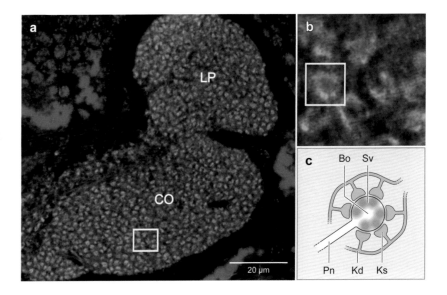

tiful anatomical analyses of the nervous systems of flies, bees, locusts, and dragonflies,[87] never seemed to have tried his hand at Golgi impregnating an ant's brain. Only several decades later, three 'ant brain pioneers' made important contributions to solving the mushroom body puzzle in insects. These pioneers are David Vowles, whom we have just mentioned, and two greatly underappreciated neuroanatomists, who coincidentally but unknown to each other made their discoveries while working on their doctoral theses in the late 1960s: Wolfgang Goll from Stuttgart-Hohenheim, Germany, and Ulrich Steiger from Zürich, Switzerland.

To appreciate the contributions of the three ant brain pioneers, let us first have a look at the mushroom body ground pattern. In Figures 6.25 and 3.31b and c, the cataglyphs' mushroom bodies are shown in total view and cross section, respectively. Peripherally, in the paired cup-shaped calyces, two major anatomically distinct subdivisions—the lip and the collar—constitute the main input regions for olfactory and visual information, respectively. In these calyces the large dendritic trees of tens of thousands of intrinsic Kenyon cells form elaborate neuropils composed of dense arrays of synaptic microcircuits, the microglomeruli (Figure 6.26). Each microglomerulus is composed of a central presynaptic bouton—the axon terminal

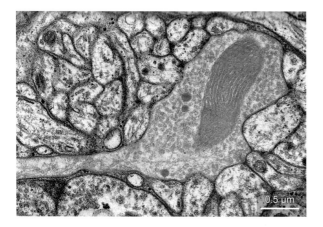

6.27 Ulrich Steiger's early electron micrograph of a microglomerulus of an ant, *Formica rufa*

In the calyx neuropil the presynaptic bouton of a sensory projection neuron *(highlighted in yellow)* is surrounded by numerous small postsynaptic profiles of Kenyon cell dendrites. In his 1967 study, Steiger correctly interpreted this microglomerular complex as a "Divergenzsynapse." The presynaptic bouton contains a large mitochondrium, a few dense core vesicles *(three black dots),* and a plethora of light core vesicles *(gray profiles).*

of a visual (in the collar) or olfactory (in the lip) projection neuron—and numerous surrounding postsynaptic profiles. These profiles are mostly dendritic spines of Kenyon cells, but also endings of recurrent inhibitory and modulatory neurons. The number of these microglomeruli is huge. In the collar alone *Cataglyphis* possesses about 400,000 of them.[88] Based on his superb electron microscopic studies of the mushroom bodies of wood ants, Ulrich Steiger was the first to unravel the functional anatomy of the microglomeruli in any insect (Figure 6.27).[89]

As sketched out in Figure 6.28a, the axon-like processes of the Kenyon cells run in parallel through the stalk of the mushroom bodies, the peduncle, and finally bifurcate to form the vertical and medial lobes (the bifurcation is not included in Figure 6.28a). Different sets of modality-specific Kenyon cells occupy different areas in the calyces, form distinct axon bundles (slabs subdivided into laminae) in the peduncle and lobes, and may be functionally further partitioned into different domains batched along the lengths of peduncle and lobes.[90] The parallel arrays of Kenyon cell axons are orthogonally intersected by axon arrays of various sets of extrinsic neurons such as sensory projection neurons, modulatory (dopaminergic and octopaminergic) neurons, and output neurons. This way, the peduncle and lobes are connected to the surrounding protocerebrum by a variety of different input and output pathways accompanied by significant feedback circuits even from the mushroom body output neurons to the calyx

6.28 Sketch of mushroom body ground pattern

(a) Schematic circuit diagram. In the peduncle and lobes the parallel fibers *(black)* of the Kenyon cells are part of local networks with input and output connections depicted here in a general and highly schematized way. *Open and filled squares* indicate pre- and postsynaptic sites, respectively. *INH,* inhibitory neurons (involved in feedback circuits); *MG,* microglomeruli; *MOD,* modulatory neurons; *OUT,* output neurons; *SENS,* sensory interneurons. (b) Wolfgang Goll's demonstration that the cell bodies of morphologically distinct types of Kenyon cells fill and surround the bowls of the paired calyces *(CA),* where they occupy different subdivisions *(black): 1,* basal ring; *2,* collar; *3,* lip. *PD,* peduncle. (c) A modern view of the general pattern of mushroom body compartmentalization as devised by Nicholas Strausfeld.

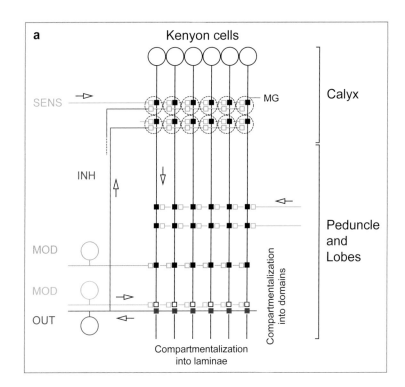

input side. Moreover, the mushroom bodies strike by a pronounced divergence-convergence flow of information. While extrinsic sensory input neurons project onto a large number of intrinsic Kenyon cells (divergence ratio at least 1:100), this large number of Kenyon cells finally projects onto relatively few extrinsic output neurons (convergence ratio at least 100:1).[91] On the one hand, the high degree of divergence at the input side is considered a means of transforming broadly tuned sensory inputs into sparse representations within large sets of narrowly tuned neurons, which—e.g., in the case of visual inputs—are able to encode a multitude of visual patterns. On the other hand, the high degree of convergence at the output side may allow for multimodal integration of behaviorally relevant sensory inputs. For example, in *Cataglyphis* navigation this could mean to associate visual memories with olfactory, tactile, proprioceptive and other sensory inputs that have been shown to contribute to route-following behavior, and thus to provide the ants with some compound information about frequently traveled routes.

After these short remarks on mushroom body circuitry, let us rejoin our little historical tour. Based mainly on his work in wood ants, David Vowles was the first to challenge the long-held view that the calyces of the mushroom bodies constitute the sole afferent input region, while the lobes function merely as sites of efferent output. In the mid-1950s he used intricate methods of producing localized lesions in the ant's brain—high-frequency radio waves emitted through a glass electrode and slivers of razor blade—for studying the effect of these lesions on subsequent neural degenerations in the brain as well as on the animal's behavior. From his ingenious, though limited experiments, he concluded that the calyces are not the sole input site. At least the lobes—he referred to the vertical lobe—receive sensory information as well. It was only much later that the input synapses that he had concluded to occur in the lobes of ants were discovered morphologically, this time in crickets. After all, Vowles had realized that the mushroom bodies play a central role in a system of feedback loops between various sensory, motor, and—as we now must add—modulatory neuropils, and thus anticipated the intricacies of mushroom body networks.[92]

A decade later, Wolfgang Goll again performed local lesions in the ant's brain, and in addition provided the first and as yet most beautiful Golgi staining study in ants. He showed that discrete subsets of Kenyon cells occupy different subregions of the calyx (Figure 6.28b) and project to the peduncle and the lobes in a highly ordered way. Owing to his limited degeneration techniques, he did not get the spatial details of this projection pattern fully right, but he was the first to demonstrate that in the insect's mushroom body the regionalization of the calyx is transformed into an equally well-structured compartmentalization of the peduncle and the lobes.[93] For several decades following Steiger's and Goll's decisive discoveries, no one else had taken a further deep look into an ant's forebrain, until in the mid-1990s Wulfila Gronenberg, then at the University of Würzurg, Germany, started his extensive morphometric work on the mushroom bodies of various species of ants. A decade later, modern techniques of immunofluorescent tracing, confocal microscopy, and 3-D imaging began to be used in fine-grain analyses of these neuropils, again at the University of Würzburg but now by Wolfgang Rössler's team, and focused on *Cataglyphis* as the model organism.[94]

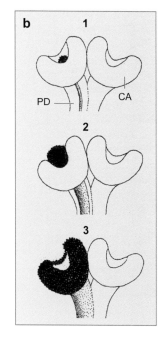

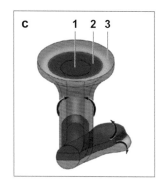

One question might have constantly hovered in the background throughout the previous considerations. Why at all do we refer to the ant's mushroom bodies in the context of the acquisition, storage, and use of landmark information? Almost all early investigators of insect brains have pointed at the obvious correlation between enlarged mushroom bodies and the social lifestyle of their bearers. Hence, it is not astounding that the two traits have been considered to be causally related ('social brain hypothesis'). However, as comparative morphological and phylogenetical analyses show, that is not the case. The expansion and elaboration of the mushroom bodies, and the substantial increase of input they receive from the optic lobes, are correlated most significantly with the lifestyle of parasitoid solitary wasps.[95] Recall that in provisioning their larvae with paralyzed prey insects, the parasitoid wasps must learn and memorize the locations of their often multiple underground nests, which they regularly replenish. Seen in this light, the evolution of mushroom body expansion in the higher hymenopterans might have been driven primarily by the behavioral challenges and computational demands associated with central place foraging and spatial learning, and thus may predate the advent of sociality by almost 100 million years of evolutionary history. This conclusion is corroborated by the fact that large, complex mushroom bodies also occur in some nonsocial insects—e.g., in certain scarab beetles and nymphalid butterflies—that exhibit navigationally demanding foraging behaviors.

In fact, the neural architecture and computational style of the mushroom bodies seem well suited to process and store the large amount of landmark information—the multitude of goal-centered and route-aligned panoramic views—that the cataglyphs are able to acquire and use in visual homing. To this end, let us return to Figure 3.31c and consider the major visual pathway that conveys visual information from the optic lobes to the mushroom bodies: the MB pathway, the anterior superior optic tract, which runs directly from the medulla to the visual input areas (the collars) of the mushroom body calyces in either hemisphere of the brain. Very pronounced in *Cataglyphis,* this MB pathway is smaller in ants that are guided by pheromone trails, and minor if not absent in many non-hymenopteran species studied so far.[96]

Barbara Webb and her collaborators have used the mushroom body architecture sketched out above as an associative learning network to account for the view-based route-following behavior in ants. The model, which was originally developed by the authors in the context of the simpler olfactory associative learning behavior of fruit flies—a behavior in which the mushroom bodies are decisively involved—is in principle also sufficient to let the simulated ants accomplish their seemingly advanced route-following tasks.[97] The key circuit properties of the mushroom body network, especially its divergence-convergence structure and sparse encoding of sensory inputs within large arrays of neurons, enable the simulated ants to separate and store a multitude of panoramic views experienced while the ants travel visually guided routes, and to distinguish these stored views from different, though highly similar, views, which the ants would encounter if they looked off-route or traveled along the route in the wrong direction.

However, what is the reinforcement signal with which the stored patterns must be associated? In olfactory association tasks the relevant signal is provided by the food reward. In the cataglyphs' landmark learning tasks it could consist in some kind of internal reward associated with the state of the ant's incessantly running path integrator. Successfully progressing along a familiar route is inevitably correlated with a 'rewarding' decrease of the travel vector, which finally reaches its zero state when the ant has arrived at the goal. Similarly, in the learning walks around the nest entrance, the reward signal could consist in the ant's alignment with the home vector during the turn-back-and-look episodes. The panoramic image experienced then could be tagged with a positive value.

The visual memory model based on realistic mushroom body circuitries offers the opportunity to put the abstract 'familiarity model' introduced in the last section for simulating the ant's route-following behavior, on realistic neurobiological footing, and to do so in a computationally even less demanding way. We now have a model at hand that in contrast to all former, mostly algorithmic modeling attempts is based on the computational capabilities of a particular neural circuit to simulate one of the ants' basic navigational behaviors. At present, however, this is primarily a promising idea, because we do not yet have much direct evidence at hand that the mushroom bodies actually handle and store landmark information. But there are some promising hints. Cockroaches, in which the medial lobes have been lesioned by meticulously inserting thin slivers of aluminum foil into the brain, fail in local visual homing tasks. In particular mushroom body neurons of honeybees, the insect homolog of an immediate-early gene involved in spatial learning in mammals is rapidly upregulated, after the bees have performed just a single learning flight or a re-learning flight in novel territory.[98] Moreover, as also first observed in honeybees by Gene Robinson and his coworkers from the University of Illinois at Urbana-Champaign, the behavioral transition of workers from performing indoor nursing to outdoor foraging tasks is associated with a remarkable volumetric expansion of the mushroom bodies, especially the calyces. In the wake of this discovery, such increases of mushroom body size during the animal's indoor-outdoor transition have been observed also in wasps and ants including the cataglyphs (Figure 6.29).[99] An avalanche of similar studies followed.

The increase of mushroom body size is not caused by adult neurogenesis. Never have active neuroblasts and the formation of new Kenyon cells been detected in the mushroom bodies of adult bees or ants.[100] The increase is caused by remarkable synaptic reorganizations occurring within the microglomeruli of the visual input centers.[101] The number of microglomeruli decreases (pruning), but this decrease is

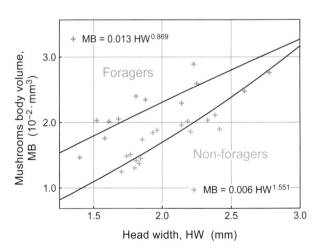

6.29 Task-related volume changes of the cataglyphs' mushroom bodies

Foragers *(orange)* have significantly larger mushroom bodies (relative to body size) than age-matched non-foragers *(blue)*. *Cataglyphis bicolor.*

associated with a massive increase in the outgrowth and branching of postsynaptic Kenyon cell dendrites.[102] Synaptic efficiency and the divergence from visual input neurons to Kenyon cells increase. Obviously, these synaptic reorganizations adapt the microglomerular synaptic complexes to the bright outdoor environment and thus may transform them from a 'default state' to a 'functional state' ready for encountering the forthcoming flood of visual information.

Contrary to what one might have expected, this neural rewiring does not result from a strict developmental program, but—as shown in the cataglyphs—is triggered by light. This most likely occurs when the ants, shortly before they start their learning walk routine, appear above ground for carrying out soil particles. In these 'digging walks' they rush straight out of the nest for only a few centimeters, deposit their load on the ground, perform a sharp U-turn, and hurry directly and with high speed back to the nest. Experimentally, the entire sequence of synaptic reorganization events mentioned above can be elicited by exposing the ants to light at any time of their indoor life. It can be elicited even in one-day-old, freshly hatched workers, so-called callows, which normally would undergo this transition only four weeks later, or in animals that have been reared in total darkness for several months (Figure 6.30). This means that the ants' visual microcircuits must have remained plastic for that long a time. Figure 6.30 also shows that a slow age-related increase of the number of microglomeruli is running in the background independently of the changes caused by light exposure. For the cataglyphs this means quite some metabolic investment, because the mushroom body calyces with their dense arrays of microglomerular synaptic complexes are energetically expensive neuropils.[103] The presynaptic boutons contain large mitochondria and contact more than 100 malleable postsynaptic spines (Figure 6.26c). Hence, in contrast to what usually happens in animals—and humans—during biological aging processes, in the cataglyphs neuronal plasticity and storage capacity seem to be maintained, and even increased, as the animals grow older.

Note that the structural reorganizations, which occur within the microglomeruli upon the animals' first outdoor appearance, are not based on any form of associative learning. Triggered by sensory exposure alone, they prepare the synaptic microcircuits for the visual

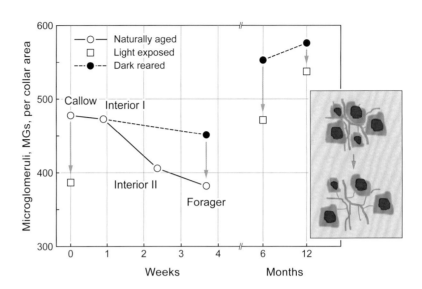

6.30 Age-related and light-triggered plasticity in the visual input area (collar) of the mushroom body calyces

In the dark the number of synaptic complexes (microglomeruli, *MGs*) slowly increases. In naturally aged as well as dark-reared animals (*open and closed circles,* respectively) light exposure leads to some pruning of MGs and considerable MG expansion (see *inset*), resulting in a decrease of MG density and an increase in overall calyx volume. The transition of MG structure from the 'default state' in the dark to the light-triggered 'functional state' is highlighted by *orange arrows. Callow, Interior I,* and *Interior II* denote different pre-foraging stages in the development of adult workers. *Cataglyphis fortis.*

navigation tasks that the ants must accomplish during their foraging lives. One of the first of these tasks is to acquire the landmark information later used in homing. As we have argued above, it is during the learning walks following the digging walks that the cataglyphs acquire stable, long-term memories of multiple landmark views from various vantage points around the nest, and as recent results show, it is during these very learning walks performed under natural skylight conditions that the number and density of the microglomerular complexes in the collar of the mushroom bodies increase. Such structural changes associated with the formation of long-term memory have previously been found in the olfactory neural circuits of honeybees and leaf-cutter ants, and have further supported the long-held view that the mushroom bodies are the neural substrates for learning and memory processes.[104] Moreover, they have been found to require at least two days to provide the animal with stable long-term memories. This is also the time span over which the cataglyphs perform their learning walks.

With these remarks, let us finally return from learning landmark views to using them. As concluded from the behavioral experiments, view matching—matching current views to learned ones—is the most likely strategy employed by the ants. Do the mushroom bodies contain the right neural circuitry to perform this view matching task? A similar 'adaptive-filter' function has been ascribed, though in a different sensory-motor context, to the cerebellum in the hindbrain of vertebrates.[105] When more than a century ago Frederick Kenyon performed his pioneering Golgi stainings in the insect brain, he already pointed toward some remarkable similarity in the neural wiring schemes characterizing the two neuropils: rectilinear networks of hundreds of thousands of thin axon-like processes running in parallel and being intersected orthogonally by sets of extrinsic nerve fibers (Figure 6.28a). More recently and independently of Kenyon, several authors have referred to these neuroarchitectural and possibly functional similarities between mushroom bodies and the cerebellum.[106] One important cerebellar function consists in disregarding predictable (reafferent) sensory inputs, as they are generated by the animal's own motor activities.[107] Neural circuitries performing this role could have been co-opted for comparing current landmark views with stored views, which have been acquired by directed learning processes. The ants have learned what to predict. Seen in this light, it may well be that comparing the actual with the desired is a common computational task accomplished by the mushroom body network. Indeed, as the following chapter will show, this capability plays a major role in the insect's navigational tool set.

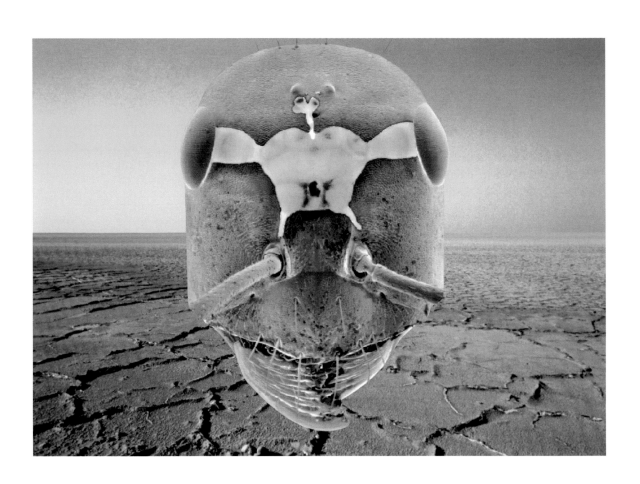

7 Organizing the Journey

"A worm is only a worm," said Diderot. "But this only means that the marvelous complexity of his organization is hidden from us by his extreme smallness."[1] By substituting Diderot's worm for a cataglyph, let us now address the "marvelous complexity" of a small insect and its navigational tool kit and ask how this complexity is generated by the interaction of various guidance systems and spatial memories.

As described and discussed in the previous chapters, the cataglyph navigator draws upon multiple sources of sensory information ranging from polarized skylight and the earth's magnetic field to the direction of the wind, from self-induced image flow to proprioceptive measures of leg movement, from time-compensated celestial cues to terrestrial views. All this information is integrated into a number of navigational routines (often referred to as systems or modules). Many questions abound. How are these separate navigational modules knitted together? Are they organized in a strictly hierarchical way such that at any one time only the highest-ranking system is used, and only if this system does not work—e.g., because of insufficient sensory input—it is switched off and the next subordinate system comes into play? Or do the systems operate in a sequential order such that at the beginning of a homeward trip one system is used, followed by other ones on subsequent stages of the trip? Would this mean that while one system is switched on, the previously active one is switched off? Or do the systems work in parallel and incessantly compete for expression, so that in situations with conflicting cues intermediate courses

7.1 The desert navigator

Desert habitat and frontal view of a cataglyph's head with the brain *(yellow)* superimposed.

may be steered? Or, last but not least, is all the sensory information employed in navigation fed into a central platform where positions of familiar places are represented topographically, so that the navigator is able to rely on some kind of internalized map—a term that will crop up time and again in this chapter? These are the kinds of question that we now shall address step-by-step.

Compromise Trajectories

In normal foraging life all navigational tools available to the animal act in unison and complement and support each other. Particularly designed cue-conflict experiments are necessary to shed some light on their relative significance and kind of interaction, and to create situations in which the navigational mechanisms can be studied in more detail. We start by letting path-integration and landmark-guidance routines compete with each other. In a basic experiment a single artificial landmark, a black cylinder, is located directly behind the nest entrance, so that the ants while returning from a familiar feeder will always keep the image of this beacon in their frontal field of view (Figure 7.2a). In subsequent tests performed in a distant test area that is free of natural landmarks, the black cylinder is placed at different positions alongside the vector route. Upon release, full-vector ants turn slightly toward the landmark without approaching it directly, but finally resume their updated path integration vector course. Turning toward the landmark gets the more pronounced, the later the ants experience the landmark on their journey. Rather than switching from one guidance system to another, the ants steer a compromise course between the bearings indicated by either system, and thus must incessantly accept information derived from both systems.[2] How do the ants weight these different sources of information? We hypothesize that at the beginning of the homeward journey the ants' decisions are biased toward the direction indicated by the path integrator, because then the travel vector is still large and hence its directional uncertainty small. The more the travel vector shrinks, the more its directional uncertainty increases, so that landmark guidance progressively dominates the ants' directional choices.

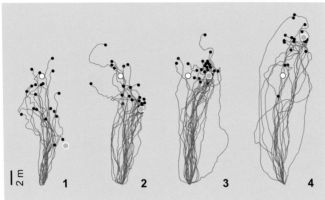

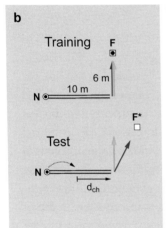
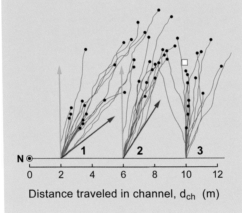

Distance traveled in channel, d_ch (m)

7.2 Compromise courses: conflict between path integration vectors and (a) landmark views or (b) local vectors

(a) In the critical tests a goal-defining landmark (green circle) appears at earlier stages and sideways of the constant path integration vector. Open red circle, fictive position of nest. The control test (C) corresponds to the training situation. In the critical tests (1–4) the landmark has been shifted. The small black dot at the end of an ant's trajectory marks the break-off point (start of search).

(b) Intermediate courses steered between path integration (global) and local vectors. Left: Outline of the experiment. Two-legged training route (top) and test situation (bottom), in which the distances run by the ants in the channel (d_{ch}) are shortened. Red and green arrows, directions indicated by the global and local vectors, respectively. Right: Results. N, nest; F, feeder; F* (open red square), fictive position of feeder. Cataglyphis fortis.

In a second type of 'cue conflict' experiment the global path integration vector is set in competition with a site-based local vector (Figure 7.2b). As described in Chapter 6, local vectors define habitual heading directions taken by the ants once a familiar site has been reached. Now the ants are trained to travel along a two-legged route from the nest to the feeder. After they have walked for a fixed distance in a channel, they leave the channel, make a 90° turn to the left and walk across open ground to the feeder. In the critical test, the channel—and with it the channel component of the global path integration vector—is shortened, so that at the channel exit the direction indicated by the global vector conflicts with that of the local vector.

7.3 Compromise courses: conflict between path integration vectors and landmark views

In the critical tests a constant goal-defining panoramic view (course to steer depicted by the *green open arrow*) is set in conflict with path integration vectors of different lengths (course to steer depicted by the *red open arrow*). *Left:* Experimental setup. *N,* nest; *F,* food sites at different distances *d* from the nest; *dotted lines,* displacements to release site, *R. Right:* Mean courses (and 90% confidence intervals) steered by the ants after displacement to *R. Cataglyphis velox.*

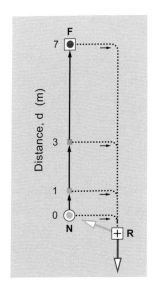
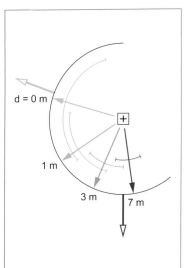

Again, as shown in the previous experiment, the ants head in intermediate directions.[3]

Let us now move from the open salt pan to the visually cluttered environment inhabited by *Cataglyphis velox,* and to an experimental paradigm, which in some way is converse to the one described in Figure 7.2a. In the present case, the landmark cues—the natural panoramic cues around the nest site—are kept in place, and the length of the global vector is varied (Figure 7.3).[4] This is achieved by capturing the ants on their outbound journey at different distances from the nest and displacing them to a location at which the directions of travel according to path integration (feeder-to-nest direction) and panoramic view (release-point-to-nest direction) differ. As expected, upon release the ants select intermediate homeward courses. The new finding is that the intermediate courses are predictably biased toward the path integration direction, the more the length of the home vector and with it the directional certainty of the path integration system increases. Antoine Wystrach and his collaborators indeed found that the weight ascribed to the path integrator quantitatively depends on its reliability, just as Bayesian reasoning implies.[5]

In the three cue-conflict experiments described so far two navigational systems were set in competition by indicating different direc-

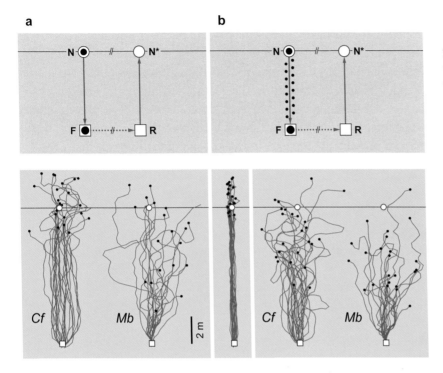

7.4 Paying out the path integration vector in the absence of familiar landmark cues

Top: Training and test situations within an open desert plain. *F,* feeder; *N,* nest; *N* (open red circle),* fictive position of nest in the test field; *R,* point of release. *Bottom:* The ants' trajectories after the ants have been trained **(a)** without and **(b)** with an alley of cylindrical landmarks (indicated by *black dots in double column*). *Middle panel:* Landmark alley present in the training and test field (recorded for *Cf*). *Cf, Cataglyphis fortis; Mb, Melophorus bagoti.*

tions to steer. Now we turn to a test situation in which one system is deprived of the cues present during training. At their salt-pan nest site, the cataglyphs are trained to walk through a long alley of cylindrical landmarks as if they headed along a tree-lined road (Figure 7.4b).[6] When in the critical tests the alley of landmarks is absent, the ants follow their vector course, but do not fully run it off. The significant, though small, amount of undershooting indicates that the two systems—path integration and landmark guidance—simultaneously contribute to the computation of the final steering course. The weight ascribed to the two systems may be subject to species-specific biases. *Melophorus bagoti,* which inhabits the cluttered environment of the Australian outback, follows the dictates of its global path integration vector for much smaller distances, even if its nest comes to lie in open terrain (Figure 7.4).[7] Forest-dwelling species such as the neotropical giant ant *Gigantiops destructor* and the wood ants of the genus *Formica* may do so even less.[8]

Beyond such interspecific propensities, it largely depends on environmental contingencies to what extent landmark guidance contributes to the overall decision process. This contribution is dynamic. When in cluttered environments, in which the ants establish familiar foraging routes, the novel test area is cluttered with natural landmarks as well, the distance traveled by path integration will depend on a suite of factors. Both the visual discrepancy between the familiar and the novel scene and the forager's individual familiarity with the landmark scene of its habitual route may play major roles. For example, when *Melophorus* ants are released in unfamiliar terrain, those individuals that have traveled their familiar route already dozens of times run less far by path integration than naive ants, which have followed their training route only once. Obviously, a distant scene looks more unfamiliar to an experienced forager than it does to a naive one, and hence is distinguished more readily from a familiar scene. Moreover, a particular route memory gets less reliable when the ants are deprived of following that route for some period of time by being induced to forage at another place. The memory of the original route has remained unchanged, but the weight assigned to it in conflict situations with path integration is weakened.[9] The marked effects of these various spatial and temporal parameters on the fraction of the home vector that the ants pay out before they start to search reflect the high flexibility of the ants' decision process and the extreme care with which even minute details of experimental design must be considered in interpreting the ants' behavior.

We end this discussion with an especially exciting experiment. The cataglyphs again forage along an alley of landmarks, this time not on bare open ground but in a straight channel (Figure 7.5). As expected, when displaced to a test channel, full-vector ants head off in the direction of the home vector (but due to the lack of the familiar landmarks in the test channel do not fully run it off), while zero-vector ants search symmetrically about the point of release. Now comes the decisive part of the experiment. If the zero-vector ants are displaced back to the feeder, they unhesitatingly travel along the familiar landmark route once again and return to the nest for a second time. This confirms what we had already seen in natural environments: Ants can recapitulate habitual visually guided routes indepen-

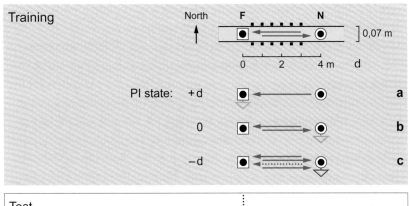

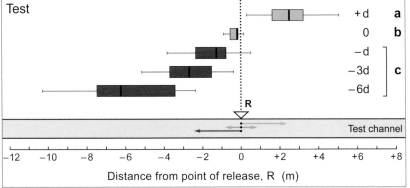

7.5 Rewinding the path integrator: how to make a homing ant run away from home

Training along a landmark alley (*d, training distance*). After training the ants are displaced into a landmark-free test channel *(a)* from the feeder (full-vector ants), *(b)* upon arrival at the nest (zero-vector ants), and *(c)* after arrival at the nest followed by passive displacement *(dotted arrow)* back to the feeder and an subsequent landmark-guided second return run to the nest ('negative-vector ants'). The latter procedure is repeated up to six times (denoted by −6*d*). For further explanations see text. *Cataglyphis bicolor.*

dently of the state of their path integrator. Upon their second arrival at the nest they are captured again and displaced to the test channel. What happens? Still loaded with prey and thus in full homing mood, the released ants move in the direction away from home.[10]

As amazing as this behavior might appear at first sight, it is to be expected. Recall that the path integrator is running all the time while the ants are outside the nest. Once they have arrived there for the second time, and this time guided by landmarks alone, their path integrator informs them that they have overshot this site considerably, namely, by exactly the feeder-to-nest distance. This means that their path integrator got wound up in the reverse direction relative to the direction of the original home vector. One can repeat this procedure of capturing the ants upon arrival at the nest and displacing them back to the feeder time and again. The ants consistently follow their habitual landmark route, so that the state of their travel vector gets increasingly negative. When transferred to the test channel, they run away from the fictive position of the nest for larger and larger distances.

Moreover, one may wonder how the ants 'feel' when they repeatedly travel a familiar homebound route without being able—due to our experimental interference—to enter the nest. Do they then then lose confidence in this landmark-defined route? Very recently Antoine Wystrach, whom we have already met and who is now at the Centre de Recherches sur la Cognition Animale in Toulouse, France, and his collaborators addressed this question in the Australian *Melophorus* ants, and answered it in the affirmative. The more often the animals have run their homebound route without getting home, the more they show signs of disorientation by meandering from side to side and performing scanning behavior. In Chapter 6 we have seen that the ants exhibit this kind of behavior when they have veered off a familiar route and experience novel views. Now they exhibit this behavior while experiencing familiar views, in which their confidence is reduced. In both cases, though for different reasons, the visual information available to them has become less reliable.

Let us return to the original rewinding experiment depicted in Figure 7.5. In these situations, when the ants are provided with an abnormally negative state of their travel vector, they often slow down, even stop walking and drop their food item. While it is normally not always easy to capture these extremely fast and agile desert runners, it gets easy now. It is as if by providing the ants with an extremely unusual state of their path integrator, we had functionally transformed a cataglyph's brain in a way that an ant's brain might have never experienced before during the more than a hundred million years of ant evolution. Apparently aberrant behavior results.

This is a vivid demonstration that the ant's navigational system does not act in isolation but interacts with the animal's motivational state and overall behavior. Similarly, it can affect the state of aggressiveness by which a homebound forager responds to a foreign conspecific. During such aggressive encounters, threatening with open mandibles may finally lead to escalating fights with fierce biting and spraying of formic acid. The state of aggressiveness is the higher, the more a homing ant after displacement to a distant test field has run off its home vector, i.e., the closer it is to the location of the fictive nest. Moreover, it is only then that the ant responds to the carbon dioxide odor plume emanating from an ant's nest, and uses this plume as a guide.[11] In both cases the functional significance of the observed interplay between path integration and other modes of behavior is obvious. Aggressiveness toward members of foreign colonies should be highest at the nest itself, which a foreigner might intend to enter, and a carbon dioxide marker of a nest should be effective only when the path integrator tells the homing ant that it is already close to its goal. Otherwise, the ant would be attracted by foreign colonies encountered on its way home, and would fall victim to the aggressive behavior of the foreigners.

At the end of this section, one conclusion seems inevitable. However diverse the preceding set of experiments has been—it should have been diverse—it clearly shows that at any one time path integration and landmark guidance uninterruptedly contribute to the animal's navigational decisions. The hypothesis now is that the contributions of the two systems are weighted according to their certainties. The dominance of one system is still possible when its weight is extremely high, but usually several systems blend their outputs, so that in situations of conflicting cues intermediate courses are steered. We have developed a decentralized model architecture, the 'Navinet,' in which the navigator's optimal directional choices result from the summed contributions of the estimates provided by the separate systems. This architecture is sufficient to simulate the results of the rather diverse set of cue-conflict experiments described above. We shall turn to it, after we have considered some additional aspects of the cataglyphs' navigational tool set.

Error Compensation

In organizing a journey, reducing positional uncertainty is one of the aims and goals, in fact the essence, of all navigational endeavors. Due to the imperfections inherent in any system of navigation, goals are surrounded by areas of uncertainty. Inadvertent disturbances by external factors further increase the risk of veering off course and finally missing the goal. In dealing with these problems, the cataglyphs employ what looks like a number of backup and error compensation strategies.

Humans have practiced such strategies at least since Micronesian and Polynesian seafarers had settled the Pacific islands. There they steered their outrigger canoes across hundreds of miles of open ocean and finally made landfall on islands, which were nothing but tiny specks of land in the vast expanse of sea.[12] In trying to travel along straight courses from one island to another, they used a celestial star compass and relied on dead reckoning as their principal navigational routines. However, as on their long interisland voyages the navigators were inevitably bound to their error-prone path integration system and subjected to passive displacements by wind and water currents, the positional errors, which had accumulated during the voyage, made the localization of a tiny goal island an uncertain and hazardous exercise. In order to reduce this unavoidable positional uncertainty, the navigators used 'expanded goal' strategies by relying on cues such as seabirds, clouds, and swell patterns that indicated the presence of land. These land indicators, which could be detected at distances of 30–40 kilometers offshore, enlarged the goal and hence reduced the search time considerably.

The desert woodlice, the cataglyphs' path-integrating companions, which we got to know in Chapter 5, take advantage of the expanded-goal strategy as well. Their burrows are surrounded by roughly circular embankments of fecal pellets, which bear the family-specific chemical badge and are thus recognized by a homing woodlouse as a sign of being close to the goal, the family-owned burrow.[13] In a similar vein, when homebound *Cataglyphis* foragers have arrived in the vicinity of the nest entrance, they are often contacted by nestmates, which they recognize by colony-specific cuticular hydrocarbon cues. The silver ants may even place nestmates at some meters' distance around the nest entrance. These outpost ants remain motionless on the ground, but quickly approach and contact returning foragers.

Another strategy of dealing with inevitable navigation errors is employed by the cataglyphs when they return to a familiar food site. As they detect and localize a food source by the odor plume emanating from it, they exploit the fairly constant wind direction prevailing in

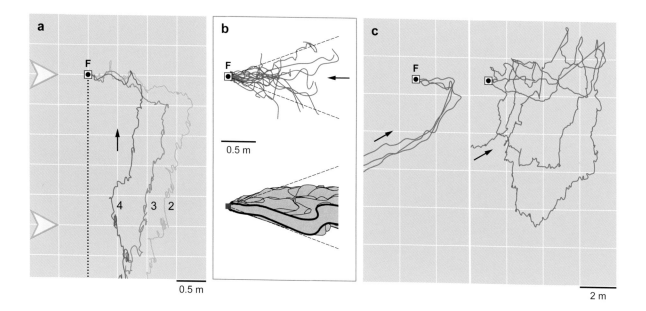

their desert habitats to embark on a two-leg journey. Rather than traveling directly from the nest to the goal, they make a detour by steering some distance downwind of the goal, picking up the odor plume, following it upwind, and arriving at the goal quickly and reliably (Figure 7.6). During their early visits of a newly detected profitable food source, they gradually decrease their upwind approach distances, but from the fourth or fifth visit onward they keep this distance constant. When the food is removed, they perform elaborate crosswind searches within an area just leeward of the feeder (Figure 7.6c, right). The same behavior is observed when the ants' antennal flagella—the seat of the olfactory receptors—have been clipped. The operated animals (Figure 7.7) approach the feeder more or less in the normal detour way, but then are unable to recognize the food. They handle the food items with their mouthparts for extended periods of time, often for several minutes, and finally drop them and continue searching. They never bring any food back to the nest.[14]

On the contrary, when the flagella are kept intact but the basal parts of the antennae are immobilized by applying small drops of insect wax to the basal joints—to the head-scape and scape-funiculus joints, where the sensors for air currents are located—the ants can no longer

7.6 Bipartite outbound paths to a familiar food source: downwind and upwind segments

(a) The video-tracked final segments of an ant's 2nd, 3rd, and 4th visit of a familiar food source (F). (b) Comparison of 19 final upwind approaches of ants to the feeder (top) and a simulated odor source (red square, bottom) visualized by the spread of a TiCl$_4$ smoke plume near the desert floor. One smoke filament is highlighted. (c) Three foraging trajectories of an individual ant to a familiar food source (left) and one trajectory of the same individual after the food has been removed (right). Note the extensive crosswind walks. Finally, the feeder is replaced. Wind direction is always from the left (green arrowheads). Black arrows mark the direction of travel. Cataglyphis fortis.

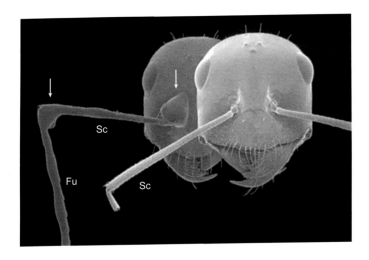

7.7 Surgical manipulation of the cataglyph's olfactory and mechanoreceptive system

In the antennae of the *front animal* the funiculus *(Fu)* has been clipped, so that the animal is deprived of olfactory input, while the scapus *(Sc)* has remained freely movable. In the *rear animal* the antennae have been immobilized by applying small drops of insect wax to their basal joints (the head-scapus and scapus-funiculus joints marked by *arrows*), so that the mechanoreceptors located at these joints no longer provide the animal with information about the direction of the wind. *Cataglyphis fortis.*

perceive the direction of the wind. After this operation they resume their normal approach to the feeder (by employing their sky compass to steer their downwind detour course), but when they encounter the odor plume, they do not walk upwind toward the source but immediately start searching and may reach the food source only after lengthy and wide-ranging search movements.

To understand the strategic rationale underlying the kind of error compensation just described, let me allude to an event that occurred in June 1931 when pilot Wiley Post and navigator Harold Gatty were on their famous eight-day flight around the world. This record-setting flight was a courageous adventure. For example, in the late afternoon of their fourth day Post and Gatty flew their Lockheed aircraft *Winnie Mae* across eastern Siberia toward a small airfield located on the Amur River near Blagoveshchensk, where they needed to refuel. There was "never a sign of habitation"; there were "no towns, no roads." It was raining heavily, darkness set in early, and gas was beginning to run low.[15]

Under these difficult circumstances, Gatty applied the "method of the intentional error, the method of aiming askew of the final target" to decrease the danger of missing the target as much as possible. In his book *Nature Is Your Guide,* he later described and explained how he advised Post to steer the *Winnie Mae* toward the airfield on the bank of the Amur River: "We knew that we were unlikely to land much before dusk and that the airfield was not lighted. We could not afford to waste any time looking for it. I purposely set a course to hit the Amur river ten miles to the left of our destination. When we reached the river I told Wiley Post to turn right. We rounded a bend and there lay the airport. We made a landing with just enough light. Wiley Post could not understand why I was so definite about turning to the right until I later explained it to him. There had been no check points for the last part of our route; and our maps were very questionable. If we had set a course directly to the field and had missed it when we reached the Amur river, we could have used up all our daylight by turning in the wrong direction."

Some passages later he continues: "This shows the value of the calculated intentional error—a simple method which can bring the traveler to the base line of his desired goal in no doubt whatever (unless he has been guilty of a really gross miscalculation) as to which way to turn. Any normal unconscious deviation is swallowed up in this intentional margin: by purposely aiming for a point either to the right or to the left of his objective, the navigator eliminates this uncertainty and is sure of his homefall."[16] Certainly, Harold Gatty would have been amazed to realize that the cataglyphs had discovered the 'method of the intentional error' long before human navigators started to consider it.

By transforming their food vector into a two-leg vector course, the cataglyphs behave in a similar way (Figure 7.8a).[17] If they traveled directly to the goal from the very beginning (what they usually do under windless or tailwind conditions), the uncertainty range caused by the noisy path integration system, and thus the search area reflecting an ant's estimation of its navigation error, would be centered about the location of the goal and consequently result in a 50% chance of ending up in an area that lacked familiar olfactory cues. Search time would substantially increase. By steering downwind detour courses from the very beginning, the ants account for this positional uncertainty by letting the uncertainty range and the odor plume overlap. As we have seen in Chapter 5, the uncertainty range, and with it the width of the ants' search density profile, increases with increasing nest-feeder distance (see Figures 5.11 and 5.13), and so does the distance of the ants' final upwind approach (Figure 7.8a*). The linear relationship between this upwind approach distance and the distance to the goal means that the detour angle, which lies in the range of 3°–8° is independent

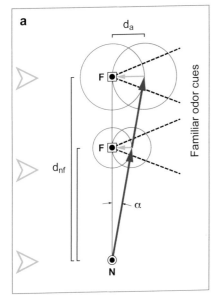

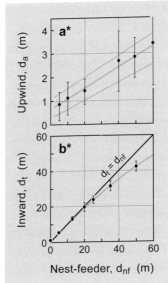

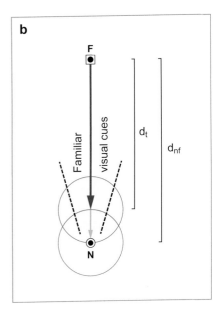

of the nest-feeder distance and thus can be regarded as the ants' estimate of their navigation uncertainty.

As an alternative to this 'error compensation strategy' one could hypothesize that the ants employed the kind of 'goal expansion strategy' described earlier for the ancient seafarers. In the case of the cataglyphs, the expanded goal would be spatially defined by the range of the odor plume. However, as this range is independent of the nest-feeder distance, the ants' upwind approach distance should be independent of this distance as well, so that the downwind detour angle should be the smaller, the larger the distances are that the ants have to travel to the goal. As we have seen, this is not the case. Obviously, the cataglyphs embark on a downwind course not to hit an expanded target but to compensate for navigational errors, which they are able to assess.

In the same vein, for pinpointing a goal the cataglyphs could also take advantage of the visual information provided by the landmark corridor along which they habitually return to their nesting site. By concentrating their searches on the corridor close to the nest rather than on the nest itself, they would avoid spending half of their search time beyond the goal in an area in which they might not have acquired

7.8 Error compensation: systematic angular and linear deviations from the ants' nest-feeder path integration course

(a) Angular deviation. Steering downwind courses *(red arrows)* from the nest (N) to a familiar feeder *(F)* and short upwind courses *(green arrows)* within the odor plume emanating from the feeder (delimited by *dashed lines*). Wind direction is indicated by *green arrowheads.* (b) Linear deviation. Underestimating the linear component of inward courses *(red arrow).* The *dashed lines* mark the familiar landmark corridor close to the nest site. d_a, upwind approach distance; d_{nf}, nest-feeder distance; d_p inward travel distance (recorded in test situation); α, downwind detour angle. The *insets* (a*) and (b*) depict how in the ants' behavior d_a and d_t change with nest-feeder distance. *Cataglyphis fortis.*

appropriate landmark information (Figure 7.8b). Indeed, when they are trained in straight channels to cover larger and larger outbound distances to an artificial feeder, they increasingly underestimate their inbound distances (Figure 7.8b*) as if they employed a leaky integration mechanism. Functionally, this systematic underestimation concentrates the animals' systematic search in the part of the landmark corridor that is next to the goal. The systematic inward errors observed after one-sided turns (see Figures 5.15 and 5.16) may fulfill a similar function.[18] These observations raise the question of whether and to what extent the leakiness of the integrator—or what we describe as leakiness—is tuned to the structure of the ant's environment and the reliability of cues that the animals can derive from it.

The ants depend so strongly on landmark-based navigation that if they happen to get lost, e.g., by being inadvertently displaced off a habitual route, they employ a number of fallback strategies. These emergency plans enable them to return as quickly as possible to visually familiar terrain. For example, when they face the danger of getting blown away by a strong gust of wind, they stop walking, lower their body close to the floor, clutch the surface, and, while clutching, record the celestial compass direction of the wind.[19] Having finally been displaced—either by the wind or by the experimenter (who has released them in unfamiliar terrain and under windless conditions)—they immediately head opposite to the wind direction experienced during clutching. Another fallback system, which they may use in cluttered environments, is 'backtracking.' This behavior can be observed when the ants have nearly reached the goal and hence have already experienced the familiar landmarks surrounding it, but for one reason or another have strayed from the familiar route, have arrived at the unfamiliar far side of the nest, or have been displaced by the experimenter to unfamiliar terrain. In such situations it is a useful option to backtrack, i.e., to walk in the reverse direction of the just run-down home vector and increase the chance of recovering familiar terrain. Thus, recent visual experience (having already encountered the visual surroundings close to the goal) can influence the ant's navigational decisions.[20] This behavior may add another level of complexity to the insect's navigational repertoire.

Situations of high uncertainty may also occur when the cataglyphs drag heavy food items backward to the nest. Then their memorized visual world is inverted. On the one hand, they are still able to read compass information from the sky and use it for steering their courses nearly as well as during forward motion.[21] On the other hand, when walking backward and thus facing the wrong way, they can no longer match the currently unfamiliar scene with their retinotopically stored route memories. To illustrate this failure, let us consider a backward-walking zero-vector ant that is displaced to a location along a familiar route and, after being released, happens to face in a wrong direction (Figure 7.9). It then displays a remarkable 'peeking' behavior.[22] It stops dragging, drops its heavy load, and performs some brief inspection loops, which finally result in peeking, i.e., in taking a few steps forward in the correct route direction. Then it turns, grasps its booty, and walks backward in the route direction. Most likely, during the forward peeks the ant has aligned itself with the familiar panoramic scene, read the corresponding compass direction from the sky, and now uses this celestial information to guide its path during backward walking.

The story that this peeking behavior tells may also shed light on what happens during normal forward walking along a familiar route. The ants may use their view matching routine only intermittently, and in between rely on the sky compass information associated with the matched views. Moreover, recordings of backward moving ants have shown that the animals can maintain their direction of travel independently of their current body orientation, be this backward, forward, or even sideways. Such active decoupling of travel direction from body orientation has often been observed in flying insects under various

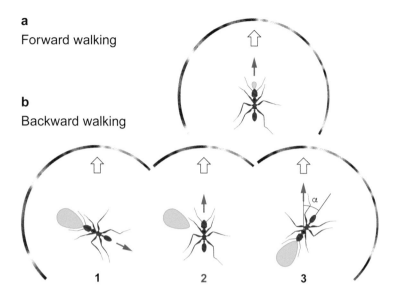

a
Forward walking

b
Backward walking

1   2   3

7.9 'Peeking' during backward homing: ants use familiar landmark views to update celestial compass courses

Zero-vector ants (**a**) carrying a small prey item and walking forward in the home direction, and (**b**) dragging a large item backward. The ant *(1)* is disoriented, *(2)* drops its item and 'peeks' in the forward direction, and *(3)* grabs the item again and moves backward in the home direction. Body orientation can deviate from the direction of backward travel (see α). *Open arrows:* home direction; *blue arrows:* travel direction. The *shaded circular arc* indicates the distant landmark panorama. *Cataglyphis velox.*

conditions, e.g., in the *Schiebeflug*—the transverse sweeping flights— of flies and wasps,[23] but for walking insects the cataglyphs provide the first example.

The high behavioral flexibility exhibited by the ants in their navigational performances raises the exciting question of whether the animals come programmed with a bunch of what we have called— maybe a bit too hastily—backup strategies, well adapted to particular situations, or whether they are endowed with more general problem-solving abilities. Can they use various means to arrive at the same end? Such questions are difficult to answer not least because the ants have certainly encountered the particular situations mentioned above—e.g., being blown off by a gust of wind, or having overshot the mark—quite regularly during their evolutionary history, and thus could have evolved particular solutions to these problems. What one would like to encourage is to present the animals with novel problems, which they and their ancestors have most likely not faced before. At present, we must leave such questions open, and turn to the model already announced above. In this model different navigational modules can flexibly interact. It might well be that what we have called backup strategies can finally be implemented in this model by varying the animal's internal motivational states and adapting the uptake and processing of sensory information accordingly.

The Navinet

The previous considerations may have given the impression that the cataglyphs' navigational gadgets come into play when particular circumstances arise. However, impressions can be deceiving. What appears as a set of separate mechanisms operating in a hierarchical order or in temporal succession is now considered as the graded expression of some basic navigational routines operating incessantly all the time. The Navinet model (Figure 7.10), which is based on a large body of experimental work previously outlined in this chapter, is an attempt to reconcile various aspects of navigation behavior and to provide a common network architecture mediating that behavior.[24] In essence, it combines spatial information used by the animal in the context of two major modes of navigation: path integration (PI, Chapter 5) and landmark guidance by view matching (VM, Chapter 6). In the latter case, let us stretch the meaning of the word a bit and use the term 'VM routine' irrespectively of whether visual panoramas, skylines, ground patterns, static or dynamic images, or other landmark cues are involved.

In the model, path integration (PI) and view matching (VM) guidance routines operate independently but concurrently. They are modulated in a context-dependent way by a network of motivation units. This is the stage where decisions are taken for future actions. Most generally, this winner-takes-all motivation network indicates whether the animal switches on its foraging mood, starts its outbound (foodward) or inbound (homeward) trip, decides to what food place to travel, and so forth. While research has mainly focused on the structure of the guidance routines (the red and green networks in Figure 7.10), less attention has been paid to the more detailed motivational conditions, under which these routines come into play (the yellow network in Figure 7.10). For example, as we have seen in Chapter 4, in estimating travel distance the cataglyphs employ their optic-flow integrator when they actively walk, or when they actively invite themselves to be carried by a nestmate from one nest of

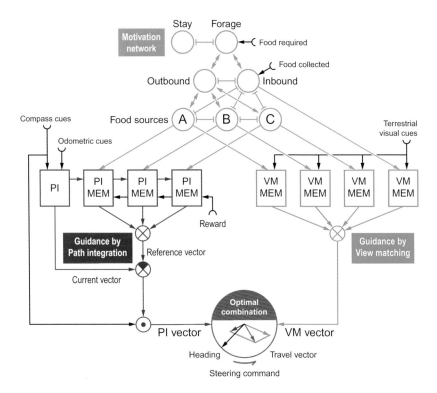

7.10 The Navinet: a model of optimal multi-guidance integration

Red and green boxes, two banks of procedural memories: path integration *(PI MEM)* and view *(VM MEM)* memories, respectively; *PI,* path integrator; *yellow circles,* motivation units. For explanation, see text.

their colony to another one (see Figures 4.6 and 4.7), but when they are passively transported across the ground in the way shown in Figure 4.8, they do not switch on their flow integrator, i.e., do not decide to path-integrate. One could reasonably ask what would happen if the passively transported ants were mounted on the transporting vehicle such that they were able to move their legs, and could even do so in their usual tripod pattern. Such detailed experimental dissection of behavior will help to unravel the ant's decision processes.

Once activated by the proper motivation units, the PI and VM network units continually compute travel vectors on their own. The lengths of these vectors are not metric dimensions. They are proportional to the certainties of the directional estimates. Both vectors are optimally combined at a downstream processing stage. What do we mean by 'optimally combined'? For various reasons, all the multimodal information acquired and processed by the animal's sensory and nervous systems is noisy. Hence, the navigator might have every interest

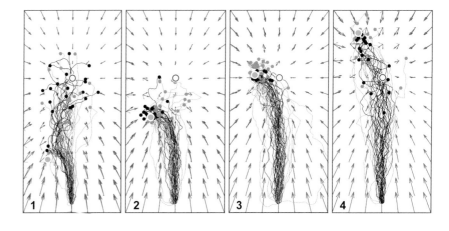

7.11 The Navinet: simulation of a cue-conflict experiment

Experimental setup as in Figure 7.2a, but this time with the landmark *(green circle)* to the left rather than the right of the path integration route. *Open red circle,* fictive position of nest; *small red* and *green arrows,* PI (path integration) and VM (view matching) vectors, respectively; *gray arrows,* resultant travel vectors; *black* and *yellow lines,* simulated and experimental ant trajectories, respectively. For clarification, see *detail at lower left.*

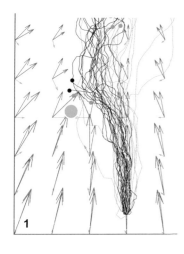

to maximize the certainty, i.e., reliability, of the combined information. We had already mentioned that multiple sources of information are combined optimally when they are weighted in inverse proportion to their variances: the smaller the variance, the higher the reliability. The PI and VM vectors computed on the basis of this premise reflect the limited knowledge that the navigator has at any one place about the direction of its goal. When we now make reasonable quantitative assumptions about how this limited knowledge varies with distance to the goal, the sum of the two vectors provides the animal's best bet of the course to steer.

How well is the Navinet hypothesis able to account for the ants' behavior in the cue-conflict experiments described above? Simulating the results depicted in Figure 7.2a may serve as a telling example (Figure 7.11). Recall that in this case the homebound full-vector ants experience a nest-defining landmark in unfamiliar positions sideways along the route indicated by the path integrator. In Navinet terms this means that at any point, which the animal may reach on its journey, the PI and VM vectors are misaligned. Their sum yields the optimal travel vector. A vector field results. As Figure 7.11 shows, the model replicates the ants' trajectories rather well indeed. Both real and virtual ants drift toward the landmark the more, the closer the landmark is placed to the fictive position of the nest, and both break off their homeward runs and start searching in the same areas. The results of other cue-conflict experiments can be simulated by the Navinet approach equally well.

At this juncture, it is important to avoid a misunderstanding that may all too easily crop up: A vector field may look like a map, yet it is anything but a map. At any one time and place the animal has access only to information corresponding to its current location, where it compares it with information stored in memory. It is 'blind' to the rest of the field. Even when the animal happens to be at the same place sometime later in its foraging life, the VM vector may differ, as in the meantime the animal's view reliability may have changed due to learning effects, and the path integrator may be in a different state. Hence, the vector fields are dynamic.

Even though the Navinet is a functional model that in principle need not imply hypotheses about neurobiological realizations, it is compatible with the computational role that the two major neuropils in the insect's forebrain play. As outlined in Chapter 5, the central complex with its recently uncovered neural wiring schemes houses the path integrator,[25] while the mushroom bodies most likely encode long-term memories of panoramic views (see Chapter 6).[26] How and in what form the information about landmark memories is conveyed from the mushroom bodies to the central complex is not known. Nor have direct connections between the two neuropils been found yet. Whatever computational steps are involved in combining the two streams of information, the cataglyphs must finally compare their current heading with the one that leads to their intended goal, and then use the difference signal to set their courses. The central complex with its highly organized sets of multiple columns and layers seems to be the place where these computations are finally done. As Nicholas Strausfeld once remarked, the central complex is the region in the brain "that supervises walking."[27]

A Map in the Ant's Mind?

"The bee is like mammals and birds in that its brain constructs an integrated, metric cognitive map"—"a common spatial memory with geometric organisation . . . as in other animals and humans." This is the conclusion drawn by Randolf Menzel and his collaborators at the

Free University of Berlin, Germany, from their extensive studies of honeybee navigation.[28] In this view, an internalized metric map provides the bee with an earthbound (geocentered) frame of reference, which "stores spatial relations between multiple landmark features,"[29] so that the animal can freely navigate between various places. In contrast, in the Navinet model, landmark information is stored as sets of egocentric views, which are not integrated into the geocentric frame of reference in which the global path integration vectors are represented and stored. Even though the Navinet approach can sufficiently well account for the cataglyphs' navigational performances tested so far, it would be unwise to rule out the appealing possibility that the ants were finally able to use all this information for acquiring and employing a map of their nest environs. As noted above, a cognitive map as considered here is an internal topographical representation of the animal's foraging environment (analogous to a city map), but in a weaker sense, one could also surmise that the animals used topological maps, in which directions and distances are preserved only approximately rather than metrically (analogous to a city subway map).

Before a potential map can be used, it must be constructed. In this respect it is important to examine how the cataglyphs acquire and increasingly improve their spatial knowledge during their foraging careers. The day will come that all journeys performed by individual cataglyphs during their entire lifetimes will be recorded, so that we will have full access to the space use patterns of individual ants. Then we will be in a better position to understand how the ants' navigational memories develop. Yet, we already have sufficient data at hand to draw a fairly clear picture. When the cataglyphs start their outdoor activities by performing the learning walks described in Chapter 6, they can do nothing but read out their path integrator in order to take nest-centered panoramic images of the landmark surroundings. Path integration is an inborn navigational routine that works already during the ants' first learning walks. For example, in the adult transport described in Chapter 4 the animals that are carried by nestmates from one nest of their colony to another (see Figure 4.6) are in most cases—as judged from the developmental status of their internal organs (Figure 7.12)—'newcomers,' which have not yet started their foraging

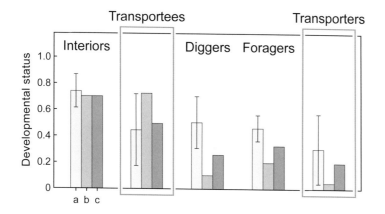

7.12 Temporal worker castes: developmental status of some internal organs

(a) Ovarioles, *(b)* fat body, *(c)* labial gland. In adult transport, individual ants (transporters) carry nestmates (transportees) from one nest of a colony to another one. Transporters tend to be experienced foragers. *Cataglyphis bicolor.*

lives, and yet are able to return to the nest of departure by path integration. From one learning walk to the next, the ants acquire landmark views from larger and larger distances and almost all directions around the nest. Usually they continue walking in various directions also during their first foraging runs (Figure 7.13), but depending on environmental factors, mainly on prey density and thus previous foraging success, each individual develops a certain degree of sector fidelity. The size of the foraging range differs substantially across species, and related to local food densities may also differ within species.[30]

How often a *Cataglyphis* forager will return to a site visited on previous foraging trips depends on the amount and quality of the food available there as well as on the reliability of encountering the food at this place on subsequent visits.[31] Usually, a foraging cataglyph—an individual forager and single-prey loader—depletes a food site in a single visit. Nevertheless, a rather simple behavioral heuristic describes how the observed sector fidelity can arise even within an isotropic food environment. The rules are as follows: Continue to forage in the direction of the preceding foraging trip, whenever this trip has been successful. If it has been unsuccessful, switch randomly to a new direction, but decrease the probability of doing so as the number of previously successful runs increases.[32] When unsuccessful at a former finding site, the cataglyphs travel beyond it, expand their searches, and may next find food within or outside the former foraging sector. Thus from time to time they may establish new foraging areas, switch back

7.13 The first foraging runs performed in the life of an individual cataglyph

At the beginning of its foraging career within a loosely cluttered, scrub-desert environment (indicated by *gray borderline*), the ant has ventured out in various directions, before discovering its first food item on run number 9. The two subsequent trips lead in the same direction. During the ant's first 22 foraging trips, only 2 of them (numbers 9 and 19, the latter path not recorded) have been successful. *Cataglyphis bicolor.*

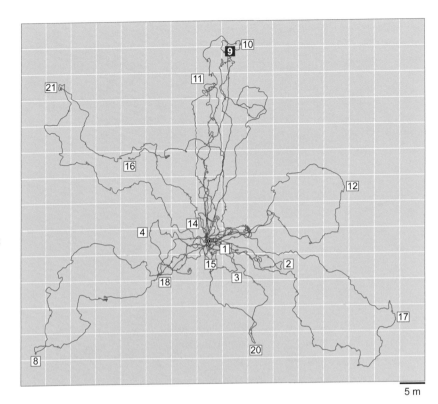

5 m

to former ones, travel to and return from different sites, and even move from one site to another.

At rich and hence repeatedly visited food sites, the cataglyphs perform orientation walks that resemble the learning walks performed around the nest entrance, but are more linearly oriented in the nest-feeder direction. They remind us of the 'turn-back-and-look' behavior observed in bees and wasps upon leaving newly discovered profitable food sources.[33] In addition, the ants perform relearning walks around the nest entrance whenever they have experienced changes in the nest-surrounding landmark panorama during their preceding home-bound journey. In many respects these relearning walks correspond to the original learning walks in the beginning of the ants' outdoor activities, and coincide with them in their effect of subsequently relocating the goal.[34] Moreover, as shown in Chapter 6, the ants continually gain landmark information about inbound and outbound routes while they travel these routes, and store this route-based information together with site-based information in long-term memory. Whereas memories of path integration vectors slowly decay over periods of 24–36 hours, landmark memories of places and routes last for the cataglyphs' entire lifetimes.[35] Except for these learning endeavors, the cataglyphs do not seem to survey their terrain in one way or another independently from performing their foraging journeys.[36]

Nevertheless, by linking information acquired during successive forays, they could in principle lay the foundation for mapping their foraging terrain, i.e., for placing landmarks within a metric framework. Let us assume that the cataglyphs were able to associate the state that their path integrator has reached at a familiar food site, say, at place A, with the landmark memory acquired at this place, and that they stored this association in long-term memory. Further suppose that they would do the same at place B as well (see Figure 7.14, but for the time being disregard all other details of this figure). When the animal is later displaced to A or B—without having actively traveled to either location—it could recognize these places on the basis of memorized landmark views, retrieve the path integration coordinates previously acquired there, compute the difference between them, and thus determine the direct path from A to B. In this way landmark memories of a set of places, all provided with metric coordinates, would finally enable a cataglyph to perform cartographic operations independently of the instructions given by its incessantly running path integrator.[37] Metaphorically speaking, the animal would 'externalize' path integration coordinates by attaching them to locations in its environment, and would thus be able to perform vector subtraction in geocentric space.

The first to claim that bees have maps of this kind—"that they draw miniature maps of where they have been"—was James Gould from Princeton University. He trained bees from the hive N to a feeding station A that was 160 meters apart from N. Later the trained foragers were transported before departure from the hive—hence as zero-vector bees—to another site, B, which was at the same distance from the hive but in a different direction, with N, A, and B forming an equilateral triangle. Upon release at B, they flew directly off to site A. From this result the author concluded that the bees must have acquired

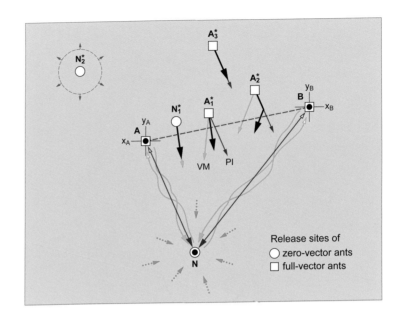

7.14 Maplike behavior resulting from flexibly combining information from path integration (PI) and view matching (VM) routines: examples of observed behaviors

Path integration vectors *(large red arrows)* and habitual routes *(green lines)* between nest *(N)* and two feeding sites *(A and B)*. *x, y* denote the path integration coordinates at *A* and *B* (0, 0 at *N*). The *dotted green arrows* mark nest-centered views taken during learning walks. The *dashed red line* indicates a novel route traveled by path integration. For the behavior of full-vector and zero-vector ants released at novel sites *(open squares and circles,* respectively), see text. The *small red* and *green arrows* mark the directions indicated by the ant's PI (path integration) and VM (view matching) vectors, respectively. The *black arrows* are the courses actually steered. The *radially expanding dashed circle* at the *upper left* indicates the ants' search pattern.

and used a landmark-based, local-area mental map to compute the novel course from B to A. However, when Fred Dyer from Michigan State University in East Lansing carefully reconsidered and repeated Gould's experiment in a landscape in which the bees' visual experience at A and B could be better controlled, the bees selected the novel course only when at B they could see distant landmarks associated with A or with the familiar route N→A. If in the reciprocal experiment (food source at B, release site at A) such views were obscured by landscape structures, the bees displaced from the hive did not head toward the food site, but flew in the compass direction they would have taken from the hive. Hence, rather than having learned the geometrical relationships between the home base N and the sites A and B, the hallmark of a mental map, the bees seemed to have relied on landmark views associated with previously traveled routes.[38] This is fully in accord with our current understanding that in landmark-based navigation bees and ants use egocentric view matching strategies.

A further claim that honeybees possessed a mental locale map of their foraging terrain was based on the often cited 'lake experiment.' When inside the hive the recruitment (waggle) dances of successful foragers advertised an 'implausible' location—a feeder on a boat in the middle of a lake—the recruits reading this dance information were supposed to consult their map, realize that the information would lead them to a location most unlikely to yield food, and thus reject this information. No recruits were observed at the feeder.[39] However, videotaping the bees' behavior inside the hive shows that the frequencies with which the recruits leave the hive is completely independent of whether a lake station or a comparable land station has been advertised, and that the arrival rate at the lake feeder depends on a number of close-range orientation factors. If the latter are controlled, there is no difference between the frequencies with which recruits arrive at either station.[40] Hence, the long-standing controversy revolving around the lake experiment has been resolved by subtly controlling experimental conditions and carefully evaluating various hypotheses that could account for the results.

Finally, as mentioned above, Randolf Menzel and his collaborators have performed thorough sets of well-designed experiments by recording the flight trajectories of foraging honeybees with harmonic radar devices. Full-vector and zero-vector bees have been displaced to various locations around the hive, and several experimental setups have been applied. From all these experiments the authors conclude that the bees acquire and use a metric map.[41] As the metrics of this map have not been deduced from the behavioral results, it is difficult to imagine what the computational operations are by which such a map is created and read. Neither do we have an idea, let alone a formalized concept, of how a metric map could be implemented in the bee's brain. Moreover, the Navinet hypothesis suffices to simulate the results even of those honeybee experiments that have deliberately been designed to strengthen the evidence for a metric map.[42]

Some light on these questions may also be shed by the spatial behavior of the honeybees' larger cousins, the bumblebees. Like many other pollinating hymenopterans, bumblebees visit multiple feeding sites and develop stable foraging circuits, so-called traplines, between these sites. The supreme trapliners are the orchid bees, which in their neotropical rain forest habitat may travel several kilometers along habitual routes to visit dozens of flower sites in a row,[43] but the best-studied insects in this case are indeed the bumblebees. Lars Chittka and his group at Queen Mary University in London designed an elegant series of large-scale field experiments to study how bumblebees develop and optimize such stable trapline routes.[44] In an open pastureland with tree lines and isolated trees providing global and local landmarks, they installed sets of artificial feeding stations, and tracked the bees' flight paths with harmonic radar. The renewable sucrose rewards at the feeders were adjusted such that the bees could fill their crops only when they had visited many feeders. After the bees had discovered all feeding sites, they reduced their travel time and costs by gradually developing efficient routes, near-optimal traplines. It is as if they tried to solve the classical 'traveling salesman problem' (to find the shortest route between a set of locations by visiting each location only once and finishing at the point of departure). Be that as it may, the bees' routing behavior can be simulated by a rather simple learning (optimization) heuristic without the need of assuming that the animals resort to a maplike representation of the feeder array. In the simulation they just need to rely on a set of flight routines—a set of site-based local vectors, as we had discussed them above for the cataglyphs. This view is supported by the observation that when one feeding station is removed from the familiar route, the bees continue, at least for a while, to follow the familiar trapline including the empty location in the same sequence as before. Hence, a rather demanding spatial problem can be accounted for by the navigational routines previously unraveled in walking ants.

Nevertheless, let us still stay with the concept of the cognitive map and finally test whether some potential preconditions for constructing such a map are fulfilled. For example, can the cataglyphs reset their path integrator by views of familiar landmarks? In the following we consider two experimental paradigms addressing this question.[45] In the first paradigm a row of conspicuous artificial landmarks is attached to the final section of the ants' habitual foraging route. In the critical tests this landmark array is installed in a distant test area, but rotated through an angle of 45° relative to the training situation (Figure 7.15a). Full-vector ants, i.e., ants that have been displaced from the feeder to this set of landmarks, occasionally disregard the landmarks and head in the vector-defined homeward direction *(1)*, but in most cases they follow the instructions of their landmark memory and travel along the rotated array of landmarks. Having arrived at the end of the array, they

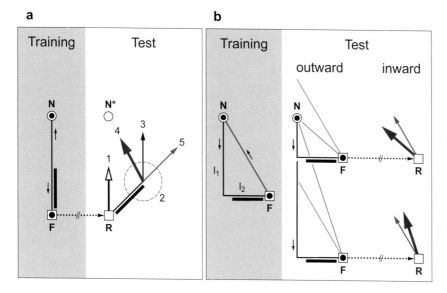

7.15 Landmark memories do not prime the recall of path integration coordinates

The last part of the training route from nest *(N)* to feeder *(F)* is provided with conspicuous landmarks (schematically depicted by *heavy black bars*). In the tests the ants are displaced from *F* to the test field (*R*, release point), where the spatial layout of the training routes has been changed. *Heavy red arrows,* directions actually taken by the ants; *blue arrows and lines,* directions to be taken, if the path integrator had been reset to the state previously associated with the landmarks. **(a)** Training array of test landmarks *(left)* is rotated through 45° *(right).* Numbers *1–5* denote possible behavioral decisions (see text). **(b)** Training along a two-leg route *(l₁, l₂),* of which l_2 is associated with landmarks. In the tests, l_1 is either shortened *(top)* or lengthened *(bottom).* For further explanations, see text. *Cataglyphis fortis.*

could adopt one of at least four different strategies. On the basis of what we already know, the first two of these options can be dismissed immediately, but for the sake of completeness let me mention them anyway. When recalling their landmark memories the ants could have switched off their path integrator, so that upon arrival at the end of the landmark array they would either start their systematic search routine *(2)* or again switch on their path integrator *(3)*. Nothing of this kind happens. Instead, having continually updated their path integrator while following the divergent landmark line, the ants head directly for the fictive position of the nest *(4)*. They do not reset their path integrator to the state previously experienced at the end of the landmark array *(5)*.

In the second experimental paradigm, the cataglyphs are trained in channels along an L-shaped route to the feeder. Again, a set of conspicuous landmarks is associated with the final part of the route (Figure 7.15b). In the critical tests, the ants perform their foodward runs in an L-shaped channel, in which the first leg is either shortened or lengthened relative to the training situation. If the view of the landmarks close to the feeder triggered the recall of the home vector, which during training had been associated with the landmarks, the animals displaced to the test field should select this very home vector

7.16 Homing from novel sites

(a) Homing trajectories of zero-vector ants *(left)* and full-vector ants *(right)* displaced to a release point *(R)* 10m from the feeder *(F)*. Cluttered steppe-like environment near Alice Springs, Australia. An alley of 0.6m tall black cylinders *(black dots)* combines nest *(N)* and feeder. (b) Homing trajectories of full-vector ants displaced from the feeder *(F)* to 12 release stations *(open squares)* more than 10m from the nest. Parkland environment near Canberra, Australia. *Green shading* indicates treetops. In (a) and (b) the *red* and *green* arrows mark the directions indicated by the ants' path integration and view matching routines, respectively. (a) *Melophorus bagoti*, (b) *Myrmecia croslandi*.

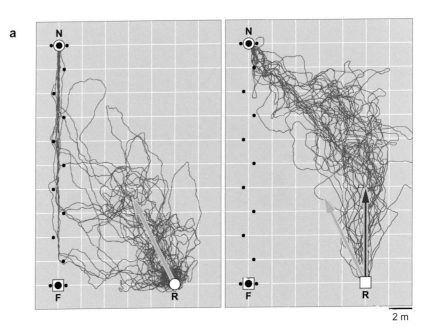

under either test condition, but they do not. Instead, they always compute their homeward courses on the basis of the immediately preceding outward journey performed in the altered channel devices. These and other experiments do not provide any evidence that the cataglyphs form long-term memories of previous path integration states attached to visually familiar sites. Of course, besides employing path integration there are other ways of designing a maplike representation of foraging space, e.g., by behaving like a human surveyor, but such methods are both time-consuming and computationally demanding. When desert ants perform their learning and relearning walks around the nest entrance, they do not survey the nest environs but instead take egocentric nest-centered views of the surrounding panorama from various locations around the nest.[46]

Later in foraging life the cataglyphs may exhibit what looks like map-based behavior, but may in fact result from the interplay of the ants' path integration (PI) and view matching (VM) routines. Just for illustration, some scenarios based on various experiments described above are sketched out in Figure 7.14. There we make the parsimonious assumption that in their past foraging lives the ants have visited two familiar feeding sites, A and B. The spatial information they

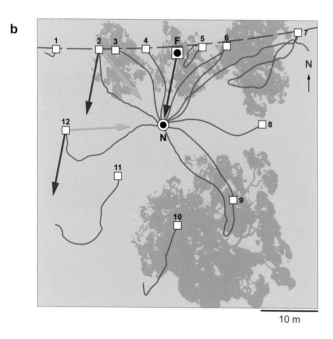

10 m

have acquired consists at least of a set of goal-centered views around the nest (N), the path integration vectors combining the nest with sites A and B, and the landmark routes between the nest and the two feeding sites (routes N→A, A→N, N→B, and B→N). Note that, as mentioned before, the memories of views are long-lasting and may be retained over a forager's entire lifetime, but the memories of path integration vectors are short-lived and gradually decay. In the critical tests the ants are displaced as either zero-vector ants (from N to sites N*) or full-vector ants (from A to sites A*) into novel territory. In this situation they may benefit from the fact that the goal-centered panoramic views can also be used at some distance from where they have been taken. How effective this way of 'view extrapolation' actually is will depend on the visual structure of the local landmark environment—how panoramas vary with distance from a known place—and where the ants have originally taken their goal-centered views. In some situations, displaced zero-vector ants are able to directly select their homeward courses (e.g., in Figure 7.16a, left panel), while in others they first engage in their characteristic search behavior until view matching strategies become successful (for examples, see Figure 6.16a).[47] Of course, the latter is also the case when the displacement distance gets so large that the ants are released in completely unknown terrain (at N^*_2). After extended searches the zero-vector ants might then reach areas in which their VM strategy starts to work. Full-vector ants displaced from A to A*, may steer intermediate courses between the directions indicated by PI and VM. Depending on the reliability of either system, the courses may lie closer to, or even coincide with, the direction of one or the other system. An interesting example is given in Figure 7.16a.[48] While displaced zero-vector ants select their landmark-guided VM course based on distant landmark cues, full-vector ants first follow their still highly reliable global PI vector until with decreasing length and thus increasing directional uncertainty of this vector, and with concomitantly increasing reliability of the landmark-based VM routine, the latter assumes control (symbolized as the behavior at A^*_2 in Figure 7.14). At distant release sites far off A and B (e.g., at A^*_3), PI completely governs the behavior of full-vector ants. These are just a few exemplary cases in which the ants successfully move toward a goal, in this case home, from places at which they have not been before.

The Navinet model suffices to account for these behaviors. As argued in Chapter 5, the PI system is geocentered with the PI vector anchored at the nest site, but the egocentrically learned views are not incorporated into this geocentric frame. They are bound to the navigator.[49] Even within the insect's brain, the view memories do not have to be arranged and stored in the sequential order in which they have been obtained. The panoramic view that matches the current one best can be used for guidance. In principle, one can map the range over which this view matching strategy works, i.e., over which it provides a heading direction, by taking systematic sets of panoramic images within the ants' nest environs, as well as recording the spatial extent over which the ants perform their learning walks and thus acquire their reference views. Whenever this has been done, the behavior of displaced zero-vector ants can be explained most parsimoniously by the view matching strategy.

As the decentralized Navinet model system implies, independent streams of navigational information optimally integrate their directional estimates only at a distant downstream site. However, as future research may modify or extend this model, we should always be open to new interpretations. For example, there could be weaker associations between familiar views and the PI coordinate system than firmly integrating the former into the latter, but conclusive evidence for such assumptions is lacking as yet. Conversely, we should not stick too strictly to the map metaphor. As the topographical-topological map distinction mentioned above already implies, maps may come in various guises. The issue that is at stake first and foremost is to unravel the insect's navigational stratagems and study their ways of interaction. The model proposed here is our current best bet.

In fact, path integration and view matching—PI and VM as proposed in the model—complement each other by balancing their strengths and weaknesses. While PI reliability is the higher, the farther the animal is away from the goal, the opposite holds for the reliability of VM. This renders the combined system a quite robust way of navigation. It is as if PI pushed the animal out of unfamiliar terrain, and VM pulled it to a familiar site, both acting in the same direction. Hence, whenever foraging in the fields, a cataglyph seems to rely on navigation vectors, which tell the navigator how reliable it is to move in a particular direction.

The separate processing of information employed by the two navigational subroutines starts already in the sensory periphery. Consider the visual system (Figure 7.17). Separate input channels are employed for picking up polarized light (and spectral) cues from the sky and panoramic views from the landmark surroundings. By polarization sharpening and low-resolution wide-field vision, either channel extracts from its visual world the most relevant and robust information that is required for accomplishing its task. In addition, ventral parts of the eye sense the self-induced optic flow and thus are involved in recording travel distance. On its way to the central brain, all information that in the periphery has been picked up by small-field visual units gets more and more integrated across large parts of the ant's omnidirectional field of view. Coarse encoding and panoramic vision might be an adequate way of handling navigational information obtained from both the ant's celestial and terrestrial world. Along these processing lines both streams of information—path integration and view matching—remain separate for quite some downstream distance (see

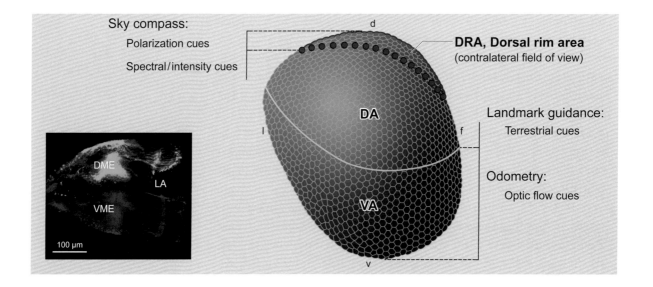

7.17 Compartmentalization of the *Cataglyphis* compound eye

The equator of the eye *(curved yellow line)* is defined as the borderline between differently structured retinulae in the dorsal and ventral retina. In running ants (for common pitch angle of the head, see Figure 3.11a), it looks at the horizon. Functional specializations are indicated. *Inset:* Injections of differently colored dyes into the dorsal and ventral medulla reveal a pronounced separation between these two parts. *d,* dorsal; *DA,* dorsal area of compound eye; *DME,* dorsal medulla; *f,* frontal; *l,* lateral; *LA,* lamina; *v,* ventral; *VA,* ventral area of compound eye; *VME,* ventral medulla. *Cataglyphis bicolor* and *(inset) C. nodus.*

the central complex and mushroom body pathways in Figure 3.31c). When the navigator is on its way to a goal, either subroutine computes a difference signal between a desired state—a reference path integration vector or a reference landmark view—and the current state. The animal performs an action to attain an expected sensory input. This is somewhat similar to what happens in olfactory nestmate recognition, in which individuals contact—'antennate'—nestmates to compare learned chemical cues (templates) with cues currently detected in these other individuals.[50] In the navigation network Navinet, both difference signals give rise to guidance estimates, the PI and VM vectors. When optimally combined at the final processing stage, the central complex, they provide the navigator with the information about the direction to steer.

It would be a challenging task for future investigation to selectively block neural activity in certain parts of the cataglyph's brain, especially in the central complex and the mushroom bodies, or parts thereof, and study the effects of these experimental interferences on the animal's navigational behavior. In the 1950s and 1970s two isolated attempts have been made to use slivers of razor blade for producing localized lesions in the mushroom bodies of wood ants, and to use thin needles to cool parts of the mushroom bodies of honeybees and study the effect of these interventions on olfactory learning.[51] In addition to pursuing such microsurgical interventions, it might be a bold flight of imagination to apply modern genetic editing tools for silencing and activating targeted sets of neurons in the ant's two major forebrain neuropils. In general, the peculiarities of the ant's reproductive life-cycle impede a straightforward application of the neurogenetic tools that have so successfully been used in fruit flies, e.g., in studies on the central complex.[52] However, there is hope. Recently, Daniel Kronauer from the Rockefeller University in New York and Claude Desplan from New York University and their collaborators have pioneered gene editing in ants by knocking out an olfactory co-receptor in a clonally reproducing doryline ant, and in a ponerine ant, in which workers can be converted to become sexually reproductive. In either case, olfaction-mediated behavior is severely impaired.[53] With the wide range of reproductive strategies exhibited by the cataglyphs (see Chapter 2), and with the increasing efficiency of genetic editing approaches, the potential of neurogenetic modifications may be *ante portas* also in our desert navigators.

Returning to the Navinet approach and its wider implications, it is important to make one point unequivocally clear. The question is not whether an insect's miniature brain would be capable at all to acquire and use a more complex, say, maplike representation of space. The much championed argument that the ant's brain lacked the required brain power to accomplish highly advanced navigational tasks is certainly misleading. Even though an ant's brain is small—smaller than a vertebrate's brain even relative to body size (Figure 7.18a),[54] "not so large as the quarter of a small pin's head," as Darwin remarked—its neuroarchitectural intricacies and computational capabilities render it anything but 'simple.' As we have seen, computational models bearing on what we know about the insect's major forebrain centers are able to mimic the principal navigational routines exhibited by the desert navigators, and the Navinet model describes how these routines can optimally be combined to generate the ant's spatial behavior.[55] After all, from an evolutionary point of view, the question is whether a maplike representation of space is the most efficient way of dealing with the navigational tasks that these central place foragers must accomplish during their short foraging lives. A cataglyph might consider

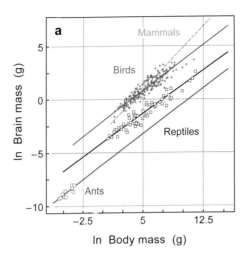

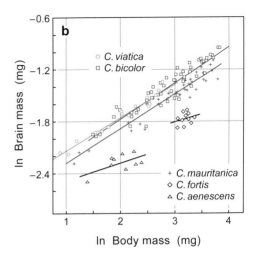

In Body mass (g)

In Body mass (mg)

7.18 Ant brain allometry

Brain mass versus body mass **(a)** in ants (based on 247 specimens of 10 species of the genera *Cataglyphis, Formica, Camponotus,* and *Lasius*) in comparison with terrestrial vertebrates (data assembled from various literature sources) and **(b)** in five *Cataglyphis* species.

it more appropriate, so to speak, in terms of time and energy expenditure[56] to constantly probe the environment and sample the visual scenes, in other words, to be online all the time by continuously integrating paths and comparing views—in a nutshell, to know where to go without knowing where to be. It acquires this knowledge in rapid, spatially, and temporally prefigured learning behaviors, which are tailored to particular navigational needs (recording and storing goal-centered panoramic views) and guarantee that the ant can accomplish its navigational tasks in fast, frugal, and robust ways without facing the need of building up some internal representation of the lay of its foraging land.

In the last decades, this very internal representation has intensively been targeted in the mammalian cortex, where it has become the focus of one of the most exciting research areas in the neurosciences. Rodents, especially rats but also mice, hamsters, and gerbils, are the champion organisms in these studies. A cursory look at them—maybe a bit 'through the cataglyphs' eyes'—shall end our journey to the desert navigator.

The Rodent's Map and the Cataglyph's View

"The incoming impulses are worked over and elaborated in the central control room into a tentative, cognitive-like map of the environment." In his emphatic Annual Faculty Research Lecture held in 1947 at the University of California, Berkeley, the psychologist Edward Tolman stated that a large body of behavioral experiments performed in rats could be understood best by assuming that while the animals explored their mazes they gradually formed an internal representation—or cognitive map—of their environment. This statement, though "brief, cavalier, and dogmatic," as Tolman himself conceded, was a forceful criticism of the then prevailing 'stimulus-response' school of behaviorists, which considered all spatial behavior as sequences of mere associations between external stimuli and behavioral responses, and thus foreshadowed the cognitive turn to follow in the behavioral sciences. Tolman's proposal received its strongest support and elaboration when three decades later John O'Keefe and Lynn Nadel published their milestone account *The Hippocampus as a Cognitive Map*. In this book Tolman's metaphorical "central control room" is defined more specifically as an internal representation of unitary Euclidian space, in which all angles and distances between places are encoded in absolute (geocentered) ways.[57]

With the discovery of the 'place cells' in the rat's hippocampus, and another 30 years later of the 'grid cells' in the entorhinal cortex, the main cortical input area to the hippocampus, the map concept received its exciting neurophysiological underpinning. In 2014 it was honored by the Nobel Prize awarded to John O'Keefe from University College London and Edvard Moser and May-Britt Moser from the Kavli Institute at Trondheim, Norway. Both types of spatial cell fire in location-dependent ways: the place cells whenever the rat is at a certain location, but not anywhere else (Figure 7.19a), while the grid cells have multiple firing (place) fields. All firing fields of one grid cell form a periodic hexagonal grid that tessellates the rat's entire environment in a strikingly regular way (Figure 7.19b). In other words, whenever the animal crosses the points of an imaginary hexagonal grid in its environment, a particular grid cell fires maximally. Different grid cells have their hexagonal grid lattices shifted relative to each other, so that at least theoretically hippocampal place fields can be computed from such lattices. These and some other types of functionally specialized 'spatial cells'—especially the location-independent head direction cells—are considered to be part of a metric navigation system, the

7.19 Place cells and grid cells in the rat's brain

Location-dependent firing activity of **(a)** a hippocampal place cell and **(b)** an entorhinal grid cell of a rat, which foraged in a square enclosure. *Left figures:* The rat's paths *(gray lines)* and the places where the cell fired *(black dots)*. *Right figures:* Spatial distribution of firing activity (so-called firing fields).

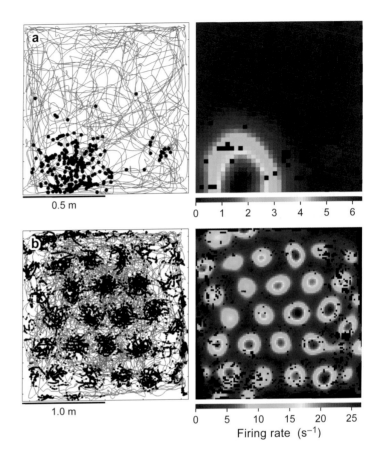

'hippocampal-entorhinal space circuit.'[58] Work in recent years has shown that the spatial role played by this space circuit in navigation is more diversified than sketched out above. Spatial cells may also encode goal vectors leading from a home base to a frequently visited goal, may encode the directions and distances to objects in the environment, may mediate information about the passage of time ('time cells') and the animal's speed ('speed cells'), may represent multiple maps for different sensory modalities and for different spatial scales, or may even map nonspatial sensory stimuli.[59] In flying mammals, Egyptian fruit bats, Nachum Ulanovsky and his coworkers from the Weizmann Institute in Rehovot, Israel, have recorded from subpopulations of hippocampal cells that encode the direction, or the distance, or even the entire vector to a goal.[60] The common tenet holds that all these types of spatial cells contribute to a metric representation of space, to Tolman's metaphorical map, but their ways of interaction in finally mapping the animal's environment remain to be elucidated.

How does the rodent's map compare with the cataglyphs' spatial knowledge? As emphasized throughout this chapter, at any one time during a journey the ants know where to go rather than where they are. Goal vectors are determined directly by combining information from path integration and landmark guidance in the central complex (see Navinet in Figure 7.10). Neither have systems of location-dependent cells been found yet in the insect's brain,[61] nor are they required to account for the ants' navigational performances studied so far. Apparently, the cataglyph's view is mainly to rely on 'views,' on the memory of scenes perceived from particular perspectives, and on vectors derived from them.

However, in this comparison one aspect should not be belittled: the way of how the experiments on the animals' spatial behavior are performed in rats and ants. While the ants are subjected to a variety of large-scale experimental setups in the field, rats are studied while they are freely moving around in small-scale laboratory enclosures. In nature, rats widely roam over hundreds of meters, travel habitual routes, repeatedly return to previously visited sites,[62] and thus experience the very navigational challenges that the cataglyphs tackle in their foraging areas. In fact, just like the cataglyphs, rats are central place foragers. Even in the spatially restricted laboratory environments they establish a home base, to which they regularly return. Early in their lives they perform exploration walks, which may be functionally similar to the ants' learning endeavors. As soon as the juveniles open their eyes, they start to move about in their close environment by exhibiting what have been dubbed "premonitions of curiosity" or the expression of an "exploratory drive," and what finally leads to "latent learning" of spatial features of the nest surroundings.[63] In the explorative walks performed by rats, slow outward journeys including frequent stops are followed by fast inward journeys,[64] and thus share at least some formal properties with the cataglyphs' learning walks. Finally, in a series of well-devised arena studies carried out by the late Ariane Etienne from the University of Geneva, Switzerland, hamsters were trained to learn landmark cues while they performed idiothetic path integration, but whether and how these two types of information are interlinked has been investigated only preliminarily.[65] Or another question: What role does external compass information play in the rodent's foraging behavior?

Using larger laboratory arenas or even natural home range areas should render it possible to subject rats and other mammalian central place foragers to the very behavioral test paradigms that have been applied in insect foragers and described in detail in this book. I would not be surprised if from such studies—potentially complemented by wireless electrophysiological recordings—some navigational routines emerged that shared essential properties with those investigated in the cataglyphs.[66] It is rather unlikely indeed that in processing navigational information a complete hiatus had occurred when evolutionary lines diverged.[67] However, what are the peculiarities that have evolved along the diverging lines? With this in mind, I wonder whether in designing relevant test paradigms there might not be lessons from the cataglyphs even for the *rationalists*.[68] We cataglyphologists are open to surprises as well—all the more, as over the years our little desert navigators have never ceased to amaze us, and certainly will continue to amaze future generations of desert ant aficionados.

Epilogue

The ant, like any higher animal, combines for its guidance all the sensory data at its disposal . . . in association with earlier impressions preserved by memory.

Santiago Ramón y Cajal, 1937

How ants find their way has fascinated scientists and scholars for centuries. Even Santiago Ramón y Cajal, the leading neuroanatomist of his time, and perhaps of all times, finally turned to studying the orientation behavior of ants. He was preceded by Lord Avebury, president of the Institute of Bankers, vice-chancellor of the University of London, and, above all, a young friend of Charles Darwin. John Lubbock, the untitled name he was known by, became so fascinated by "the power and flexibility of the ant's mind" that whenever his "professional occupations and parliamentary duties" allowed him to indulge in his spare-time studies, he performed some pioneering experiments on insect orientation. For example, he let *Lasius* ants carry their larvae across a circular tabletop, which he had transformed into an exquisite laboratory arena. By shifting the position of a candle and observing how the ants changed their courses, he was the first to describe, at the end of the nineteenth century, a light compass response in any animal species.[1]

Let us now move from Victorian England and Lord Avebury's tabletop to a North African desert plain and the celestial hemisphere vaulting it. There we enter the '*Cataglyphis* arena.' We see how researchers paint gridworks of white lines on the barren ground and use special gadgets to record the paths of ants, which have particular parts of their eyes covered with removable light-tight lacquer sheets, or which perform their home runs under an experimental trolley provided with optical devices of all kinds. Some ants have their legs lengthened by stilts attached to their tibial and tarsal segments, others have their antennae manipulated, and yet others walk in curious arrays of two- and three-dimensional open-topped channels erected in the desert. Novice ants, which are just starting their foraging lives, perform learning walks within sets of artificial landmarks, or may even do so on a platform surrounded by Helmholtz coils, which alter the earth's magnetic field in predictable ways. A visitor of the *Cataglyphis* arena may even experience a tethered forager performing its virtual home run on top of a spherical treadmill, or may observe an insect-inspired robot moving across the desert floor.

Over the years, this whole scenario—'studying *Cataglyphis* in the test field'—has become a dynamic experimental system, in which new gadgets, devices, and procedures have been developed at accelerating rates. These developments have largely been facilitated by the amazing readiness with which the cataglyphs cooperate in such endeavors, even if one intervenes in their sensory and locomotor systems quite substantially and exposes them to the most outlandish test situations and material arrangements. No wonder that they have attracted an increasing number of researchers and have induced many of them to turn from investigating the spatial behavior of bees and wasps to focusing on the bees' and wasps' cursorial cousins. Desert ants have indeed become model organisms, true research generators, in the study of animal navigation. The intricate familiarity with this organism has continually opened up new scientific fronts in unprecedented and unanticipated directions. While researchers have manipulated the cataglyphs in various experimental approaches, the cataglyphs have also succeeded in "manipulating the researcher," in shaping the researcher's attitudes, concepts, and experimental approaches in various ways.[2] Let us finally turn to some of their lessons.

First, while readily unfolding their rich navigational repertoire in front of the beholder's eyes, the ants save us from proceeding in a top-down way and relying on overarching concepts often drawn from human introspection and indiscriminately applied across animal taxa.[3] Note that it was already Niko Tinbergen who advised us that "one should not use identical experimental techniques to compare two species because they would almost certainly not be the same to them."[4] Depending on a species' natural history, the functional significance of a given experimental task will vary across species and higher taxa, and even if the behavioral outcomes are similar, the neural circuitries mediating these outcomes might vary across species and taxa as well. The cataglyphs have taught us to consider the navigational problems in the ecophysiological context within which they must solve them. On the one side, this means reconstructing the kind and amount of visual information that a visual navigator with a low-resolution, wide-angle visual system is able to derive from its natural environment. On the other side, there is the need to record the ant's behavior within its environmental settings in as much structural detail as possible, and try to discover peculiarities and behavioral patterns. The increasing availability of optical tracking devices and suitable high-speed cameras may render this feasible in the fields.[5] On the basis of such observations and recordings, hypotheses and proper experimental programs can be developed.

As we have realized on our journeys to the desert navigator, such experiments are open-ended. Their results may depend on minute variations in experimental strategies. Take the following example. When in our first experiments on landmark guidance the cataglyphs' homing behavior was studied with arrays of artificial landmarks installed in the open desert, the animal was assumed to take a snapshot of the landmark scene surrounding the goal. However, later observations showed that the ants perform distinct learning walks, during which they take several goal-directed views from all directions around the goal rather than acquire a single snapshot at the goal itself. This finding has significantly changed our way of understanding the animal's navigational tool set: the insect stores multiple views leading to the goal rather than a single view defining the goal. The overall lesson is to let the animal guide your way of investigation. Start out as naturalist, and then try to understand the animal's solutions by strategically oscillating between developing hypotheses and designing experiments.

Second, the research strategy entertained above immediately leads behavioral biologists and neuroscientists on the evolutionary track, and lets them wonder how the various navigational routines described in this book, their neural underpinnings and interconnections, might have evolved for particular needs. Surely, the bottom-up approach pursued throughout this book is justified as a general research strategy—avoiding higher-order explanations as long as lower-order mechanisms suffice for explaining the observed behavior—but also, and more importantly, it is rooted in the evolutionary process, which with all its vicissitudes has eventually led to the present system of navigation. The modular structure of this system is the direct result of its evolutionary history.

In the nested hierarchy of phylogenetic clades, the cataglyphs are not only ants but also hymenopterans, insects, and arthropods, and thus look back at a long line of ancestors and ways of life. Due to this heritage their navigational repertoire has most likely evolved by co-opting and modifying more basic behavioral mechanisms, i.e., by adopting preexisting neural circuits to meet new needs. For example, the ants' path integration routine may be related to the common counterturning tendency, by which many arthropods subjected to a forced detour maintain an initially adopted heading.[6] Similarly, the local search behavior of flies walking around a food spot and repeatedly returning to it, even in the dark, can be regarded as a small-scale idiothetic path integration maneuver.[7] By recruiting external compass systems, it could have laid the foundation for the advanced large-scale path integration routines observed in ants and bees. Moreover, the capacity to acquire and store a multitude of goal-centered panoramic views may have evolved from the short-term visual place memories studied in flies and other insects.[8] What we observe at present is an elaborate set of strategies strung together in a collective system of multiple interacting guidance routines that constitute the ant's current navigational tool set.

As distant relatives of us, the cataglyphs have followed a different historical path, and hence have a different story to tell. They might tell that their arthropod ancestors had hardly ever been encouraged to behave like geometers, to develop a "position sense" or "map sense," and thus to acquire and use a unified internal representation in which all spatial information is combined.[9] Seen in the light of the animal's present behavioral requirements, this is not at all a drawback. Rather, it is a phylogenetically plausible and economically efficient way of accomplishing the real-life tasks that the cataglyphs face when navigating within their foraging grounds. As it appears, the landmark knowledge that the ants acquire in rapid learning events during well-organized sequences of behavior consists in sets of panoramic views. The insect navigator later uses these memorized views, together with information from an incessantly operating path integrator, to find its way in its nest environs. The particular weights it assigns to the acquired views will depend on the species' evolutionary background and on current environmental and situational contingencies. The question still is whether, to what extent and to what effect, the memorized views are combined with information derived from path integration, and whether this way the cataglyphs can acquire some more elaborate spatial knowledge. The historical paths that have led to the ant and the mouse—to mention just two representatives of evolutionary lines that separated about 550 million years ago—have most likely also led to particular navigational solutions. If so, what are the peculiarities of these solutions? The jury is still out.

Third, to inquire about the ant's navigational repertoires at the level of the brain has hitherto been hampered by the cataglyph's small body size and reproductive strategy, which impede electrophysiological and neurogenetic approaches, respectively. However, with the rise of advanced genetic editing tools, the latter situation may change.[10] Moreover, as the major forebrain neuropils involved in navigation are rather conserved across insect taxa, neurobiological data obtained in larger insects (locusts) or in genetically more tractable species (fruit flies) can be drawn upon. As our present knowledge suggests that there is not one distinct network uniquely dedicated to navigation, the cataglyphs may guide us to various regions and subregions of their brain, and advise us to investigate the computational capabilities of these particular neural networks.[11] One or another of these capabilities might have been co-opted by the navigator for accomplishing particular modes of spatial behavior. This way, the neural jigsaw puzzle will be pieced together step-by-step.

Cataglyphis brain research has recently gathered momentum and proved its high potential in various ways: for combining neural and behavioral approaches, for unraveling remarkable age-dependent and behaviorally mediated neural rewiring processes that occur in the brain during the ant's lifetime,[12] and for considering species-specific differences. Let us finally turn to the latter aspect. Given the high energetic costs of maintaining nervous tissues,[13] one can certainly assume that the species-specific relative size of a given brain region is a reliable measure of how important a role this region plays in the animal's overall behavioral repertoire. In this respect, allometric investigations of taxa-specific differences in the relative sizes of various brain regions are essential. On a grand scale, they show that the mushroom bodies are relatively larger in ants than in bees, but that the reverse is true for the optic lobes. On a smaller scale, exceptions to such rules are especially revealing. For example, the cataglyphs have exceedingly larger collar (visual input) regions in the calyces of their mushroom bodies than expected for the sizes of their eyes and optic lobes.[14] This trait may be correlated with the computational tasks that these visual navigators must perform in learning, retaining, and using view-based landmark information.

Even within the cataglyphs the relative size of the brain varies across species (see Figure 7.18b).[15] We do not know yet what functional differences may be hidden behind these variations, nor do we know how specific subcompartments contribute to the observed differences in brain size. This is a promising field for future research all the more, as some behavioral studies indicate that the weight ascribed to particular navigational routines may vary across species as well. Such neuroarchitectural differences might be slight, but as they may be correlated with types of habitat and thus navigational requirements, locomotor agility, and even colony structure and associated life history patterns, alongside a suite of other ecological factors, they could provide a direct means of tracing evolutionary adaptations to particular needs, and thus may shed light on how the corresponding neural circuits have gradually evolved. The wide species spectrum of the cataglyphs will offer a large potential for such investigations.

The journey to the desert navigator continues. Conceptually and technologically, modern methods are developing at a staggering rate in all fields of the molecular, organismal, and evolutionary life sciences, but they do so mainly in a handful of decontextualized model organisms such as the fly and the mouse in the laboratory. The time has come to increasingly apply this plethora of tools, in a joint venture, in an organism 'in the wild.' Our desert navigator superbly lends itself for such future endeavors. On the one hand, one cannot help but marvel at the complexity and richness of the navigational tasks that a cataglyph is able to accomplish—by "combining for its guidance all the sensory data at its disposal" and associating them "with earlier impressions preserved by memory." On the other hand, one should never underrate the efficacy and versatility of the navigator's miniature brain, which after all contains almost half a million neurons and reaches a degree of miniaturization of cells and circuits that by far exceeds that of the human brain. To increasingly unravel the computational power of these miniaturized circuitries in the ant's cockpit is an intellectually challenging task that will continue to drive future research agendas. Seen in this light, we can give the last word again to Darwin, who argued that the ant's brain "is one of the most marvelous atoms of matter in the world, perhaps more so than the brain of a man."

NOTES

ACKNOWLEDGMENTS

ILLUSTRATION CREDITS

INDEX

Notes

References, which have already been cited in a previous Note, are referred to as *nX* (meaning, see note X). Work on insect—especially ant—navigation is presently increasing at an accelerating pace at various fronts. Papers that appeared after fall 2018 could be included only if I had access to them prior to publication.

Prologue

Epigraph: Darwin C (1871): The Descent of Man, and Selection in Relation to Sex. *London, J. Murray.* Reprint by *Princeton University Press,* 1981, *quot.* p. 145.

1. Krogh A (1929): The progress of physiology. *Am J Physiol 90, 243–251, quot.* p. 247.

2. Wehner R, Fukushi T, Isler K (2007): On being small: ant brain allometry. *Brain Behav Evol 69, 220–228.*

3. Schaffner W (2015): Enhancers, enhancers—from their discovery to today's universe of transcription enhancers. *Biol Chem 396, 311–327.*

4. For an emphasis on natural history, see Edward O. Wilson's foreword in Canfield MR (2011): Field Notes in Science and Nature. *Cambridge MA, Harvard University Press:* "The wellspring of the new biology is scientific natural history," *quot.* p. xi.

5. Excellent books have recently appeared about the superorganisms of social insects, especially of ants and honeybees: Seeley TD (1995): The Wisdom of the Hive. *Cambridge MA, Harvard University Press.* Hölldobler B, Wilson EO (2009): The Superorganism. *New York, Norton.* Seeley TD (2010): Honeybee Democracy. *Princeton NJ, Princeton University Press.*—For treating superorganisms from the computer scientist's point of view, see Dorigo M, Stützle T (2004): Ant Colony Optimization. *Cambridge MA, MIT Press.*

6. Edinburgh International Festival 1957: An Exhibition of Paintings of Claude Monet. *Edinburgh, Royal Scottish Academy, quot.* p. 7.

1. Setting the Scene

1. Taxonomically, *Cataglyphis fortis* was first described as *Myrmecocystus albicans* var. *fortis:* Forel A (1902): Les fourmis du Sahara Algérien. Récoltées par M. le Professeur A. Lameere et le Dr. A. Diehl. *Ann Soc Entomol Belgique 46, 147–158.*—The description was based on a few worker specimens collected in 1896 by the neurologist August Diehl from Lübeck, Germany, in a small salt pan near Touggourt, Algeria. Diehl's personal notes have been published by Auguste Forel: Forel A (1903): Die Sitten und Nester einiger Ameisen der Sahara bei Tugurt und Biskra. *Mitt Schweiz Entomol Ges 10, 453–459.*—In 1982 the author and his wife visited this first finding site in Algeria and collected animals from there and later from many other North African locations; see figure 6 in Wehner R, Wehner S, Agosti D (1994): Patterns of biogeographic distribution within the *bicolor* species group of the North African desert ant, *Cataglyphis*

Foerster 1850. *Senck biol 74, 163–191.*—Based on morphological studies of the workers and the newly discovered and described males and reproductive females, *C. fortis* could be clearly separated from *C. albicans* and raised to species rank: Wehner R (1983): Taxonomie, Funktionsmorphologie und Zoogeographie der saharischen Wüstenameise *Cataglyphis fortis* (Forel 1902) stat. nov. (Insecta: Hymenoptera: Formicidae). *Senck biol 64, 89–132.*

2. In social hymenopterans the concept of central place foraging was first outlined for honeybees: Kacelnik A, Houston AI, Schmid-Hempel P (1986): Central-place foraging in honey bees: the effect of travel time and nectar flow on crop filling. *Behav Ecol Sociobiol 19, 19–24.*—For more recent work in ants, see Dornhaus A, Collins EJ, Dechaume-Moncharmont FX, Houston AI, Franks NR, McNamara JM (2006): Paying for information: partial loads in central place foragers. *Behav Ecol Sociobiol 61, 151–116.* Schmolke A (2009): Benefits of dispersed central-place foraging: an individual-based model of a polydomous ant colony. *Am Nat 173, 772–778.*

3. Stein G (1937): Everybody's Autobiography. *New York, Random House, quot.* p. 289.

4. As the ant's navigational tool set developed over evolutionary time, so did the story of Ariadne's strategy over historical time. In the first published version (Diodoros: *Bibliotheca Historica* 4, 61, 4) the thread is not mentioned at all. We only learn that Ariadne informed Theseus about the general location of the exit door. The trick of using the thread as a safety line is told only later, and in slightly different ways, by Plutarch (*Theseus* 19,1f), Catullus (*Carmina* 64, 112f), and Ovid (*Metamorphoses* 8, 172).

5. Wehner R (1982): Himmelsnavigation bei Insekten. Neurophysiologie und Verhalten. *Neujahrsbl Naturforsch Ges Zürich 184, 1–132.* Collett M, Collett TS, Wehner R (1999): Calibration of vector navigation in desert ants. *Curr Biol 9, 1031–1034.* Collett M, Collett TS (2000): How do insects use path integration for their navigation? *Biol Cybern 83, 245–259.*

6. The rule of thumb of not using a moving landmark for navigation has been derived from experiments in rats. These animals use a landmark as a location cue only if the landmark maintains a stable position within a geometric frame of reference. Biegler R, Morris RGM (1993): Landmark stability is a prerequisite for spatial but not discrimination learning. *Nature 361, 631–633.*—For neurophysiological evidence, see Knierim JJ, Kudrimoti HS, McNaughton BL (1998): Interactions between idiothetic cues and external landmarks in the control of place cells and head direction cells. *J Neurophysiol 80, 425–446.* Jeffery KJ (1998): Learning of landmark stability and instability by hippocampal place cells. *Neuropharmacol 37, 677–687.*

7. Dahmen H, Wahl VL, Pfeffer SE, Mallot HA, Wittlinger M (2017): Naturalistic path integration of *Cataglyphis* desert ants on an air-cushioned lightweight spherical treadmill. *J Exp Biol 220, 634–644.*—Erich Buchner, then at the Max Planck Institute for Biological Cybernetics in Tübingen, Germany, was the first to design an air-suspended spherical treadmill device (for *Drosophila*). Buchner E (1976): Elementary movement detectors in an insect visual system. *Biol Cybern 24, 85–101.*—Recordings of neuronal activity in the brains of *Drosophila* flies walking on a trackball in a virtual reality arena have been performed by Johannes D. Seelig and Vivek Jayaraman. Seelig JD, Jayaraman V (2015): Neural dynamics for landmark orientation and angular path integration. *Nature 521, 186–191.*

8. For optical characteristics as determined by microophthalmological (deep pseudopupil) measurements in the compound eyes of *Cataglyphis,* see Wehner R (1982), *n5.* Zollikofer CPE, Wehner R, Fukushi T (1995): Optical scaling in conspecific *Cataglyphis* ants. *J Exp Biol 198, 1637–1646.*—Given the rather small size and only slightly arched shape of the cataglyphs' compound eyes, the huge visual fields may be surprising. They result from the fact that the optical axes of the ommatidia are normal to the corneal surfaces only in a small dorsal part of the eye and deviate systematically the more toward the periphery, the more lateral the ommatidia are positioned in the eye. In *Melophorus* optical resolution data are derived from histological sections: Schwarz S, Narendra A, Zeil J (2011): The properties of the visual system in the Australian desert ant *Melophorus bagoti. Arthropod Struct Dev 40, 128–134.*

9. Mainly due to the animals' small body and thus eye sizes, the interommatidial divergence angles, $\Delta\varphi$, of compound eyes, as defined by the density of ommatidia per unit visual angle, cannot decrease beyond a certain limit. They are rarely smaller than 1°–2°. The smallest ones ($\Delta\varphi = 0.24°$

and 0.30°) are found in zones of acute vision in the large compound eyes of dragonflies (with about 30,000 ommatidia) and fiddler crabs (with about 8,000 ommatidia), respectively: Sherk TE (1978): Development of the compound eyes of dragonflies (Odonata). III. Adult compound eyes. *J Exp Zool 203, 61–80.* Smolka J, Hemmi JM (2009): Topography of vision and behaviour. *J Exp Biol 212, 3522–3532.*

10. Wystrach A, Dewar ADM, Philippides A, Graham P (2016): How do field of view and resolution affect the information content of panoramic scenes for visual navigation? A computational investigation. *J Comp Physiol A 202, 87–95.* Wystrach A, Beugnon G, Cheng K (2011): Landmarks or panoramas: what do navigating ants attend to for guidance? *Front Zool 8, 21.* See also Philippides A, Baddeley B, Cheng K, Graham P (2011): How might ants use panoramic views for route navigation? *J Exp Biol 214, 445–451.*

11. For matched sensory filters, see Wehner R (1987): "Matched filters"—neural models of the external world. *J Comp Physiol A 161, 511–531.* Krapp HG, Hengstenberg R (1996): Estimation of self-motion by optic flow processing in single visual neurons. *Nature 384, 463–466.* Von der Emde G, Warrant E, eds. (2016): The Ecology of Animal Senses: Matched Filters for Economical Sensing. *Heidelberg, Springer.*

12. In two *Cataglyphis* species behavioral repertoires of 40–50 acts have been recognized: Retana J, Cerdá X (1991): Behavioral repertoire of the ant *Cataglyphis cursor* (Hymenoptera: Formicidae): is it possible to elaborate a standard specific one? *J Insect Behav 4, 139–155.* Cerdá X, Retana J, Carpintero S (1996): The caste system and social repertoire of *Cataglyphis floricola* (Hymenoptera Formicidae). *J Ethol 14, 1–8.*—Myrmecologists have tried hard to inventory the types of behavioral acts that individual workers perform as they age. For task specifications as defined in harvester ants, *Pheidole dentata,* and fire ants, *Solenopsis invicta,* see Wilson EO (1976): Behavioral characterization and the number of castes in an ant species. *Behav Ecol Sociobiol 1, 141–154.* Fagen RM, Goldman RN (1977): Behavioural catalogue analysis methods. *Anim Behav 25, 261–274.* Tschinkel WR (2006): The Fire Ants. *Cambridge MA, Harvard University Press.* See also Herbers JM, Cunningham M (1983): Social organization in *Leptothorax longispinosus. Anim Behav 31, 759–771.* Cole BJ (1985): Size and behavior in ants: constraints on complexity. *Proc Natl Acad Sci USA 82,*

8548–8551.—Honeybee workers have been reported to display nearly 60 distinct and (at least partly) hardwired behavior patterns: Chittka L, Niven J (2009): Are bigger brains better? *Curr Biol 19, R995–R1008.*

13. Retana J, Cerdá X (1991): Behavioural variability and development of *Cataglyphis cursor* ant workers (Hymenoptera: Formicidae). *Ethology 89, 275–286.* Schmid-Hempel P, Schmid-Hempel R (1984): Life duration and turnover of foragers in the ant *Cataglyphis bicolor* (Hymenoptera, Formicidae). *Insect Soc 31, 345–360.* Nowbahari, E, Fénéron R, Malherbe MC (2000): Polymorphism and polyethism in the fomicine ant *Cataglyphis niger* (Hymenoptera). *Sociobiology 36, 485–496.*—In general, concentrically arranged task allocations in social insect colonies have been discussed in Beshers SN, Fewell SH (2001): Models of division of labor in social insects. *Annu Rev Entomol 46, 413–440.*

14. Even though not all individuals are strictly bound to this scheme, and especially smaller individuals tend to start foraging later in life than their larger nestmates, the majority of the cataglyphs does not exhibit distinct size-related divisions of labor (morphological polyethism). An exception to this rule is the extremely dimorphic *Cataglyphis bombycina* (see Chapter 2 and Figure 2.5 therein).

15. Nowbahari E, Scohier A, Durand JL, Hollis KL (2009): Ants, *Cataglyphis cursor,* use precisely directed rescue behavior to free entrapped relatives. *PLoS ONE 4(8), e6573.* Nowbahari E, Hollis KL, Durand JL (2012): Division of labor regulates precision rescue behavior in sand-dwelling *Cataglyphis cursor* ants: to give is to receive. *PLoS ONE 7(11), e48516.*

16. In *Cataglyphis* chemical recognition based on cuticular hydrocarbons is described in various contexts and species: Nowbahari E, Lenoir A, Clément JL, Lange C, Bagnères AG, Joulie C (1990): Individual, geographical and experimental variation of cuticular hydrocarbons of the ant *Cataglyphis cursor* (Hymenoptera: Formicidae): their use in nest and subspecies recognition. *Behav Syst Ecol 18, 63–74.* Dahbi A, Cerdá X, Hefetz A, Lenoir A (1996): Social closure, aggressive behaviour, and cuticular hydrocarbon profiles in the polydomous ant *Cataglyphis iberica* (Hymenoptera, Formicidae). *J Chem Ecol 22, 2173–2186.* Dahbi A, Hefetz A, Cerdá X, Lenoir A (1999): Trophallaxis mediates uniformity of colonial odor in *Cataglyphis iberica* ants (Hymenoptera,

Formicidae). *J Insect Behav 12, 559–567.* Lahav S, Soroker V, Hefetz A, Vander Meer RK (1999): Direct behavioral evidence for hydrocarbon as ant recognition discriminators. *Naturwissenschaften 86, 246–249.* Lahav S, Soroker V, Vander Meer RK, Hefetz A (2001): Segregation of colony odor in the desert ant *Cataglyphis niger. J Chem Ecol 27, 927–943.*—In *Cataglyphis niger* it has been shown that individuals become less aggressive against non-nestmates, when they have encountered these non-nestmates repeatedly before: Nowbahari E (2007): Learning of colony odor in the ant *Cataglyphis niger* (Hymenoptera: Formicidae). *Learn Behav 35, 87–94.* Foubert E, Nowbahari E (2008): Memory span of heterospecific individual odors of an ant, *Cataglyphis cursor. Learn Behav 36, 319–326.*—For task-related cuticular cue composition and recognition by nestmates in other ant species, see Bonavita Cougourdan A, Clement JL, Lange C (1993): Functional subcaste discrimination (foragers and brood tenders) in the ant *Camponotus vagis:* polymorphism of cuticular hydrocarbon patterns. *J Chem Ecol 19, 1461–1477.* Wagner D, Tissot M, Gordon DM (1998): Task-related environment alters the cuticular hydrocarbon composition of harvester ants. *J Chem Ecol 27, 1805–1819.* Greene MJ, Gordon DM (2003): Social insects: cuticular hydrocarbons inform task decisions. *Nature 423, 32.* Bos N, d'Ettorre (2012): Recognition of social identity in ants. *Front Psychol 3, 83.*—See also Lenoir A, Fresneau D, Errard C, Hefetz A (1999): The individuality and the colonial identity in ants: the emergence of the social representation concept. In: Detrain C, Deneubourg JL, Pastells J, eds., Information Processing in Social Insects, pp. 219–237. *Basel, Birkhäuser.*

17. Life duration, life history schedules, and turnover rates in *Cataglyphis* workers with their temporal polyethism were determined in the ants' natural habitats by recording resighting frequencies of animals that had been individually marked at different times of their life history: Harkness RD (1977): The carrying of ants *(Cataglyphis bicolor)* by others of the same nest. *J Zool 183, 419–430.* Wehner R, Harkness RD, Schmid-Hempel P (1983): Foraging Strategies in Individually Searching Ants, *Cataglyphis bicolor* (Hymenoptera: Formicidae). *Mainz, G. Fischer.* Schmid-Hempel P, Schmid-Hempel R (1984), *n13.*—For *Melophorus,* see Muser B, Sommer S, Wolf H, Wehner R (2005): Foraging ecology of the thermophilic Australian desert ant, *Melophorus bagoti. Austr J Zool 53, 301–311.*

18. When in the second half of the nineteenth century biologists realized that bees and ants were able to accomplish quite sophisticated tasks of spatial orientation, object detection, and decision making, they endowed these insects with "intelligence" (Dujardin 1853, Emery 1893); "understanding" (Romanes 1882); "mental faculties" (Brandt 1879); "rationality" (Forel 1886–1888); and "higher psychological functions" (Viallanes 1893, Wasmann 1909). References: Brandt EK (1879): Vergleichend-anatomische Skizze des Nervensystems der Insekten. *Horae Soc Ent Ross 15, 3–19.* Dujardin F (1853): Quelques observations sur les abeilles et particulièrement sur les actes, qui chez ces insectes peuvent être rapportés à l'intelligence. *Ann Sci Nat Sér (Zool Biol Anim) 18, 231–240.* Emery C (1893): Intelligenz und Instinkt der Tiere. *Biol Cbl 13, 151–155.* Forel A (1886–1888): Expériences et remarques critiques sur les sensations des insectes. *Rec Zool Suisse, Tome 4, Nos 1–4.* Romanes G (1882): Animal Intelligence. *London, Kegan Paul and Trench.* Viallanes MH (1893): Études biologiques et organologiques sur les centres nerveux et les organes de sens des animaux articulés. *Ann Sci Nat Sér 7 (Zool Paléont) 14, 405–456.* Wasmann E (1909): Die psychischen Fähigkeiten der Ameisen. Mit einem Ausblick auf die vergleichende Tierpsychologie. 2nd ed. *Stuttgart, Schweizerbart.*

Crediting ants and bees with these top-class mental faculties was also favored by the leading neuroanatomists of the time, who at the end of the nineteenth century did their pioneering work on the hymenopteran forebrain. Dujardin F (1850): Mémoire sur le système nerveux des insectes. *Ann Sci Nat Sér 3 (Zool Biol Anim) 14, 195–206.* Flögel JHL (1878): Über den einheitlichen Bau des Gehirns in den verschiedenen Insektenordnungen. *Z Wiss Zool Suppl 30, 556–592.* Kenyon FC (1896): The meaning and structure of the so-called "mushroom bodies" of the hexapod brain. *Am Nat 30, 643–650.*—For *Formica rufa* Franz Leydig argued in the same way: Leydig F (1864): Vom Bau des thierischen Körpers. Handbuch der vergleichenden Anatomie. Vol. 1, part 1. *Tübingen, Laupp.*

Soon the tide turned. By trying to get rid of any form of "mentalism," physiologists such as Jacques Loeb and Albrecht Bethe regarded animal behavior as the mere outcome of stimulus-dependent forced actions and thus became forerunners of John B. Watson's radical behaviorism. Bethe A (1898): Dürfen wir den Ameisen und Bienen psychische Fähigkeiten zuschreiben? *Pfügers Arch Ges Physiol 70, 15–100.* Loeb J (1899): Einleitung in die vergleichende Gehirnphysiologie und vergleichende Psychologie mit besonderer Berücksichtigung der wirbellosen Tiere. *Leipzig, J. A. Barth.* Watson JB (1913): Psychology as the behaviorist views it. *Psychol Rev 20, 158–177.*—However, Bethe's mechanistic view of insect behavior was already criticized by Hugo Berthold von Buttel-Reepen, a German businessman and gentleman scientist, to whom we owe careful studies of honeybee behavior: von Buttel-Reepen H (1900): Sind die Bienen Reflexmaschinen? *Biol Zbl 20, 97–109, 130–144, 177–193, 209–224, 289–304.*

19. Abbott RL (1931): Instinct or intelligence in the great golden digger? *Proc Iowa Acad Sci 38, 255–258.* Wooldridge DE (1931): The Machinery of the Brain. *New York, McGraw-Hill.*—For "sphexish" behavior, see Hofstadter DR (1982): Can inspiration be mechanized? *Scient Amer 247(3), 18–31.* Dennett DC (1996): Elbow Room. *Cambridge MA, MIT Press.*—The original description of the wasp's apparently inflexible chain of behavioral acts is given in Fabre JH (1879): Souvenirs Entomologiques. Sér 1, pp. 87–92. *Paris, Delagrave.*

20. Keijzer F (2013): The *Sphex* story: how the cognitive sciences kept an old and questionable anecdote. *Phil Psychol 26, 502–519.*—It was already Jean-Henri Fabre, *n19,* who provided evidence that the wasp's prey retrieval behavior could be modulated by experience.

21. Fabre JH (1879), *n19.* Lubbock J (1882): Ants, Bees, and Wasps: A Record of Observations on the Habits of the Social Hymenoptera. *London, Kegan Paul, Trench.*—Fabre, who has often been considered a hard-nosed antievolutionist, cautioned against deducing natural phenomena from a single unifying framework; see the critical account in Yavetz I (1991): Theory and reality in the work of Jean Henri Fabre. *Hist Phil Life Sci 13, 33–72.*

22. Ritter H, Cruse H, Dean J, eds. (2000): Prerational Intelligence: Adaptive Behavior and Intelligent Systems without Symbols and Logic. Vol. 2. *Dordrecht, Kluwer Academic Publishers.*—At present, the tide *(n18)* has turned again. Heisenberg M (1983): Initiale Aktivität und Willkürverhalten bei Tieren. *Naturwissenschaften 70, 70–78.* Brembs B (2011): Towards a scientific concept of free will as a biological trait: spontaneous actions and decision-making in invertebrates. *Proc R Soc B 278, 930–939.* Heisenberg M (2013): Action selection: The brain as a behavioral organizer. In: Menzel R, Benjamin PR, eds., Invertebrate Learning and Memory, pp. 9–13. *Amsterdam, Elsevier.*

2. The Thermophiles

1. For insect thermoregulation in general, see Bernd Heinrich's magnum opus: Heinrich B (1993): The Hot-Blooded Insects. *Berlin, Springer.* The author has condensed this broad overview of insect-temperature relationships into a wonderful small book: Heinrich B (1996): The Thermal Warriors. *Cambridge MA, Harvard University Press.*—The term 'thermophiles' usually refers to thermophilic single-celled microorganisms, especially to the hyperthermophilic archaea; see, e.g., Stetter KO (2006): Hyperthermophiles in the history of life. *Phil Trans R Soc B 361, 1837–1843.* Li FL, ed (2015): Thermophilic Microorganisms. *Norfolk UK, Caister Academic Press.*

2. The ants depicted in *Description de l'Égypt (Paris, L'Imprimerie Impériale,* 1809–1813) were described as *C. bombycina* and *C. savignyi* only several decades later by Julius Roger and Léon Dufour, respectively. Roger J (1859): Beiträge zur Kenntnis der Ameisenfauna der Mittelmeerländer. *Berliner Ent Z 3, 225–259.* Dufour L (1862): Notices entomologiques. 3. Notice sur la *Formica savignyi. Ann Soc Ent France, Sér 4, 2, 131–148.*

3. The size of these ants, "which very much resemble in form the ants found in the land of the Hellenes" (most probably *C. nodus*), has been poetically exaggerated ("smaller than dogs but larger than foxes," Herodotus), and so has their "ferocity and passion for gold" (Pliny): Herodotus (ca. 450 BC): The History of Herodotus. Book 3, chapter 102. *London, Macmillan, 1904, quot.* p. 260. Plinius C. Secundus (ca. 100): The Natural History of Pliny. Book 11, chapter 36. *London, Bohn, 1855, quot.* p. 39.—In the alluvial sands of the area described by the ancient writers, gold particles can indeed be found. Ball V (1881): A Manual of the Geology of India. Part 3. *Calcutta, Office of the Geological Survey of India,* p. 213. The "ferocity" could refer to *C. bellicosa, n4.* See also

Harkness RD, Wehner R (1977): *Cataglyphis. Endeavour NS 1, 115–121.*

4. Agosti D (1990): Review and reclassification of *Cataglyphis* (Hymenoptera, Formicidae). *J Nat Hist 24, 1457–1505.*—Currently, there are 93 valid species names: Bolton B (2019): An online catalog of the ants of the world. Available from http://antcat.org (accessed 8 May 2019).

The actual number of species will certainly be larger. For example, the present species *C. bicolor* and *C. savignyi* most likely include several separate species, which in the time to come might be recognized by genetic methods. See also Bolton B (1995): A New General Catalogue of the Ants of the World. *Cambridge MA, Harvard University Press.*—As mentioned in the text, the cataglyphs do not sting (all formicine ants have evolutionarily lost their sting apparatus). However, at least one species *(C. bellicosa)* is strongly aggressive. When Vladimir Karavaiev described this species from a stone pit south of Teheran, he already noted that the ants "furiously attacked" the collector by "approaching him in troops," so that he had to retreat. Karavaiev V (1924): Zur Systematik der paläarktischen *Myrmecocystus* (Formicidae), nebst einigen biologischen Notizen. *Konowia 3, 301–308, quot.* p. 308.

5. At this juncture, a note about the proper gender of the *Cataglyphis* species names may be in order. Agosti (1990) and Bolton (1995), see *n4,* use the masculine form throughout. Other authors use either the feminine or the masculine form—see remark in Lenoir et al. (2009), *n35*—mostly in accord with the usage by former researchers. As due to its Greek origin the genus name *Cataglyphis* is feminine (*glyphís,* genitive *glyphídos,* see *n24*), we consistently use the feminine form. Often, when *Cataglyphis* species have originally been described as members of the genus *Myrmecocystus* (e.g., *M. albicans ibericus* Emery 1906, *M. bicolor bellicosus* Karavaiev 1924) or *Monocombus,* and have later been transferred to *Cataglyphis,* the masculine form has been conserved. Note that in *Cataglyphis nodus* the species name is a masculine noun rather than an adjective, and hence must remain masculine.

6. It was already William Morton Wheeler who realized that the ammochaetae reached their greatest development in species inhabiting the driest deserts and that by occurring in several unrelated genera, they represented a striking example of convergent evolution. Wheeler WM (1907): On certain modified hairs peculiar to the ants of arid regions. *Biol Bull 13, 185–202.*—For functional significance in sand-carrying behavior, see Santschi F (1909): Sur la signification de la barbe des fourmis arénicoles. *Rev Suisse Zool 17, 449–458.* Spangler HG, Rettenmeyer CW (1966): The function of the ammochaetae or psammophores of harvester ants, *Pogonomyrmex* spp. *J Kansas Ent Soc 39, 739–745.* Dlussky GM (1981): The Ants of the Desert. *Moscow, Academia Nauk USSR* [in Russian], p. 72.

7. In most *Cataglyphis* species the workers exhibit substantial variations in body size. For that reason alone, they are often called 'polymorphic,' e.g., in Cerdá X, Retana J (1997): Links between worker polymorphism and thermal ecology in a thermophilic ant species. *Oikos 78, 467–474.* Cerdá X (2001): Behavioural and physiological traits to thermal stress tolerance in two Spanish desert ants. *Etologia 9, 15–27.*—However, polymorphism represents the phenomenon that different forms (morphs) occur within a species or caste; see already Wheeler WM (1910): Ants. *New York, Columbia University Press,* p. 86.—The worker caste of *C. bombycina* is a case in point.

8. The only other *Cataglyphis* species possessing a soldier caste is *C. kurdistanica,* a member of the *C. altisquamis* species group. Pisarski B (1965): Les fourmis du genre *Cataglyphis* en Irak (Hymenoptera, Formicidae). *Bull Acad Pol Sci Cl II 13, 417–422.*—In *C. bombycina* there are indications that genetic factors are involved in the determination of the minor workers and soldiers: Leniaud L, Pearcy M, Aron S (2013): Sociogenetic organization of two desert ants. *Insect Soc 60, 337–344.*—As demonstrated by Christian Peeters, the development of the marked dimorphism within the worker caste could result from variations in growth rules. Molet M, Maicher C, Peeters C (2014): Bigger helpers in the ant *Cataglyphis bombycina:* increased worker polymorphism or novel soldier caste? *PLoS ONE 9(1), e84929.*—For the defense function of soldiers, see Délye G (1957): Observations sur la fourmi saharienne *Cataglyphis bombycina. Insect Soc 4, 77–82.* Wehner R, Marsh AC, Wehner S (1992): Desert ants on a thermal tightrope. *Nature 357, 586–587.*—A role in nest construction has been proposed in Bernard F (1951): Adaptation au milieu chez les fourmis sahariennes. *Bull Soc Hist Nat Toulouse 86, 88–96.*—Even a function as

food-storing repletes has been suggested: Molet M, Maicher C, Peeters C (2014), *loc. cit.*

9. *Vanessa cardui* is a regular spring migrant from West and North Africa northward to as far as Britain and southern Scandinavia. Stefanescu C, Páramo F, Åkesson S, Alarcón M, Brereton T, et al., Chapman JW (2013): Multigeneration long-distance migration of insects: studying the painted lady butterfly in the Western Palaearctic. *Ecography 36, 474–486.*

10. Most necromones are fatty acids, which are common compounds of insect cuticles. In bees and ants they have been mainly studied in the context of the removal of dead nestmates from the colony; e.g., Diez L, Moquet L, Detrain C (2013): Post-mortem changes in chemical profile and their influence on corpse removal in ants. *J Chem Ecol 39, 1424–1432.*—A large range of necromones has been tested for plume-following behavior: Bühlmann C, Graham P, Hansson BS, Knaden M (2014): Desert ants locate food by combining high sensitivity to food odors with extensive crosswind runs. *Curr Biol 24, 960–964.*

11. For first accounts on diet composition in *Cataglyphis,* see Délye G (1968): Recherches sur l'écologie, la physiologie et l'éthologie des fourmis du Sahara. *PhD Thesis, Université d'Aix-Marseille.* Harkness RD, Wehner R (1977), *n3.* Wehner R, Harkness RD, Schmid-Hempel P (1983): Foraging Strategies in Individually Searching Ants, *Cataglyphis bicolor* (Hymenoptera: Formicidae). *Mainz, G. Fischer.* Schmid-Hempel P (1983): Foraging ecology and colony structure of two sympatric species of desert ants, *Cataglyphis bicolor* and *Cataglyphis albicans. PhD Thesis, University of Zürich.* Cerdá X (1988): Food collection by *Cataglyphis iberica* (Hymenoptera, Formicidae). *Ann Zool Polsk Akad Nauk 41, 515–525.* Cerdá X, Retana J, Bosch J, Alsina A (1989): Daily foraging activity and food collection of the thermophilic ant *Cataglyphis cursor* (Hymenoptera, Formicidae). *Vie Milieu 39, 207–212.*—Plant material generally accounts for much less than 10% of all collected items. For the unique petal foraging of *C. floricola,* see Cerdá X, Retana J, Carpintero S, Cros S (1992): Petals as the main resource collected by the ant *Cataglyphis floricola* (Hymenoptera, Formicidae). *Sociobiology 20, 315–319.*—In *Ocymyrmex* the collection of plant material may occur to a larger extent, e.g., in *O. picardi.* Prins AJ (1965): African Formicidae: De-

scription of a new species. *S Afr J Agricult Sci 8, 1021–1024.*—In Australia, depending on season *Melophorus bagoti* can collect large amounts of seeds especially of one species of grass. Schultheiss P, Nooten SS (2013): Foraging patterns and strategies in an Australian desert ant. *Austr Ecol 38, 942–951.*

12. In *C. albicans* and *C. bicolor* the distance range that includes 90% of the total search density roughly coincides with the nearest neighbor distance between nests of different colonies. Hence, the foraging ranges of adjacent colonies largely overlap. Wehner R, Harkness RD, Schmid-Hempel P (1983), *n11.* Schmid-Hempel P (1983), *n11.* Wehner R (1987): Spatial organization of foraging behaviour in individually searching desert ants, *Cataglyphis* (Sahara Desert) and *Ocymyrmex* (Namib Desert). *Experientia Suppl 54, 15–42.*

13. For definition and discussion of 'diffuse foraging,' see Oster GF, Wilson EO (1978): Caste and Ecology in the Social Insects. *Princeton NJ, Princeton University Press.* Traniello JFA (1989): Foraging strategies of ants. *Annu Rev Entomol 34, 191–210.*

14. For the development of sector fidelity, see Wehner R (1987), *n12.* Wehner R, Meier C, Zollikofer CPE (2004): The ontogeny of foraging behaviour in desert ants, *Cataglyphis bicolor. Ecol Entomol 29, 240–250.*—Site fidelity has been studied in Bolek S, Wittlinger M, Wolf H (2012): What counts for ants? How return behaviour and food search of *Cataglyphis* ants are modified by variations in food quantity and experience. *J Exp Biol 215, 3218–3222.*

15. For data on the structure and dynamics of the forager force of *C. albicans, C. bicolor,* and *C. nodus,* see *n11* and Schmid-Hempel P (1987): Foraging characteristics of the desert ant *Cataglyphis. Experientia Suppl 54, 43–61.* Wehner R, Rössler W (2013): Bounded plasticity in the ant's navigational tool kit. In: Menzel R, Benjamin PR, eds., Invertebrate Learning and Memory, pp. 514–529. *Amsterdam, Elsevier.*—Studies performed on laboratory colonies: Retana J, Cerdá X (1991): Behavioral variability and development of *Cataglyphis cursor* ant workers (Hymenoptera, Formicidae). *Ethology 89, 275–286.*

16. B. Hölldobler, *in litt;* J. Billen, *in litt.*

17. For recruitment topics in *Cataglyphis,* see especially Schmid-Hempel P (1983), *n11.* Wehner R, Harkness RD, Schmid-Hempel P (1983), *n11.* Wehner R (1987), *n12.* Lenoir

A, Nowbahari E, Quérard L, Pondicq N, Delalande C (1990): Habitat exploitation and intercolonial relationships in the ant *Cataglyphis cursor* (Hymenoptera, Formicidae). *Acta Oecol 11, 3–18.* Amor F, Ortega P, Cerdá X, Boulay RR (2010): Cooperative prey-retrieving in the ant *Cataglyphis floricola:* an unusual short-distance recruitment. *Insect Soc 57, 91–94.*

18. For remarks on the functional roles of the postpharyngeal gland in *Cataglyphis,* see *n20;* the cloacal gland, see *n22;* the mandibular gland, see *n53.*

19. Several aspects of the 'dispersed central-place foraging' that are associated with the cataglyphs' polydomous colony structure are discussed in Schmid-Hempel P, Schmid-Hempel R (1984): Life duration and turnover of foragers in the ant *Cataglyphis bicolor* (Hymenoptera, Formicidae). *Insect Soc 31, 345–360.* Cerdá X, Retana J, de Haro A (1994): Social carrying between nests in polycalic colonies of the monogynous ant *Cataglyphis iberica* (Hymenoptera: Formicidae). *Sociobiology 23, 215–231.* Cerdá X, Dahbi A, Retana J (2002): Spatial patterns, temporal variability, and the role of multi-nest colonies in a monogynous Spanish desert ant. *Ecol Entomol 27, 7–15.* Dillier FX, Wehner R (2004): Spatio-temporal patterns of colony distribution in monodomous and polydomous species of North African desert ants, genus *Cataglyphis. Insect Soc 51, 186–196.* Dahbi A, Retana J, Lenoir A, Cerdá X (2008): Nest-moving by the polydomous ant *Cataglyphis iberica. J Ethol 26, 119–126.*—Modeling attempts show that polydomy increases foraging success in randomly scattered food source distributions: Schmolke A (2009): Benefits of dispersed central place foraging: an individual-based model of a polydomous ant colony. *Am Nat 173, 772–778.* Debout G, Schatz B, Elias M, McKay D (2007): Polydomy in ants: what we know, what we think we know, and what remains to be done. *Biol J Linn Soc 90, 319–348.*

20. Soroker V, Vienne C, Hefetz A, Nowbahari E (1994): The postpharyngeal gland as a 'gestalt' organ for nestmate recognition in the ant *Cataglyphis niger. Naturwissenschaften 81, 510–513.* Dahbi A, Cerdá X, Hefetz A, Lenoir A (1997): Adult transport in the ant *Cataglyphis iberica:* a means to maintain a uniform colonial odour in a species with multiple nests. *Physiol Entomol 22, 13–19.* Dahbi A, Hefetz A, Cerdá X, Lenoir A (1999): Trophallaxis mediates uniformity of colonial odor in *Cataglyphis iberica* ants (Hymenoptera,

Formicidae). *J Insect Behav 12, 559–567.* Lahav S, Soroker V, Vander Meer RK, Hefetz A (2001): Segregation of colony odor in the desert ant *Cataglyphis niger. J Chem Ecol 27, 927–943.*

21. Schmid-Hempel P (1983), *n11.*

22. Laboratory experiments indicate that secretions of the cloacal gland *(C. nigra)* or fecal deposits *(C. cursor)* may serve as nest-site markings: Mayade S, Cammaerts MC, Suzzoni JP (1993): Home-range marking and territorial marking in *Cataglyphis cursor* (Hymenoptera: Formicidae). *Behav Processes 30, 131–142.* Wenseleers T, Billen J, Hefetz A (2002): Territorial marking in the desert ant *Cataglyphis niger:* does it pay to play bourgeois? *J Insect Behav 15, 85–93.*—For factors determining levels of nest-site aggressiveness, see Knaden M, Wehner R (2003): Nest defense and conspecific enemy recognition in the desert ant *Cataglyphis fortis. J Insect Behav 16, 717–730.* Knaden M, Wehner R (2004): Path integration in desert ants controls aggressiveness. *Science 305, 60.*—For aggressive behavior in *C. bellicosa,* see Karavaiev V (1924), *n4.*

23. Attacks of *Zodarion* spiders on *Cataglyphis* species have been described by Harkness RD, Wehner R (1977), *n3.* See also Cushing PE, Santangelo RG (2002): Notes on the natural history and hunting behavior of an ant-eating zodariid spider (Arachnida, Araneae). *J Arachnol 30, 618–621.* Pekár S (2004): Predatory behavior of two European ant-eating spiders (Araneae, Zodariidae). *J Arachnol 32, 31–41.*—For the ecological prey-predator interactions between desert lizards and silver ants, see Wehner R (1989): Strategien gegen den Hitzetod. Thermophilie und Thermoregulation bei Wüstenameisen *(Cataglyphis bombycina). Jb Akad Wiss Lit Mainz 89, 101–112.* Wehner R, Marsh AC, Wehner S (1992), *n8.*—Ants are mentioned as a food source of the Lake Eyre Dragon by Mitchell FJ (1973): Studies on the ecology of the agamid lizard *Amphibolurus maculosus. Trans R Soc South Austr 97, 47–76.* Pedler R, Neilly H (2010): A re-evaluation of the distribution and status of the Lake Eyre Dragon *(Ctenophorus maculosus):* an endemic South Australian salt lake specialist. *South Austr Naturalist 84, 15–29.*—The attacks of parasitoid wasps of the genus *Kollasmosoma* on *Cataglyphis* workers are described in Huddleston T (1976): A revision of *Elasmosoma* (Hymenoptera, Braconidae) with two new species from Mongolia. *Ann Hist Nat Mus Natl Hung 68, 215–*

225. Durán JMG, van Achterberg C (2011): Oviposition behaviour of four parasitoids (Hymenoptera, Braconidae, Euphorinae, Neoneurini and Ichneumonidae, Hybrizontinae), with the description of three new European species. *ZooKeys 125.*

24. Foerster A (1850): Eine Centurie neuer Hymenopteren, zweite Decade. *Verh Nat Hist Ver Preuss Rheinl Westf 7, 485–500.*—The Greek name *Cataglyphis* (*kata*, κατα: down there, caudal; glyphis, γλυφίς: groove) is related to the fact that the original description was based on a male. It refers to the marked segmental constrictions identifying the male's gaster: "omnibus apice transversim impressis"; see Figures 2.14a and 2.30. In Greek architecture, compare the triglyph frieze of Doric temples.—Santschi F (1929): Étude sur les *Cataglyphis. Rev Suisse Zool 36, 25–70.* Agosti D (1990), *n4.*—Agosti's synopsis was complemented by a study by Alexander Radchenko, who used morphological characters of the workers to provide a species-group classification. Radchenko AG (2001): The phylogeny and faunogenesis of the genus *Cataglyphis* (Hymenoptera, Formicidae). *Ent Oboz 80, 885–895* [in Russian].—For references to species numbers, see *n4.*

25. Collingwood CA (1960): The 3rd Danish expedition to Central Asia. Zoological results 27. Formicidae (Insecta) from Afghanistan. *Vidensk Medd Dansk Nat For Kjobenhavn 123, 51–79.* Schneider P (1971): Vorkommen und Bau von Erdhügelnestern bei der Afghanischen Wüstenameise *Cataglyphis bicolor. Zool Anz 187, 202–213.* Collingwood CA (1985): Hymenoptera: Fam. Formicidae of Saudi Arabia. In: Büttiker W, Krupp F, eds., Fauna of Saudi Arabia, vol. 7, pp. 230–302. *Basel, Pro Entomologia.*—A *Cataglyphis* species belonging to the *aenescens* species group has been collected at Tashkurghan on the Silk Road in western Xinjiang at 3,010 m above sea level (E. P. Meyer and M. Frischknecht, *in litt.*). Obviously for thermoregulatory reasons, in the high-altitude gravel plains of the Dasht-e-Khoshi (Logar Province, Afghanistan) a *Cataglyphis* species of the *C. bicolor* species group constructs conspicuous gravel mounds. Schneider P (1971), *loc. cit.*

26. Moreau CS, Bell CD, Vila R, Archibald SB, Pierce NE (2006): Phylogeny of the ants: diversification in the age of angiosperms. *Science 312, 101–104.* Blaimer BB, Brady SG, Schulz TR, Lloyd MW, Fisher BL, Ward PS (2015): Phylogenomic methods outperform traditional multi-locus approaches in resolving deep evolutionary history: a case study of formicine ants. *BMC Evol Biol 15, 271.*

27. Dlussky GM (1981), *n6,* pp. 204–205; Agosti D (1990), *n4;* Radchenko AG (2001), *n24;* Agosti D (1994): The phylogeny of the ant tribe Formicini (Hymenoptera: Formicidae), with the description of a new genus. *Syst Ent 19, 93–117.*—On the basis of morphological characters *Alloformica* is considered to be phylogenetically closest to *Cataglyphis.* It was first described as a subgenus of *Proformica:* Dlussky GM (1969): Ants of the genus *Proformica* of the USSR and contiguous countries (Hymenoptera: Formicidae). *Zool J 48, 218–232* [in Russian]; but it was later raised to genus rank by Dlussky GM, Fedoseeva EB (1988): Origin and early evolution of ants. In: Pomomarenko AG, ed., Origin and Early Stages of Evolution in Ants, pp. 70–144. *Moscow, Akademiia Nauk* [in Russian].—Molecular data render *Rossomyrmex,* a slave-maker of several *Proformica* host species, the sister genus of *Cataglyphis.* Blaimer BB, Brady SG, Schulz TR, Lloyd MW, Fisher BL, Ward PS (2015), *n26.*—For arguments of the ancestral status of *C. emeryi,* see Kusnezov NN (1926): Die Entstehung der Wüstenameisenfauna Turkestans. *Zool Anz 65, 140–160.* Dlussky GM (1981), *n6,* pp. 38, 75, 82.

28. Emery C (1912): Der Wanderzug der Steppen- und Wüstenameisen von Zentral-Asien nach Süd-Europa und Nord-Afrika. *Zool Jb Suppl 15(1), 95–104.* Radchenko AG (2001), *n24.*—A Miocene origin of the *Cataglyphis* crown group is also supported by molecular data; see Blaimer BB, Brady SG, Schulz TR, Lloyd MW, Fisher BL, Ward PS (2015), *n26.*

29. Wehner R, Wehner S, Agosti D (1994): Patterns of biogeographic distribution within the *bicolor* species group of the North African desert ant, *Cataglyphis* Foerster 1850. *Senck biol 74, 163–191.*

30. The Sicilian cataglyphs reached the island from the north, from the Italian mainland. They belong to the *C. cursor* species group: Baroni Urbani C (1971): Catalogo delle specie di Formicidae d'Italia. Studi sulla mirmecofauna d'Italia, X. *Mem Soc Entomol Ital 50, 5–287.* The same most certainly also pertains to the cataglyphs of Corsica and Sardinia: Borowiec L (2014): Catalogue of ants of Europe, the Mediterranean Basin and adjacent regions (Hymenoptera: Formicidae). *Biol Silesiae 25(1–2), 56.*

31. For details on the biogeography and possible evolutionary history of *C. fortis,* see Wehner R (1983): Taxonomie, Funktionsmorphologie und Zoogeographie der saharischen Wüstenameise *Cataglyphis fortis* (Forel 1902) stat. nov. (Insecta: Hymenoptera: Formicidae). *Senck biol 64, 89–132.*

32. Knaden M, Tinaut A, Stökl J, Cerdá X, Wehner R (2012): Molecular phylogeny of the desert ant genus *Cataglyphis* (Hymenoptera: Formicidae). *Myrm News 16, 123–132.* Aron S, Mardulyn P, Leniaud L (2016): Evolution of reproductive traits in *Cataglyphis* desert ants: mating frequency, queen number, and thelytoky. *Behav Ecol Sociobiol 70, 1367–1379.*—*Alloformica* has not yet been included in the molecular phylogenies, in which *Rossomyrmex,* a slave-maker of *Proformica* ants, features as the sister group of *Cataglyphis:* Blaimer BB, Brady SG, Schulz TR, Lloyd MW, Fisher BL, Ward PS (2015), *n26.*—Earlier chemotaxonomic studies based on cuticular hydrocarbons and Dufour gland secretions have been performed in a few *Cataglyphis* species groups in some geographical regions: Dahbi A, Cerdá X, Hefetz A, Lenoir A (1996): Social closure, aggressive behavior, and hydrocarbon profiles in the polydomous ant *Cataglyphis iberica* (Hymenoptera, Formicidae). *J Chem Ecol 22, 2173–2186.* Oldham NJ, Morgan ED, Agosti D, Wehner R (1999): Species recognition from postpharyngeal gland contents of ants of the *Cataglyphis bicolor* group. *J Chem Ecol 25, 1383–1393.* Gökcen OA, Morgan ED, Dani FR, Agosti D, Wehner R (2002): Dufour gland contents of ants of the *Cataglyphis bicolor* group. *J Chem Ecol 28, 71–87.* Dahbi A, Hefetz A, Lenoir A (2008): Chemotaxonomy of some *Cataglyphis* ants from Morocco and Burkina Faso. *Biochem Syst Ecol 36, 564–572.*—A first molecular (mtDNA) phylogenetic analysis dealt with the Tunisian species of the *bicolor* species group and separated three parapatric species, which had previously been described as *C. bicolor.* Knaden M, Tinaut A, Cerdá X, Wehner S, Wehner R (2005): Phylogeny of three parapatric species of desert ants, *Cataglyphis bicolor, C. viatica,* and *C. savignyi:* a comparison of mitochondrial DNA, nuclear DNA, and morphological data. *Zoology 108, 169–177.*—Molecular mtDNA analyses were also used to determine the genetic population structures of *C. bicolor* and *C. mauritanica* in Tunisia. Knaden M, Wehner R (2006): Fundamental differences in life history traits of two species of *Cataglyphis* ants. *Front Zool 3, 21.*

33. The cataglyphoid ants were first described as *Camponotus constrictus* by Mayr GL (1868): Die Ameisen des baltischen Bernsteins. *Beitr Naturk Preuss Königl Phys Ökon Ges Königsberg 1, 1–102;* later as *Formica constricta* by Wheeler WM (1915): The ants of the Baltic amber. *Schr Phys Ökon Ges Königsberg 55, 1–142;* finally as *Cataglyphoides constrictus* by Dlussky GM (2008): Ants of the tribe Formicini (Hymenoptera, Formicidae) from Late Eocene amber of Europe. *Paleont J 42, 500–513.*—For micro-CT analyses, see Wehner R, Rabenstein R, Habersetzer J: Long-leggedness in cataglyphoid Baltic amber ants, *Palaeobiodiv Palaeoenviron, in press.*

34. Ward PS (2014): The phylogeny and evolution of ants. *Annu Rev Ecol Evol Syst 45, 23–43.*

35. The variety of sociogenetic systems that sharply contrasts with the uniformity of foraging strategies has been emphasized by Serge Aron. We owe him and his collaborators a growing number of analyses and conceptualizations of the various ways of sociogenetic organization and reproductive systems occurring in *Cataglyphis* species: Leniaud L, Pearcy M, Aron S (2013), *n8.* Eyer PA, Leniaud L, Darras H, Aron S (2013): Hybridogenesis through thelytokous parthenogenesis in two *Cataglyphis* desert ants. *Mol Ecol 22, 947–955.* Darras H, Leniaud L, Aron S (2014): Large-scale distribution of hybridogenetic lineages in a Spanish desert ant. *Proc R Soc B 281, 20132396.* Aron S, Mardulyn P, Leniaud L (2016), *n32.*—For reviews, see Lenoir A, Aron S, Cerdá X, Hefetz A (2009): *Cataglyphis* desert ants: a good model for evolutionary biology in Darwin's anniversary year. *Isr J Entomol 39, 1–32.* Boulay RR, Aron S, Cerdá X, Doums C, Graham P, Hefetz A, Monnin T (2017): Social life in arid environments: the case study of *Cataglyphis* ants. *Annu Rev Entomol 62, 305–321.*

36. It came as a real surprise when, in 1973, Henry Cagniant reported that in a cataglyph the workers were able to produce female offspring by so-called thelytokous parthenogenesis—a phenomenon discovered many decades earlier by George Onions in the Cape honeybee, *Apis mellifera capensis.* First considered to be a peculiar case in *C. cursor,* this way of worker and queen reproduction has now been found also in several other *Cataglyphis* species, so that one might wonder whether it is even an ancestral trait of the genus rather than a peculiarity. Onions GW (1912): South

African "fertile worker bees." *S Afr Agricult J 1, 720–728.* Cagniant H (1973): Apparition d'ouvrières à partir d'œufs pondus par les ouvrières chez la fourmi *Cataglyphis cursor. C R Acad Sci Paris D 277, 2197–2198.*—Several thelytokous *Cataglyphis* species are listed in Rabeling C, Kronauer DJC (2013): Thelytokous parthenogenesis in eusocial hymenoptera. *Annu Rev Entomol 58, 273–292.* See also Aron S, Mardulyn P, Leniaud L (2016), *n32.*

37. Sexual selection is widely considered to be a driving force for genital divergence. In this context polyandrous mating systems are often correlated with pronounced structural divergences of genitalia. Arnquist G (1998): Comparative evidence for the evolution of genitalia by sexual selection. *Nature 393, 784–786.* Hosken DJ, Stocley P (2004): Sexual selection and genital evolution. *Trends Ecol Evol 19, 87–93.*—See also Rowe L, Arnquist G (2011): Sexual selection and the evolution of genital shape and complexity in water striders. *Evolution 66, 40–54.* Simmons LW (2014): Sexual selection and genital evolution. *Austral Entomol 53, 1–17.*—Polyandry is widespread in the cataglyphs; see, e.g., Leniaud L, Pearcy M, Aron S (2013), *n8.* Aron S, Mardulyn P, Leniaud L (2016), *n32.*—Polyandry could be beneficial for a scavenger in increasing colony resistance to pathogens. It could also promote the division of labor. For the latter, see Eyer PA, Freyer J, Aron S (2013): Genetic polyethism in the polyandrous desert ant *Cataglyphis cursor. Behav Ecol 24, 144–151.*

38. Knaden M, Wehner R (2005): The coexistence of two large-sized thermophilic desert ants: the question of niche differentiation in *Cataglyphis bicolor* and *Cataglyphis mauritanica* (Hymenoptera: Formicidae). *Myrm News 7, 31–42.*

39. For example, Knaden M, Wehner R (2006), *n32.* Timmermans I, Grumiau L, Hefetz A, Aron S (2010): Mating system and population structure in the desert ant *Cataglyphis livida. Insect Soc 57, 39–46.* Leniaud L, Hefetz A, Grumiau L, Aron S (2011): Multiple mating and supercoloniality in *Cataglyphis* desert ants. *Biol J Linn Soc 104, 866–876.* Jowers MJ, Leniaud L, Cerdá X, Alasaad S, et al., Boulay RR (2013): Social and population structure in the ant *Cataglyphis emmae. PLoS ONE 8(9), e72941.* Peeters C, Aron S (2017): Evolutionary reduction of female dispersal in *Cataglyphis* desert ants. *Biol J Linn Soc 20, 1–13.*

40. Local analyses already show that species ranges can be extremely small, as in two *Cataglyphis* sibling species in southwestern Spain, or may form band-like zones adjacent to each other along a geographical gradient; see Figure 2.11 and Wehner R, Wehner S, Agosti D (1994), *n29.* Jowers MJ, Amor F, Ortega P, Lenoir A, et al., Galarza JA (2014): Recent speciation and secondary contact in endemic ants. *Mol Ecol 23, 2529–2542.*

41. Arnold G (1916): A monograph of the Formicidae of South Africa, part 2. *Ann S Afr Mus 14, 159–270.* Bolton B, Marsh AC (1989): The Afrotropical thermophilic ant genus *Ocymyrmex* (Hymenoptera: Formicidae). *J Nat Hist 23, 1267–1308.*—The latter authors not only provide a description of, and key to, the 37 known *Ocymyrmex* species, but also treat the behavioral biology and ecology of the genus.

42. For the discovery and description of ergatoid females in *Ocymyrmex,* see Bolton B (1981): A revision of six minor genera of Myrmicinae in the Ethiopian zoogeographical region. *Bull Brit Mus Nat Hist (Entomol) 45, 307–370.* Bolton B, Marsh AC (1989), *n41.* Forder JC, Marsh AC (1989): Social organisation and reproduction in *Ocymyrmex foreli. Insect Soc 36, 106–115.*—Ergatoid (small, wingless, worker-like) queens have been described also for one *Cataglyphis* species, the Iberian *C. tartessica* (in the original paper assigned to *C. floricola*), where they occur alongside small-winged (brachypterous) queens: Amor F, Ortega P, Jowers MJ, Cerdá X, et al., Boulay RR (2011): The evolution of worker-queen polymorphism in *Cataglyphis* ants: interplay between individual- and colony-level selections. *Behav Ecol Sociobiol 65, 1473–1482.*—For a survey of the occurrence of "ergatoid queens" in ants in general, see Peeters C (2012): Convergent evolution of wingless reproductives across all subfamilies of ants, and sporadic loss of winged queens (Hymenoptera: Formicidae). *Myrm News 16, 75–91.*

43. Marsh AC (1985): Microclimatic factors influencing foraging patterns and success of the thermophilic desert ant, *Ocymyrmex barbiger. Insect Soc 32, 286–296.* Wehner R (1987), *n12.* Sommer S, Weibel D, Blaser N, Furrer A, Wenzler NE, Rössler W, Wehner R (2013): Group recruitment in a thermophilic desert ant, *Ocymyrmex robustior. J Comp Physiol A 199, 711–722.*—The kind of recruitment observed in *Ocymyrmex* comes closest to that of *Camponotus socius* and *Myrmecocystus mimicus:* Hölldobler B (1971): Recruitment behavior in *Camponotus socius* (Hym. Formicidae). *Z vergl Physiol 75, 123–142.* Hölldobler B (1981): Foraging and spatiotemporal

territories in the honey ant *Myrmecocystus mimicus* (Hymenoptera: Formicidae). *Behav Ecol Sociobiol 9, 301–314.*—For *Melophorus,* see Schultheiss P, Schwarz S, Cheng K, Wehner R (2013): Foraging ecology of an Australian salt-pan desert ant (genus *Melophorus*). *Austr J Zool 60, 311–319.* Schultheiss P, Nooten SS (2013), *n11.*

44. For the large-scale distribution of *Melophorus,* see Shattuck SO (1999): Australian Ants. *Melbourne, CSIRO Publishing.* Andersen AN (2000): The Ants of Northern Australia: A Guide to the Monsoonal Fauna. *Collingwood, CSIRO Publishing.* Andersen AN (2003): Ant biodiversity in arid Australia: productivity, species richness, and community organization. *S Austr Mus Monogr Ser 7, 79–92.*—Alan Andersen estimated about 500 species in arid Australia alone. A recent revision based on morphometric, mtDNA and ncDNA lists a total of 93 species, of which 74 species are newly described. Heterick B, Castalanelli M, Shattuck SO (2017): Revision of the ant genus *Melophorus* (Hymenoptera, Formicidae): *ZooKeys 700, 1–420.*

45. The lifestyles of a variety of *Melophorus* species have been described by Greenslade PJM (1979): A Guide to Ants of South Australia. *Adelaide, South Australian Museum.* General predators: Andersen AN (2000), *n44,* p. 78.—Specialists on termite prey: McAreavey JJ (1947): New species of the genera *Prolasius* and *Melophorus* (Hymenoptera: Formicidae). *Mem Natl Mus Victoria 15, 7–27.*—Brood raiding of other ants: Clark J (1941): Australian Formicidae. Notes and new species. *Mem Natl Mus Victoria 12, 71–94.* Agosti D (1997): Two new enigmatic *Melophorus* species (Hymenoptera: Formicidae) from Australia. *J New York Ent Soc 105, 161–169.*—Specialized seed harvesters: Briese DT, Macauley BJ (1980): Temporal structure of an ant community in semiarid Australia. *Austr J Ecol 5, 121–134.* Morton SR, Davidson DW (1988): Comparative structure of harvester ant communities in arid Australia and North America. *Ecol Monogr 58, 19–38.*

46. Keith Christian and Stephen Morton, then at the CSIRO Field Station in Alice Springs, Australia, have provided a paradigmatic study in the thermal biology of desert ants: Christian KA, Morton SR (1992): Extreme thermophilia in a central Australian ant, *Melophorus bagoti. Physiol Zool 65, 885–905.*

47. Even though the nests of most *Cataglyphis fortis* colonies are located at the edges of the salt lakes, and the entrances of some especially large colonies are surrounded by conspicuous mounds of excavated soil, many nest openings are on level ground and thus must get flooded in winter. Nevertheless the colonies survive and have been recorded at the same place for many years. Dillier FX, Wehner R (2004), *n19.*—This poses the ecophysiological question of how these colonies protect themselves from getting drowned. Could they breathe from a plastron or physical gill, as some water beetles and bugs do? Computations show that for refilling, the air bubble must be rather small, in the size order of an individual ant rather than a nest chamber.

48. Heterick B, Castalanelli M, Shattuck SO (2017), *n44.* For a first behavioral ecological study of this species, see Schultheiss P, Schwarz S, Cheng K, Wehner R (2013), *n43.*

49. *Melophorus:* "Furnace ants" in Andersen AN (2002): Common names for Australian ants (Hymenoptera: Formicidae). *Austr J Entomol 41, 285–293.* "Ituny, ituny" in Conway JR (1992): Notes on the excavation of a nest of *Melophorus bagoti* in the Northern Territory, Australia (Hymenoptera: Formicidae). *J Austr Ent Soc 31, 247–248.*—*Cataglyphis:* "Englishmen" in Harkness RD, Wehner R (1977), *n3.*

50. The maximal values of solar radiation recorded under the conditions of our measurements (Figure 2.19) come close to the solar constant, the extra-atmospheric solar radiation per square meter (1.37 $kW \cdot m^{-2}$). About one month after the measurements were taken, i.e., at the summer solstice, the sun culminated nearly in the zenith. The asymmetric shape of the surface-temperature curve is due to the fact that at night heat is radiated off the ground, and that the heat capacity of sand is larger than that of air. The shape of the soil-temperature curve results nearly exclusively from the diffusion of heat within the soil, so that in this case the 24-h temperature cycle can be directly described by the heat diffusion equation complemented by a term, which represents the daily variation of solar radiation. Given the thermal diffusivity of sand (SiO_2) of about 1 $mm^2 \cdot s^{-1}$, such computations show that the maximum of the soil temperature recorded at a depth of 0.2 m should lag behind the temperature maximum at the surface by about 8–10 h. This is roughly what we observe.

51. For some examples, see *n10* and Heatwole H, Muir R (1989): Seasonal and daily activity of ants in the pre-Saharan steppe of Tunisia. *J Arid Environ 16, 49–67.* Wehner R, Wehner S (2011): Parallel evolution of thermophilia: daily and seasonal foraging patterns of heat-adapted desert ants, *Cataglyphis* and *Ocymyrmex* species. *Physiol Entomol 36, 271–281.* Amor F, Ortega P, Cerdá X, Boulay RR (2011): Solar elevation triggers activity in a thermophilic ant. *Ethology 117, 1031–1039.*—For *Melophorus,* see Christian KA, Morton SR (1992), *n46.*

52. Marsh AC (1985), *n43.* Wehner R, Wehner S (2011), *n51.*—The latter authors found that in the northern and southern winter *Cataglyphis desertorum* and *Ocymyrmex velox* exhibited similar daily activity patterns at the Tropic of Cancer and the Tropic of Capricorn, respectively *(unpubl. data).* Schultheiss P, Nooten SS (2013), *n11.*

53. The well-developed mandibular gland of *C. bombycina* produces citronellol and geraniol, which have been found to elicit 'alarm and recruitment' behavior. Hefetz A, Lloyd HA (1985): Mandibular gland secretions as alarm pheromones in two species of the desert ant *Cataglyphis. Z Naturforsch C 40, 665–666.*

54. For the influence of light on the onset and offset of the foraging activity in ants, see Hodgson ES (1955): An ecological study of the behavior of the leaf-cutting ant *Atta cephalotes. Ecology 36, 293–304.* Curtis BA (1985): Activity of the Namib Desert dune ant, *Camponotus detritus. S Afr J Zool 20, 41–48.* Narendra A, Reid SF, Hemmi JM (2010): The twilight zone: ambient light levels trigger activity in primitive ants. *Proc R Soc B 277, 1531–1538.* Amor F, Ortega P, Cerdá X, Boulay RR (2011): Solar radiation triggers foraging activity in a thermophilic ant. *Ethology 117, 1031–1039.*—Further note that for these thermophilic scavengers prey density will increase during the course of the day, and hence might shift foraging offset to higher temperatures than foraging onset.

55. Marsh AC (1985): *Ocymyrmex robustior [barbiger]:* Thermal responses and temperature tolerance in a diurnal desert ant, *Ocymyrmex barbiger. Physiol Zool 58, 629–636.* Wehner R, Wehner S (2011), *n51.*—*Melophorus bagoti:* Christian KA, Morton SR (1992), *n46.* Schultheiss P, Nooten SS (2013), *n11.*—*Cataglyphis bombycina:* Délye G (1968), *n11.*

56. For North American harvester ants, see Whitford WG, Ettershank G (1975): Factors affecting foraging activity in Chihuahuan desert harvester ants. *Environ Ecol 4, 689–696.* Whitford WG, Johnson P, Ramirez J (1976): Comparative ecology of the harvester ants *Pogonomyrmex barbatus* and *Pogonomyrmex rugosus. Insect Soc 23, 117–132.* Mehlhop P, Scott NJ (1983): Temporal patterns of seed use and availability in a guild of desert ants. *Ecol Entomol 8, 69–85.* Cole BJ, Smith AA, Huber ZJ, Wiernasz DC (2010): The structure of foraging activity in colonies of the harvester ant, *Pogonomyrmex occidentalis. Behav Ecol 21, 337–342.*—For ants in general, see the tabulated temperature preferences in Hölldobler B, Wilson EO (1990): The Ants. *Cambridge MA, Harvard University Press,* pp. 380–381.

57. Lutterschmidt WJ, Hutchinson VH (1997): The critical thermal maximum: history and critique. *Canad J Zool 75, 1561–1574.*—The technique of thermolimit respirometry developed by John Lighton and Robbin Turner circumvents the necessity of visual observation by using the steep decline in the rate of CO_2 emission and the loss of spiracular control as endpoint criteria. Lighton JRB, Turner RJ (2004): Thermolimit respirometry: an objective assessment of critical thermal maxima in two sympatric desert harvester ants, *Pogonomyrmex rugosus* and *P. californicus. J Exp Biol 207, 1903–1913.*—For a thorough treatment of some confounding effects and their ecological relevance, see Terblanche JS, Hoffmann AA, Mitchell KA, Rako L, Le Roux PC, Chown SL (2011): Ecologically relevant measures of tolerance to potentially lethal temperatures. *J Exp Biol 214, 3713–3725.*

58. CT_{max} values determined by the same version of the dynamic method applied in *Cataglyphis* are 51.5°C for *Ocymyrmex robustior* (Marsh AC (1985), *n55*) and 56.7°C; 56.9°C for *Melophorus bagoti* (Christian KA, Morton SR (1992), *n46;* Wehner R, Wehner S (2011), *n51,* respectively). See also de Bie G, Hewitt PH (1990): Thermal responses of the semi-desert zone ants *Ocymyrmex weitzeckeri* and *Anoplolepis custodiens. J Ent Soc S Afr 53, 65–73.*

59. Lutterschmidt WJ, Hutchinson VH (1997), *n57.* At present, the dynamic method is used most frequently. In *Cataglyphis* the static method has been applied by Délye G (1968), *n11,* p. 102. Heatwole H, Harrington S (1989): Heat tolerance of some ants and beetles from the pre-Saharan steppe of Tunisia. *J Arid Environ 16, 69–77.* Cerdá X, Retana J, Manzaneda A (1998): The role of competition by dominants and temperature in the foraging of subordinate species in

Mediterranean ant communities. *Oecologia 117, 404–412.* Clémencet J, Cournault L, Odent A, Doums C (2010): Worker thermal tolerance in the thermophilic ant *Cataglyphis cursor* (Hymenoptera, Formicidae). *Insect Soc 57, 11–15.*—In *Melophorus bagoti* static and dynamic methods have been applied in parallel by Christian KA, Morton SR (1992), *n46.*

60. Wu GC, Wright JC (2015): Exceptional thermal tolerance and water resistance in the mite *Paratarsotomus macropalpis* (Erythracaridae) challenge prevailing explanations of physiological limits. *J Insect Physiol 82, 1–7.*—In a suite of superb studies Jonathan Wright and his collaborators from Pomona College, Claremont, CA, have made this mite a model organism for studies in thermal tolerance, desiccation resistance, and locomotor behavior. See also *n81* and *n97.*

61. Heatwole H, Muir R (1979): Thermal microclimates in the pre-Saharan steppe of Tunisia. *J Arid Environ 2, 119–136.* Marsh AC (1985), *n43.* Wehner R, Marsh AC, Wehner S (1992), *n8.* Wehner R, Wehner S (2011), *n51.*

62. Shi NN, Tsai CC, Camino F, Bernard GD, Yu N, Wehner R (2015): Keeping cool: enhanced optical reflection and radiative heat dissipation in Saharan silver ants. *Science 349, 298–301.* Willot Q, Simonis P, Vigneron JP, Aron S (2016): Total internal reflection accounts for the bright color of the Saharan silver ant. *PLoS ONE 11(4), e152325.*

63. For heat shock responses in general, see, e.g., Moseley PL (1997): Heat shock proteins and heat adaptation of the whole organism. *J Appl Physiol 83, 1413–1417.* Feder ME, Hofmann GE (1999): Heat-shock proteins, molecular chaperons, and the stress response: evolutionary and ecological physiology. *Annu Rev Physiol 61, 243–282.* Evgen'ev MB, Garbuz DG, Zatsepina OG (2014): Heat Shock Proteins and Whole Body Adaptation to Extreme Environments. *Berlin, Springer.* Nguyen AD, Gotelli NJ, Cahan SH (2016): The evolution of heat shock protein sequences, cis-regulatory elements, and expression profiles in the eusocial Hymenoptera. *BMC Evol Biol 16, 15.*

64. Gehring W, Wehner R (1995): Heat shock protein synthesis and thermotolerance in *Cataglyphis,* an ant from the Sahara Desert. *Proc Natl Acad Sci USA 92, 2994–2998.* Willot Q, Gueydan C, Aron S (2017): Proteome stability, heat hardening, and heat-shock protein expression profiles in *Cataglyphis* desert ants. *J Exp Biol 220, 1721–1728.* Willot Q,

Mardulyn P, Defrance M, Gueydan C, Aron S (2018): Molecular chaperoning helps safeguarding mitochondrial integrity and motor functions in the Sahara silver ant *Cataglyphis bombycina. Sci Rep 8, 9220.*—For comparable studies in other ant species (genera *Formica, Aphaenogaster, Pogonomyrmex*), see Gehring W, Wehner R (1995), *loc. cit.* Ślipiński P, Pomorski JJ, Kowalewska K (2015): Heat shock protein expression during thermal risk exposure in the xerothermic ant *Formica cinerea. Sociobiology 62, 457–459.* Nguyen AD, Gotelli NJ, Cahan SH (2016), *n63.*

65. For comparison, lizard species inhabiting the Middle Asian deserts are characterized by higher constitutive HSG expressions and higher thermal thresholds for HSG induction than species from more mesic habitats. Ulmasov KA, Shammakov S, Karaev K, Evgen'ev MB (1992): Heat shock proteins and thermoresistance in lizards. *Proc Natl Acad Sci USA 89, 1666–1670.* Zatsepina OG, Ulmasov KA, Beresten SF, Molodtsov VB, Rybtsov SA, Evgen'ev MB (2000): Thermotolerant desert lizards characteristically differ in terms of heat-shock system regulation. *J Exp Biol 203, 1017–1025.*—For ecological significance, see also Hoffmann AA, Chown SL, Clusella-Trullas S (2013): Upper thermal limits in terrestrial ectotherms: how constrained are they? *Funct Ecol 27, 934–949.*

66. Principal aspects of microclimatology dealing with the climatic conditions close to the ground are treated in Oke TR (1978): Boundary Layer Climates. *London, Methuen.* Geiger R, Aron RH, Todhunter P (2009): The Climate near the Ground. 2nd ed. *Lanham MD, Rowman and Littlefield.* Implications for the microenvironmental physiology of insects are discussed in Willmer PG (1982): Microclimate and the environmental physiology of insects. *Adv Insect Physiol 16, 1–57.* Casey TM (1988): Thermoregulation and heat exchange. *Adv Insect Physiol 20, 119–146.* Gates DM (2003): Biophysical Ecology. *Mineola NY, Dover.*

67. The operative environmental temperature (OET) is an integrative measure of all heat gains and losses due to physical (radiant and convective) thermal exchanges between animal and environment, provided that neither metabolic heat production nor evaporative heat losses occur. For a small arthropod the latter two factors are normally insignificant compared to radiative and convective heat exchanges. The concept of the OET was introduced by Bakken GS, Gates

DM (1975): Heat-transfer analysis of animals: some implications for field ecology, physiology, and evolution. In: Gates DM, Schmerl RB, eds., Perspectives in Biophysical Ecology, pp. 255–290. *New York, Springer.*—For technical aspects, see Unwin DM (1980): Microclimate Measurement for Ecologists. *London, Academic Press.*—The microenvironmental physiology of insects is comprehensively treated in Willmer PG (1982), *n66;* Casey TM (1988), *n66.*

68. Actual measurements of T_e in thermophilic desert ants have been provided by Marsh AC (1985), *n55.* Christian KA, Morton SR (1992), *n46.* Wehner R, Marsh AC, Wehner S (1992), *n8.*

69. Heat-stressed honey bees display 'tongue-lashing' behavior by extending their tongues and evaporating collected nectar; see, e.g., Heinrich B (1980): Mechanisms of body temperature regulation in honeybees, *Apis mellifera.* I. Regulation of head temperature. *J Exp Biol 85, 61–72.*

70. For general treatments and literature surveys of water relations in insects and other arthropods, see the monographs and reviews by Edney EB (1977): Water Balance in Land Arthropods. *Berlin, Springer.* Noble-Nesbitt J (1991): Cuticular permeability and its control. In: Binnington K, Retnakaran A, eds., Physiology of the Insect Epidermis, pp. 252–283. *Melbourne, CSIRO Publishing.* Hadley NF (1994): Water Relations of Terrestrial Arthropods. *San Diego, Academic Press.*

71. For measurements in *Pogonomyrmex, Messor,* and *Forelius* species, see Quinlan MC, Lighton JRB (1999): Respiratory physiology and water relations of three species of *Pogonomyrmex* harvester ants (Hymenoptera: Formicidae). *Physiol Entomol 24, 293–302.* Schilman PE, Lighton JRB, Holway DA (2005): Respiratory and cuticular water loss in insects with continuous gas exchange: comparison across five ant species. *J Insect Physiol 51, 1295–1305.* Johnson RA, Kaiser A, Quinlan MC, Sharp W (2011): Effect of cuticular abrasion and recovery on water loss rates in queens of the desert harvester ant *Messor pergandei. J Exp Biol 214, 3495–3506.*—For a general treatment of the temperature-humidity envelope in ants, see Hölldobler B, Wilson EO (1990), *n56,* pp. 379–381.

72. In several *Cataglyphis* species the cuticular water loss per unit surface area is significantly smaller and the relative content of saturated, longer-chain hydrocarbons is considerably higher than in *Myrmica rubra,* an ant inhabiting the temperate climatic zone: Lenoir A, Aron S, Cerdá X, Hefetz A (2009), *n35.*—An interspecific difference in water loss rates has been found in size-matched small specimens of *C. rosenhaueri* and *C. velox.* Cerdá X, Retana J (2000): Alternative strategies by thermophilic ants to cope with extreme heat: individual versus colony level traits. *Oikos 89, 155–163.*—In general, transpiration rates in xeric as compared to mesic ant species are given in, e.g., Lighton JRB, Feener DH (1989): Water-loss rate and cuticular permeability in foragers of the desert ant *Pogonomyrmex rugosus. Physiol Zool 62, 1232–1256.* Hood WG, Tschinkel WR (1990): Desiccation resistance in arboreal and terrestrial ants. *Physiol Entomol 15, 23–35.*—The water-proofing properties of cuticular lipids are comprehensively treated in Lockey KH (1988): Lipids of the insect cuticle: origin, composition and function. *Comp Biochem Physiol B 89, 595–645.* Buckner JS (1993): Cuticular polar lipids of insects. In: Stanley-Samuelson DW, Nelson DR, eds., Insect Lipids: Chemistry, Biochemistry and Biology, pp. 227–270. *Lincoln, University of Nebraska Press.* Gibbs AG (1998): Water-proofing properties of cuticular lipids. *Am Zoologist 38, 471–482.*

73. In *Pogonomyrmex barbatus* harvester ants, foragers have higher amounts of linear as compared to branched alkanes than ants of indoor life stages. Wagner D, Tissot M, Gordon D (2001): Task-related environment alters the cuticular hydrocarbon composition of harvester ants. *J Chem Ecol 27, 1805–1819*

74. See *n16* in Chapter 1.

75. Lighton JRB, Wehner R (1993): Ventilation and respiratory metabolism in the thermophilic desert ant, *Cataglyphis bicolor* (Hymenoptera: Formicidae). *J Comp Physiol B 163, 11–17.* In *C. albicans* a sudden increase of water loss occurs at an ambient temperature of 50°C: Délye G (1967): Physiologie et comportement de quelques fourmis (Hym. Formicidae) du Sahara en rapport avec les principaux facteurs du climat. *Insect Soc 14, 323–338.*—Desert harvester ants, *Pogonomyrmex,* lose motor coordination and die when their hydration level reaches a critical value of 52%: Lighton JRB, Feener DH (1989), *n72.* See also Feener DH, Lighton JRB (1991): Is foraging in the desert ant, *Messor pergandei* (Hymenoptera: Formicidae), limited by water? *Ecol Entomol 16, 183–191.*—For the grasshopper study, see Rourke BC (2000): Geographic and altitudinal variation in water balance

and metabolic rate in a Californian grasshopper, *Melanoplus sanguinipes*. *J Exp Biol 203, 2699–2712.*

76. A critical review is given in Chown SL, Gibbs AG, Hetz SK, Klok CJ, Lighton JRB, Marais E (2006): Discontinuous gas exchange in insects: a clarification of hypotheses and approaches. *Physiol Biochem Zool 79, 333–343.*—More recent evidence for the hygrig (water conservation) hypothesis is provided, e.g., by Schimpf NG, Matthews PGD, Wilson RS, White CR (2009): Cockroaches breathe discontinuously to reduce respiratory water loss. *J Exp Biol 212, 2773–2780, quot.* p. 2773.—For the sleep hypothesis, see Matthews PGD, White CR (2011): Discontinuous gas exchange in insects: is it all in their heads? *Am Nat 177, 130–134.*—Of course, the various hypotheses are not mutually exclusive. Matthews PGD, Terblanche JS (2015): Evolution of the mechanisms underlying insect respiratory gas exchange. *Adv Insect Physiol 49, 1–24.*

77. For the shape of propodeal spiracles in desert ants, see Dlussky GM (1981), *n6*, pp. 91–92. In *Cataglyphis*, respiration patterns have been measured independently and simultaneously in different body parts. Lighton JRB, Fukushi T, Wehner R (1993): Ventilation in *Cataglyphis bicolor*: regulation of carbon dioxide release from thoracic and abdominal spiracles. *J Insect Physiol 39, 687–699.*

78. Remarks on the extreme celerity of the thermophiles date back to the eighteenth century. The quotations are from Fabricius JC (1787): Mantissa Insectorum. Tom 1. *Hafniae, C. G. Proft, quot.* p. 308. Latreille PA (1802): Histoire naturelle des fourmis, et recueil de mémoires et d'observations sur les abeilles, les araignées, les faucheurs, et autres insectes. *Paris, Barrois Père, quot.* p. 174. Cornetz V (1910): Trajets de fourmis et retours au nid. *Mém Inst Gen Psychol, Sect Psychol Zool 2, 1–167, quot.* p. 93. Karavaiev V (1912): Ameisen aus Tunesien und Algerien, nebst einigen unterwegs in Italien gesammelten Arten. *Russk Ent Obozr 12, 1–22, quot.* p. 18. Arnold G (1916), *n41, quot.* p. 194.

79. In the cataglyphs, the highest walking speeds as revealed by high-speed video recordings (500 frames · s^{-1}) at surface temperatures $T_s = 45°C–50°C$ are 855 mm · s^{-1} (*C. bombycina*, M. Wittlinger, *in litt.*). Speeds of up to 1.0 m per second have been clocked in this species at $T_s = 55°C–60°C$ along a 1.0 m calibration distance (R. Wehner and S. Wehner, *unpubl.*).

80. Nicolson SW, Bartholomew GA, Seely MK (1984): Ecological correlates of locomotion speed, morphometrics and body temperature in three Namib Desert tenebrionid beetles. *S Afr J Zool 19, 131–134.* Full RJ, Tu MS (1991): Mechanics of rapid running insects: two-, four- and six-legged locomotion. *J Exp Biol 156, 215–231.* Kamoun S, Hogenhout SA (1996): Flightlessness and rapid terrestrial locomotion in tiger beetles of the *Cicindela* subgenus *Rivacindela* from saline habitats of Australia (Coleoptera: Cicindelidae). *Coleopt Bull 50, 221–230.*

81. Rubin S, Young MHY, Wright JC, Whitaker DL, Ahn AN (2016): Exceptional running and turning performances in a mite. *J Exp Biol 219, 676–685.*

82. For considerations on how running speed scales with body mass, see Schmidt-Nielsen K (1984): Scaling: Why Is Animal Size so Important? *Cambridge, Cambridge University Press,* p. 174. Biewener AA (2003): Animal Locomotion. *Oxford, Oxford University Press,* p. 62. Rubin S, Young MHY, Wright JC, Whitaker DL, Ahn AN (2016), *n81,* figure 7.

83. Sommer S, Wehner R (2012): Leg allometry in ants: extreme long-leggedness in thermophilic species. *Arthropod Struct Dev 41, 71–77.* M. Wittlinger, *in litt.*

84. According to the size-grain hypothesis the rugosity of the ground has been influential in the evolution of leg allometries in ants, and in insects in general. For the pros and cons of this hypothesis, see Kaspari M, Weiser MD (1999): The size-grain hypothesis and interspecific scaling of ants. *Funct Ecol 13, 530–538.* Teuscher M, Brändle M, Traxel V, Brandl R (2009): Allometry between leg and body length of insects: lack of support for the size-grain hypothesis. *Ecol Entomol 34, 718–724.*

85. Cornetz V (1933): L'homme et la fourmi. *Alger, L'Association des Écrivains Algériens, quot.* p. 52.

86. McMeeking RM, Arzt E, Wehner R (2012): *Cataglyphis* desert ants improve their mobility by raising the gaster. *J Theor Biol 297, 17–25.*—The forward and upward shift of the center of mass associated with the upward tilt of the gaster has already been proposed by Arnoldi KV (1964): The most highly specialized running ants and the phaetons of the genus *Cataglyphis* (Hymenoptera, Formicidae) in the fauna of the Soviet Union. *Zool Zh 43, 1800–1814* [in Russian].—Details of changes in the skeletal and muscular

system associated with the ability to raise the gaster are given in Dlussky GM (1981), *n6,* pp. 75–83.

87. It has been claimed that the primary function of raising the gaster is to protect the vital organs contained in it from high temperatures. Cerdá X (2001), *n7.*—Of course, both hypotheses are not mutually exclusive. However, in terms of ultimate reasoning it is more likely that the behavioral trait of elevating the gaster has evolved for increasing locomotor agility. It also occurs in several other very agile ants, e.g., in the arboreal weaver ants *Oecophylla.*

88. Two publications can be regarded as the classical accounts on inter-leg coordination in insect walking: Hughes GM (1952): The coordination of insect movements. I. The walking movements of insects. *J Exp Biol 29, 267–284.* Wilson DM (1966): Insect walking. *Annu Rev Entomol 11, 103–122.*—For later work, see, e.g., Graham D (1985): Pattern and control of walking in insects. *Adv Insect Physiol 18, 31–140.* Full RJ, Tu MS (1990): Mechanics of six-legged runners. *J Exp Biol 148, 129–146.* Ting LH, Blickhan R, Full RJ (1994): Dynamic and static stability in hexapod runners. *J Exp Biol 197, 251–269.* Cruse H, Dürr V, Schilling M, Schmitz J (2009): Principles of insect locomotion. In: Arena P, Patane L, eds., Spatial Temporal Patterns from Action-Oriented Perception in Roving Robots, pp. 43–96. *Berlin, Springer.*—On wood ants: Reinhardt L, Weihmann T, Blickhan R (2009): Dynamics and kinematics of ant locomotion: do wood ants climb on level surfaces? *J Exp Biol 212, 2426–2435.* Reinhardt L, Blickhan R (2014): Level locomotion in wood ants: evidence for grounded running. *J Exp Biol 217, 2358–2370.*

89. After the original work of Christoph Zollikofer, the walking kinematics of the cataglyphs have been investigated by Matthias Wittlinger, now at the University of Freiburg, Germany, and his students. Zollikofer CPE (1988): Vergleichende Untersuchungen zum Laufverhalten von Ameisen. *PhD Thesis, University of Zürich.* Shorter versions of this study have appeared in Zollikofer CPE (1994a): Stepping patterns in ants. I. Influence of speed and curvature. *J Exp Biol 192, 95–106.* Zollikofer CPE (1994b): Stepping patterns in ants. II. Influence of body morphology. *J Exp Biol 192, 107–118.* Zollikofer CPE (1994c): Stepping patterns in ants. III. Influence of load. *J Exp Biol 192, 119–127.*—For conclusions drawn from high-speed video recordings of walking cataglyphs, see

Wittlinger M, Wehner R, Wolf H (2007): The desert ant odometer: a stride integrator that accounts for stride length and walking speed. *J Exp Biol 210, 198–207.* Seidl T, Wehner R (2008): Walking on inclines: how do desert ants monitor slope and step length. *Front Zool 5, 8.* Steck K, Wittlinger M, Wolf H (2009): Estimation of homing distance in desert ants, *Cataglyphis fortis,* remains unaffected by disturbance of walking behaviour. *J Exp Biol 212, 2893–2901.* Wittlinger M, Wolf H (2013): Homing distance in desert ants, *Cataglyphis fortis,* remains unaffected by disturbances of walking behaviour and visual input. *J Physiol 107, 130–136.* Wahl VL, Pfeffer SE, Wittlinger M (2015): Walking and running in the desert ant *Cataglyphis fortis. J Comp Physiol A 201, 645–656.* Pfeffer SE, Wahl VL, Wittlinger M (2016): How to find home backwards? Locomotion and inter-leg coordination during rearward walking of *Cataglyphis fortis* desert ants. *J Exp Biol 219, 2110–2118.*

90. Hoyt DF, Taylor CR (1981): Gait and energetics of locomotion in horses. *Nature 292, 239–240.*—In quadruped mammals the transition from trot to gallop, and at a lower speed from walk to trot, occurs in a discontinuous way. At any one speed the energetically most efficient gait is used. Even when in some insect species at low speeds other than tripod leg coordination patterns occur, the transitions are always continuous. Schilling M, Hoinville T, Schmitz J, Cruse H (2013): Walknet, a bio-inspired controller for hexapod walking. *Biol Cybern 107, 397–419.* Wosnitza A, Bockemühl T, Dübbert M, Scholz H, Büschges A (2013): Inter-leg coordination in the control of walking speed in *Drosophila. J Exp Biol 216, 480–491.*

91. See Simon Sponberg and Richard J. Full's comprehensive treatment of locomotor control on rough terrain in a large (2 g) cockroach, *Blaberus discoidalis.* Sponberg S, Full RJ (2008): Neuromechanical response of muscular-skeletal structures in cockroaches during rapid running on rough terrain. *J Exp Biol 211, 433–446.*

92. For the development and use of highly sensitive three-dimensional, ultra-miniature force platforms, see Bartsch MS, Federle W, Full RJ, Kenny TW (2007): A multiaxis force sensor for the study of insect biomechanics. *J Microelectromech Syst 16, 709–718.* Reinhardt L, Blickhan R (2014): Ultra-miniature force plate for measuring triaxial forces in the micronewton range. *J Exp Biol 217,*

704–710. Reinhardt L, Blickhan R (2014): Level locomotion in wood ants: evidence for grounded running. *J Exp Biol 217, 2358–2370.*

93. Wöhrl T, Reinhardt L, Blickhan R (2017): Propulsion in hexapod locomotion: how do desert ants traverse slopes? *J Exp Biol 220, 1618–1625.*

94. For comparison, see the masterly studies of Walter Federle at the University of Cambridge, UK: Federle W, Brainerd EL, McMahon TA, Hölldobler B (2001): Biomechanics of the movable pretarsal adhesive organ in ants and bees. *Proc Am Acad Sci USA 98, 6215–6220.* Federle W, Endlein T (2004): Locomotion and adhesion: dynamic control of adhesive surface contact in ants. *Arthropod Struct Dev 33, 67–75.* Endlein T, Federle W (2008): Walking on smooth or rough ground: passive control of pretarsal attachment in ants. *J Comp Physiol A 194, 49–60.* Endlein T, Federle W (2015): On heels and toes: how ants climb with adhesive pads and tarsal friction hair arrays. *PLoS ONE 10, e0141269.*

95. In a partly autobiographical book originally titled *Racing the Antelope,* biologist and ultra-endurance runner Bernd Heinrich provides a lucid account on the physiology and evolution of animal and human locomotion. Heinrich B (2001): Why We Run: A Natural History. *New York, HarperCollins.*—Definition: Stride length is the distance between two successive footfall points of the same leg (see Figure 2.32b). In bipedal walking it consists of the length of the right step plus the length of the left step. Hence, stride length equals twice the step length.

96. The longest foraging trajectory described as yet for a cataglyph (*C. fortis,* 1238 m) has been recorded by GPS methods and hence will not have included all small scale details, so that the ant's actual path length and walking speed might have certainly been larger than indicated in the text. Bühlmann C, Graham P, Hansson BS, Knaden M (2014), *n10.*

97. Stride frequency scales negatively with body mass over a wide range of body sizes in vertebrates and invertebrates (with an allometric exponent of about −0.17): Full RJ (1989): Mechanics and energetics of terrestrial locomotion: bipeds to polypeds. In Wieser W, Gnaiger E, eds., Energy Transformations in Cells and Organisms, pp. 175–182. *Stuttgart, Thieme.* Full RJ, Tu MS (1991), *n80.* Ting LH, Blickhan R, Full RJ (1994), *n88.* Wu GC, Wright JC, Whitaker DL, Ahn AN (2010): Kinematic evidence for superfast locomotory muscle in two species of teneriffiid mites. *J Exp Biol 213, 2551–2556.*—Even the maximal stride frequencies ever reported for an animal (135 s^{-1} in the mite *Paratarsotomus macropalpis*) are largely in accord with scaling predictions: Rubin S, Young MHY, Wright JC, Whitaker DL, Ahn AN (2016), *n81.*

98. In terms of speed the pronghorn antelope is rivaled, and even outperformed, only by the cheetah, which was clocked to reach 104 km·h^{-1} when lured to run in a straight line over 200 m, and 93 km·h^{-1} when running freely in the wild. In contrast to antelopes, cheetahs, which prey on antelopes, reach their maximum speeds only during short, highly accelerated bursts of chasing activity. Sharp NCC (1997): Timed running speed of the cheetah *(Acinonyx jubatus). J Zool 241, 493–494.* Wilson AM, Lowe JC, Roskilly K, Hudson P, Golabek KA, McNutt JW (2013): Locomotion dynamics of hunting in wild cheetahs. *Nature 498, 185–189.*—Data on maximal mass-specific O$_2$ consumption rates of pronghorn antelopes and other mammals are given in Lindstedt SL, Hokanson JF, Wells DJ, Swain SD, Hoppeler H, Navarro V (1991): Running energetics in the pronghorn antelope. *Nature 353, 748–749.* Lindstedt SL, Schaeffer PJ (2002): Use of allometry in predicting anatomical and physiological parameters of mammals. *Lab Animals 36, 1–19.*—The current "best guess" of a pronghorn's standard (resting) metabolism is 0.2 mlO$_2$·kg^{-1}·s^{-1} (S. L. Lindstedt, *in litt.*).

99. After in the 1970s Torben Jensen and Mogens Nielsen from the University of Aarhus, Denmark, had applied micro-Warburg techniques to determine respiratory rates in a number of ant species, in the past three decades John Lighton has profoundly developed and refined techniques for measuring metabolic rates, e.g., by recording CO$_2$ production rates in individual ants by using high-resolution, flow-through respirometry systems: Jensen TF, Nielsen MG (1975): The influence of body size and temperature on worker ant respiration. *Nat Jutland 18, 21–25.* Bartholomew GA, Lighton JRB, Feener DH (1988): Energetics of trail-running, load carriage, and emigration in the column-raiding army ant, *Eciton hamatum. Physiol Zool 61, 57–68.* Lighton JRB (2009): Measuring Metabolic Rates: A Manual

for Scientists. *Oxford, Oxford University Press.* Halsay LG, Lighton JRB (2011): Flow-through respirometry applied to chamber systems: pros and cons, hints and tips. *Comp Biochem Physiol A 158, 265–275.*—Data on ventilation patterns and respiratory metabolism of *Cataglyphis bicolor* are given in Lighton JRB, Wehner R (1993), *n75.* Lighton JRB, Fukushi T, Wehner R (1993), *n77.*—The metabolic rates of *C. velox* lie in the same ballpark as the ones recorded in *C. bicolor:* Cerdá X, Retana J (2000), *n72.*

100. The hypothesis that lowered standard metabolic rates are an adaptation to arid conditions received some early support from studies on desert beetles and ants. Bartholomew GA, Lighton JRB, Louw GN (1985): Energetics of locomotion and patterns of respiration in tenebrionid beetles from the Namib Desert. *J Comp Physiol B 155, 155–162.* Lighton JRB, Bartholomew GA (1988): Standard energy metabolism of a desert harvester ant, *Pogonomyrmex rugosus:* effects of temperature, body mass, group size, and humidity. *Proc Natl Acad Sci USA 85, 4765–4769.*—A survey of all currently available data in ants including *Cataglyphis* does not lend further support to this hypothesis. Lighton JRB, Wehner R (1993), *n75.* Vogt JT, Appel AG (1999): Standard metabolic rate of the fire ant, *Solenopsis invicta:* effects of temperature, mass, and caste. *J Insect Physiol 45, 655–666.* Hou C, Kaspari M, Vander Zanden HB, Gillooly JF (2010): Energetic basis of colonial living in social insects. *Proc Natl Acad Sci USA 107, 3634–3638.* Käfer H, Kovac H, Stabentheiner A (2012): Resting metabolism and critical thermal maxima of vespine wasps (*Vespula* sp.). *J Insect Physiol 58, 679–689,* including data on some ant species.

101. A thorough analysis based on a comprehensive survey of standard and flight metabolic rates in insects has been provided by Niven JE, Scharlemann JPW (2005): Do insect metabolic rates at rest and during flight scale with body mass? *Biol Lett 1, 346–349.*—It supports the 'aerobic capacity hypothesis,' which is discussed for insects in Reinhold K (1999): Energetically costly behaviour and the evolution of resting metabolic rate in insects. *Funct Ecol 13, 217–224.*—See also Käfer H, Kovac H, Stabentheiner A (2012), *n100.*

102. For metabolic rates in walking ants, see Jensen TF, Holm-Jensen I (1980): Energetic cost of running in workers of three ant species, *Formica fusca* L. *Formica rufa* L., and *Camponotus herculeanus* L. (Hymenoptera, Formicidae). *J Comp Physiol 137, 151–156.* Lighton JRB, Feener DH (1989): A comparison of energetics and ventilation of desert ants during voluntary and forced locomotion. *Nature 342, 174–175.* Lighton JRB, Fukushi T, Wehner R (1993), *n77.* Roces F, Lighton JRB (1995): Larger bites of leaf-cutting ants. *Nature 373, 392–393.*—The energetic costs of walking differ very little between running on inclines and level running. Lipp A, Wolf H, Lehmann FO (2005): Walking on inclines: energetics of locomotion in the ant *Camponotus. J Exp Biol 208, 707–719.*—For an excellent study based on treadmill running in cockroaches, see Full RJ, Tullis A (1990): Capacity for sustained terrestrial locomotion in an insect: energetics, thermal dependence, and kinematics. *J Comp Physiol B 160, 573–581.*—The maximal aerobic capacity ever recorded in any nonflying animal is reached in leaf-cutter ants, *Atta sexdens,* during leaf cutting. In this case the ants increase their metabolic rate 31 times over their standard metabolic rate. Roces F, Lighton JRB (1995), *loc. cit.*

103. The relative importance of physiological and behavioral adaptations of small ectotherms to extremes of temperature and aridity has long occupied physiologists and ecologists; see Edney EB (1974): Desert arthropods. In: Brown GW, ed., Desert Biology, vol. 2, pp. 311–384. *New York, Academic Press.* Stevenson RD (1985): The relative importance of behavioral and physiological adjustments controlling body temperatures in terrestrial ectotherms. *Am Nat 126, 362–386.* Seely MK (1989): Desert invertebrate physiological ecology: is anything special? *S Afr J Sci 85, 266–270.*

104. Forel A (1878): Etudes myrmécologiques en 1878 avec l'anatomie du gésier des fourmis. *Bull Soc Vaud Sci Nat 15, 337–392.* Emery C, Forel A (1879): Catalogue des Formicides d'Europe. *Mitt Schweiz Ent Ges 5, 441–481.*—The situation was later resolved by Wheeler WM (1908): Honey ants, with a revision of the American Myrmecocysti. *Bull Am Mus Nat Hist 24, 345–397.*

105. Even though in *Myrmecocystus* the diurnal species (e.g., *M. mimicus,* $CT_{max} = 47.7°C–48.4°C$) exhibit higher thermal tolerances than the nocturnal ones (e.g., *M. mexicanus,* $CT_{max} = 43.7°C–44.7°C$), they cannot tolerate the extremely high temperatures at which the cataglyphs are active.

They stop foraging at $T_s = 45°C$. Kay CAR (1978): Preferred temperatures of desert honey ants (Hymenoptera: Formicidae). *J Therm Biol 3, 213–217.*

106. Diet preferences of *Myrmecocystus* species: Snelling RR (1976): A revision of the honey ants, genus *Myrmecocystus* (Hymenoptera: Formicidae). *Nat Hist Mus Los Angeles Cty Sci Bull 24, 1–163.* Conway JR (1980): Foraging behavior of the honey ant, *Myrmecocystus mexicanus,* in Colorado. *Trans Illinois State Acad Sci 72, 81–93.* Cole BJ, Haight K, Wiernasz DC (2001): Distribution of *Myrmecocystus mexicanus* (Hymenoptera: Formicidae): association with *Pogonomyrmex occidentalis* (Hymenoptera: Formicidae). *Ecol Pop Biol 94, 59–63.*—Foraging territories as studied in *M. mimicus:* Hölldobler B (1976): Tournaments and slavery in a desert ant. *Science 192, 912–914.* Hölldobler B (1981), *n43.*

107. Cole AC (1968): *Pogonomyrmex* Harvester Ants: A Study of the Genus in North America. *Knoxville, University of Tennessee Press, quot.* p. 3.—A vivid illustration of the high population densities, which *Pogonomyrmex* species can reach, is given by Diane Wiernasz and Blaine Cole, who mapped the nest sites of more than 1,300 colonies of one species in the Colorado saltbush desert. Wiernasz DC, Cole BJ (1995): Spatial distribution of *Pogonomyrmex occidentalis:* recruitment, mortality and overdispersion. *J Anim Ecol 64, 519–527.*—Foraging and recruitment: Hölldobler B (1971): Homing in the harvester ant *Pogonomyrmex badius. Science 171, 1149–1151.* Hölldobler B (1974): Home range orientation and territoriality in harvesting ants. *Proc Natl Acad Sci USA 71, 3274–3277.* Hölldobler B (1976): Recruitment behavior, home range orientation and territoriality in harvester ants, *Pogonomyrmex. Behav Ecol Sociobiol 1, 3–44.*—Even though *Pogonomyrmex* ants are seed harvesters par excellence, they include dead insects and even live termites into their diet. In most species such insect matter represents less than 5% of all collected items, e.g., Rissing SW (1988): Dietary similarity and foraging range of two seed-harvester ants during resource fluctuations. *Oecologia 75, 362–366.* Morehead SA, Feener DH (1998): Foraging behavior and morphology: seed selection in the harvester ant, *Pogonomyrmex. Oecologia 114, 548–555.* Cole BJ, Haight K, Wiernasz DC (2001), *n106.*—However, some species are quite omnivorous and may collect seeds and insect corpses in about equal amounts. Traniello JFA, Beshers SN (1991):

Polymorphism and size-pairing in the harvester ant *Pogonomyrmex badius:* a test of the ecological release hypothesis. *Insect Soc 38, 121–127.* Ferster B, Traniello JFA (1995): Polymorphism and foraging behavior in *Pogonomyrmex badius* (Hymenoptera: Formicidae): worker size, foraging distance, and load size association. *Environ Entomol 24, 673–678.*—Thermal tolerances are given in Whitford WG, Ettershank G (1975), *n56,* and Lighton JRB, Turner RJ (2004), *n57.*—Blaine Cole and his colleagues provide a detailed analysis of intercolony differences in the thermal foraging ranges of *P. occidentalis.* Cole BJ, Smith AA, Huber ZJ, Wiernasz DC (2010), *n56.* On extremely hot summer days, *P. rugosus* even switches to nocturnal foraging: Hölldobler B (1976), *loc. cit.;* R. A. Johnson, *in litt.*

108. Philippi RA (1860): Reise durch die Wüste Atacama. *Halle, E. Anton, quot.* p. 7. Rauh W (1985): The Peruvian-Chilean deserts. In: Evenari M, Noy-Meir I, Goodall DW, eds., Hot Deserts and Arid Shrublands: Ecosystems of the World, vol. 12A, pp. 239–267. *Amsterdam, Elsevier.* Gómez-Silva B (2009): On the limits imposed to life by the hyperarid Atacama Desert in northern Chile. In: Basiuk VA, ed., Astrobiology, pp. 1–13. *Stevenson Ranch CA, American Scientific Publishers.*

109. Wilhelm Goetsch, who has worked substantially on the biogeography and biology of Chilean ants, described *Dorymyrmex goetschi* as the "real desert species," extremely long-legged and fast running, even "resembling the desert solifuges." Goetsch W (1932): Beiträge zur Biologie südamerikanischer Ameisen. I. Teil: Wüstenameisen. *Z Morph Ökol Tiere 25, 1–30, quot.* p. 2. Goetsch W (1935): Biologie und Verbreitung chilenischer Wüsten-, Steppen- und Waldameisen. *Zool Jb Syst Ökol Geogr Tiere 67, 235–318, quot.* p. 239. Dlussky GM (1981), *n6,* pp. 49–50.—However, like other species of this genus (e.g., *D. antarcticus*) *D. goetschi* is not a thermophile. It restricts its foraging to $T_s < 50°C$. Hunt JH (1974): Temporal activity patterns in two competing ant species (Hymenoptera: Formicidae). *Psyche 81, 237–242.* Torres-Contreras H, Vásquez RA (2007): Spatial heterogeneity and nestmate encounters affect locomotion and foraging success in the ant *Dorymyrmex goetschi. Ethology 113, 76–86.* Wehner R, Wehner S, *unpubl. data.*—Several species of the genus, e.g., *D. exsanguis,* are active even at extremely low temperatures: Bestelmeyer BT (1997): Stress tolerance in

some Chacoan dolichoderine ants: implications for community organization and distribution. *J Arid Environ 35, 297–310.*—Furthermore, as regard to diet, insects comprise only 27%–47% of all collected items, so that *D. goetschi* has also been listed as a harvester ant, see Medel RG, Vásquez RA (1994): Comparative analysis of harvester ant assemblages of Argentinian and Chilean arid zones. *J Arid Environ 26, 363–371.*

110. Snelling RR, Hunt JH (1975): The ants of Chile. *Rev Chil Entomol 9, 63–129.* Heatwole H (1996): Ant assemblages at their dry limits: the northern Atacama Desert, Peru, and the Chott El Djerid, Tunisia. *J Arid Environ 33, 449–456.*

111. Heatwole H, Muir R (1991): Foraging abundance and biomass of ants in the pre-Saharan steppe of Tunisia. *J Arid Environ 21, 337–350.*—Granivorous ants are the dominant trophic group in all desert ant communities. Especially detailed studies on species richness and species distribution, dietary relationships, and food availability have been performed by Alan Marsh in 27 ant species of the Namib Desert gravel plains. Marsh AC (1985): Forager abundance and dietary relationships in a Namib Desert ant community. *S Afr J Zool 20, 197–203.* Marsh AC (1986): Ant species richness along a climatic gradient in the Namib Desert. *J Arid Environ 11, 235–241.* Marsh AC (1987): The foraging ecology of two Namib Desert harvester ant species. *S Afr J Zool 22, 130–136.*

112. The dominance status is usually determined by the outcome of interspecific competitive interactions at artificial baits. In ants of semiarid habitats consistent correlations have been found between thermal tolerance and submissive behavior at feeding sites. Bestelmeyer BT (1997), *n109.* Cerdá X, Retana J, Manzaneda A (1998), *n59.* Bestelmeyer BT (2000): The trade-off between thermal tolerance and behavioural dominance in a subtropical South American ant community. *J Anim Ecol 69, 998–1009.*

113. Even though meat ants are strictly diurnal foragers, they cease foraging when temperatures at ant height reach 43.5°C. Greenaway P (1981): Temperature limits to trailing activity in the Australian arid-zone meat ant *Iridomyrmex purpureus* form *viridiaenus. Austr J Zool 29, 621–630.*—In contrast, *Melophorus bagoti* commences foraging only when ant-height temperatures exceed 43.9°C. Christian KA, Morton SR (1992), *n46.*—In turn, in competition with the behaviorally dominant invasive Argentine ant, *Linepithema humile,* the meat ants are the subordinate ones. They occupy feeding sites only at surface temperatures of $T_s = 30°C–40°C$, not tolerated by the Argentine ants, which occupy the baiting stations at even lower temperatures ($T_s = 18°C–25°C$): Thomas ML, Holway DA (2005): Condition-specific competition between invasive Argentine ants and Australian *Iridomyrmex. J Anim Ecol 74, 532–542.*

3. Finding Directions

1. It is a matter of debate whether the magnetic compass used for marine navigation was invented independently in China and Europe, or whether it had been passed from China to the West. May WE (1955): Alexander Neckam and the pivoted compass needle. *J Navigation 8, 283–284.* Kreutz BM (1973): Mediterranean contribution to the medieval mariner's compass. *Techn Cult 14, 367–383.* May WE (1981): Were compasses used in antiquity? *J Navigation 34, 414–423.*—Literally, 'navigation' means to steer a ship to its destination; derived from Latin: *navis,* ship; *agere,* to drive.

2. Dominique Arago published his skylight observations two years after he had made them: Arago DFJ (1811): Mémoire sur une modification remarquable qu'éprouvent les rayons lumineux dans leur passage à travers certains corps diaphanes, et sur quelques autres nouveaux phénomènes d'optique. *Mém Cl Sci Math Phys Inst France 1, 93–134.* Brewster D (1863): On the polarization of light by rough and white surfaces. *Trans R Soc Edinburgh 23, 205–210,* quot. p. 210. Brewster D (1847): On the polarization of the atmosphere. *Phil Mag 31, 444–454.* Tyndall J (1888): The sky. *Forum 1888/2, 595–603.*

3. It has been a long-held view, derived in the 1970s from spectroscopic measurements in frog rod outer segments, that in vertebrate photoreceptors the rhodopsin molecules are completely free to rotate. This view has recently been challenged. The dimerization and higher-order oligomerization of rhodopsin molecules could form the basis of dichroism at least in some morphologically distinct types of cells, e.g., double cones in fish. Roberts NW (2014): Polarization vision of fishes. In: Horváth G, ed., Polarized Light and Polarization Vision in Animal Sciences, pp. 225–247. *Berlin, Springer.*

4. Santschi F (1923): L'orientation sidérale des fourmis, et quelques considérations sur leurs différentes possibilités d'orientation. *Mém Soc Vaud Sci Nat 4, 137–175.* Von Frisch K (1949): Die Polarisation des Himmelslichts als orientierender Faktor bei den Tänzen der Bienen. *Experientia 5, 142–148.* Von Frisch K (1965): Tanzsprache und Orientierung der Bienen. *Berlin, Springer,* pp. 384–421.

5. In chronological order: Wehner R, Duelli P (1971): The spatial orientation of desert ants, *Cataglyphis bicolor,* before sunrise and after sunset. *Experientia 27, 1364–1366.* Kirschfeld K (1972): Die notwendige Anzahl von Rezeptoren zur Bestimmung der Richtung des elektrischen Vektors linear polarisierten Lichtes. *Z Naturforsch C 27, 578–579.* Zolotov V, Frantsevich L (1973): Orientation of bees by polarized light of a limited area of the sky. *J Comp Physiol 85, 25–36.* Von Helversen O, Edrich W (1974): Der Polarisationsempfänger im Bienenauge: ein Ultraviolettrezeptor. *J Comp Physiol 94, 33–47.* Van der Glas HW (1974): Polarization induced colour patterns: a model of the perception of the polarized skylight by insects. *Netherl J Zool 25, 476–505.* Brines ML, Gould JL (1979): Bees have rules. *Science 206, 571–573.*—For a historical treatment, see Wehner R (2014): Polarization vision: a discovery story. In: Horváth G, ed., Polarized Light and Polarization Vision in Animal Sciences, pp. 3–25. *Berlin, Springer.* Wehner R (2016): Early ant trajectories: spatial behaviour before behaviourism. *J Comp Physiol A 202, 247–266.*

6. Wehner R, Bernard GD, Geiger E (1975): Twisted and non-twisted rhabdoms and their significance for polarization detection in the bee. *J Comp Physiol 104, 225–245.* Herrling PL (1975): Topographische Untersuchungen zur funktionellen Anatomie der Retina von *Cataglyphis bicolor. PhD Thesis, University of Zürich.*

7. Strutt JW (Lord Rayleigh) (1871): On the light from the sky, its polarization and colour. *Phil Mag 41, 107–120, 274–279.* Coulson KL (1988): Polarization and Intensity of Light in the Atmosphere. *Hampton VA, Deepak.*

8. Brines ML, Gould JL (1982): Skylight polarization patterns and animal orientation. *J Exp Biol 96, 69–91.* Pomozi I, Horváth G, Wehner R (2001): How the clear-sky angle of the polarization pattern continues underneath clouds: full-sky measurements and implications for animal orientation. *J Exp Biol 204, 2933–2942.*

9. Horváth G, Wehner R (1999): Skylight polarization as perceived by desert ants and measured by video polarimetry. *J Comp Physiol A 184, 1–7, 347–349.* Labhart T (1999): How polarization-sensitive interneurons of crickets see the polarization pattern of the sky: a field study with an optoelectronic model neurone. *J Exp Biol 202, 757–770.* Pomozi I, Horváth G, Wehner R (2001), *n8.*—E-vector patterns under various environmental conditions are discussed in several chapters of Horváth G, ed. (2014): Polarized Light and Polarization Vision in Animal Sciences. *Berlin, Springer.* The polarization pattern of the moonlit sky is portrayed in Gál J, Horváth G, Barta A, Wehner R (2001): Polarization of the moonlit clear night sky measured by full-sky imaging polarimetry at full noon: comparison of the polarization of moonlit and sunlit skies. *J Geophys Res 106, 22647–22653.* –It can be used by nocturnal dung beetles, *Scarabaeus zambesianus,* for steering straight courses: Dacke M, Nilsson DE, Scholtz CH, Byrne M, Warrant EJ (2003): Insect orientation to polarized moonlight. *Nature 424, 33.*

10. Rozenberg GV (1966): Twilight. A Study in Atmospheric Optics. *New York, Plenum Press.* Coulson KL (1988), *n7.* Cronin TW, Warrant EJ, Greiner B (2006): Celestial polarization pattern during twilight. *Appl Optics 45, 5582–5589.* —For the twilight experiment in *Cataglyphis,* see Figure 3.25. Wehner R, Duelli P (1971), *n5.*—For bull ants and dung beetles active at nocturnal and crepuscular times, respectively, see Reid SF, Narendra A, Hemmi JM, Zeil J (2011): Polarized skylight and landmark panorama provide night-active bull ants with compass information during route following. *J Exp Biol 214, 363–370.* Freas CA, Narendra A, Lemesle C, Cheng K (2017): Polarized light use in the nocturnal bull ant, *Myrmecia midas. R Soc Open Sci 4, 170598.* Dacke M, Nordström P, Scholtz CH (2003): Twilight orientation to polarized light in the crepuscular dung beetle *Scarabaeus zambesianus. J Exp Biol 206, 1535–1543.*

11. For different versions of the experimental trolley, with which the cataglyphs' sky compass has been unraveled behaviorally, see Duelli P, Wehner R (1973): The spectral sensitivity of polarized light orientation in *Cataglyphis bicolor* (Formicidae, Hymenoptera). *J Comp Physiol 86, 37–53.* Lanfranconi B (1982): Kompassorientierung nach dem rotierenden Himmelsmuster bei der Wüstenameise *Cataglyphis bicolor. PhD Thesis, University of Zürich.* Fent K

(1985): Himmelsorientierung bei der Wüstenameise *Cataglyphis bicolor*: Bedeutung von Komplexaugen und Ocellen. *PhD Thesis, University of Zürich.* Müller M (1989): Mechanismus der Wegintegration bei *Cataglyphis fortis. PhD Thesis, University of Zürich.* Lebhardt F, Ronacher B (2014): Interactions of the polarization and the sun compass in path integration of desert ants. *J Comp Physiol A 200, 711–720.* See also Wehner R (2019): The *Cataglyphis* Mahrèsienne. Fifty years of *Cataglyphis* research at Mahrès. *J Comp Physiol A, in press.*

12. Duelli P, Wehner R (1973), *n11.* Von Helversen O, Edrich W (1974), *n5.*—For references to other insects, see review in Wehner R, Labhart T (2006): Polarization vision. In: Warrant EJ, Nilsson DE, eds., Invertebrate Vision, pp. 291–348. *Cambridge, Cambridge University Press.*

13. See the tabulated data in Coulson KL (1988), *n7.* In addition, Lythgoe JN (1979): The Ecology of Vision. *Oxford, Clarendon Press.* Johnsen S, Kelber A, Warrant EJ, Sweeney AM, et al., Widder EA (2006): Crepuscular and nocturnal illumination and its effects on color perception by the nocturnal hawkmoth *Deilephia elpenor. J Exp Biol 209, 789–800.*

14. Brines ML, Gould JL (1982), *n8.* Pomozi I, Horváth G, Wehner R (2001), *n8.* Barta A, Horváth G (2004): Why is it advantageous for animals to detect celestial polarization in the ultraviolet? Skylight polarization under clouds and canopies is strongest in the UV. *J Theor Biol 226, 429–437.* Wang X, Gao J, Fan Z (2014): Empirical corroboration of an earlier theoretical resolution to the UV paradox of insect polarized skylight orientation. *Naturwissenschaften 101, 95–103.*

15. For the DRA as the retinal input area of the polarized-light compass, see Wehner R (1982): Himmelsnavigation bei Insekten. Neurophysiologie und Verhalten. *Neujahrsbl Naturf Ges Zürich 184, 1–132.* Fent K (1985), *n11.* The stringent necessity/sufficiency tests have been performed by M. Müller and R. Wehner *(unpubl. data).* As shown later in the text, there is a slight twist in this story. Bees and ants tested with small patches of the sky do not expect individual e-vectors to occur exactly at the positions at which they are present in the sky (Figure 3.13b). This refinement has already been taken into account in evaluating the data shown in Figure 3.9e, f.

16. Wehner R, Müller M (1985): Does interocular transfer occur in visual navigation by ants? *Nature 315, 228–229.*—After occlusion of one eye, the cataglyphs consistently deviate by a certain angular amount toward the seeing side. This deviation has been taken into account in the data analysis; for details, see Wehner R (1997): The ant's celestial compass system: spectral and polarization channels. In: Lehrer M, ed., Orientation and Communication in Insects, pp. 145–185. *Basel, Birkhäuser.*

17. Peter Duelli recorded the inclination of the ant's head and body by using a macrozoom camera mounted inside the trolley and focused on the trolley center, where the ant was steadily walking in its home direction. Duelli P (1974): Polarisationsmusterorientierung bei der Wüstenameise *Cataglyphis bicolor* (Formicidae, Hymenoptera). *PhD Thesis, University of Zürich.* Duelli P (1975): A fovea for e-vector orientation in the eye of *Cataglyphis bicolor* (Formicidae, Hymenoptera). *J Comp Physiol 102, 43–56.* Of course, in this context the term 'fovea' has not its usual meaning of a zone of acute vision.—Duelli's results are in accord with a previous set of experiments in which the ants could see the entire sky, but had parts of their eyes painted over. Weiler R, Huber M (1972): The significance of different eye regions for astromenotactic orientation in *Cataglyphis bicolor.* In: Wehner R, ed., Information Processing in the Visual Systems of Arthropods, pp. 287–294. *Berlin, Springer.*

18. In the experiments described in the text the sun was occluded, but intensity/spectral gradients were available alongside the polarization gradients and could have been used by the polarization insensitive parts of the contralateral eye.

19. For various theoretical treatments of the compass problem, see Kirschfeld K (1972), *n5.* Kirschfeld K, Lindauer M, Martin H (1975): Problems of menotactic orientation according to the polarized light of the sky. *Z Naturforsch C 30, 88–90.* Brines ML (1978): Skylight polarization patterns as cues for honey bee orientation: physical measurements and behavioral experiments. *PhD Thesis, Rockefeller University.* Brines ML, Gould JL (1979), *n5.*—All these theoretical concepts were developed before the dorsal rim area as the exclusive input channel of the ant's and bee's polarized-light compass had been discovered.

20. Müller M (1989), *n11*. Wehner R, Müller M (2006): The significance of direct sunlight and polarized skylight in the ant's celestial system of navigation. *Proc Natl Acad Sci USA 103, 12575–12579.*

21. Wehner R (1991): Visuelle Navigation—Kleinstgehirn-Strategien. *Verh Dtsch Zool Ges 84, 89–104,* table 1 therein.

22. For *Cataglyphis,* see Fent K (1985), *n11.* Fent K (1986): Polarized skylight orientation in the desert ant *Cataglyphis. J Comp Physiol A 158, 1–7.*—For *Apis,* see Rossel S, Wehner R (1984): How bees analyse the polarization patterns in the sky. *J Comp Physiol A 154, 607–615.* Wehner R, Rossel S (1985): The bee's celestial compass—a case study in behavioural neurobiology. *Fortschr Zool 31, 11–53.* Rossel S, Wehner R (1986): Polarization vision in bees. *Nature 323, 128–131.*—For the bees' waggle dance, see Couvillon MJ (2012): The dance legacy of Karl von Frisch. *Insect Soc 59, 297–306.*

23. In contrast to the cataglyphs, the dancing honeybees tested with the same natural sky window apply the convention to interpret e-vectors in the solar half of the sky as invariably lying in the antisolar half. This behavior may be a recruitment dance convention.

24. Réaumur RAF (1740): Mémoires pour servir à l'histoire des insectes. Tome 5. *Paris, L'Imprimerie Royale,* p. 287. Homann H (1924): Zum Problem der Ocellenfunktion bei den Insekten. *Z Vergl Physiol 1, 541–578.* Stockhammer K (1959): Die Orientierung nach der Schwingungsrichtung linear polarisierten Lichtes und ihre sinnesphysiologischen Grundlagen. *Ergeb Biol 21, 23–56.* Von Frisch K (1965), *n4,* p. 412–413.

25. Wilson M (1978): The functional organization of the insect ocelli. *J Comp Physiol 124, 297–316, quot.* p. 297. Warrant EJ, Kelber A, Wallén R, Wcislo WT (2006): Ocellar optics in nocturnal and diurnal bees and wasps. *Arthropod Struct Dev 35, 293–305, quot.* p. 29.

26. Hesse R (1908): Das Sehen der niederen Tiere. *Jena, G. Fischer,* pp. 44–45. Wilson M (1978), *n25.* Stange G (1981): The ocellar component of flight equilibrium control in dragonflies. *J Comp Physiol 141, 335–347.* Taylor CP (1981): Contribution of compound eyes and ocelli to steering of locusts in flight. I. Behavioural analysis. *J Exp Biol 93, 1–18.*—For an upsurge of interest in ocellar function, see Taylor GK, Krapp HG (2007): Sensory systems and flight stability: what do insects measure and why? *Adv Insect Physiol 34, 231–316.*—There are recent indications that in some flying insects ocelli might even be able to resolve images, as in these cases the focal plane of the lens does not lie too far off the retina: Berry RP, Stange G, Warrant EJ (2007): Form vision in the insect dorsal ocelli: an anatomical and optical analysis of the dragonfly median ocellus. *Vision Res 47, 1394–1409.* Warrant EJ, Kelber A, Wallén R, Wcislo WT (2006), *n25.* Hung YS, Ibbotson MR (2014): Ocellar structure and neural innervation in the honeybee. *Front Neuroanat 19(8), 6.*

27. Fent K (1985), *n11.* Fent K, Wehner R (1985): Ocelli: a celestial compass in the desert ant, *Cataglyphis. Science 228, 192–194.*—Recently, it has been shown that also in *Melophorus bagoti* the ocelli can derive compass information from the sunfree sky (with the sun occluded), but whether polarized light is the decisive cue, remains to be elucidated. Schwarz S, Albert L, Wystrach A, Cheng K (2011): Ocelli contribute to encoding of celestial compass information in the Australian desert ant *Melophorus bagoti. J Exp Biol 214, 901–906.*—In a short note William Wellington described some observations that bumblebees, *Bombus terricola,* in which the compound eyes had been occluded, could select their homeward courses, at least over short distances. Spectral / intensity gradients have not been considered, let alone excluded, as cues. Wellington WG (1974): Bumblebee ocelli and navigation at dusk. *Science 183, 550–551.*

28. Mote MI, Wehner R (1980): Functional characteristics of photoreceptors in the compound eye and ocellus of the desert ant, *Cataglyphis bicolor. J Comp Physiol 137, 63–71.* Geiser FX (1985): Elektrophysiologische Charakterisierung der Ocellen von *Apis mellifera* und *Cataglyphis bicolor.* PhD Thesis, University of Zürich.

29. Karl Kral from the University of Graz, Austria, was the first to measure the angular distribution of the axes of rhabdom cross sections in entire retinae of insect ocelli. He found random orientation in honeybees, but a significant preference orientation in a species of vespid wasps: Kral K (1978): Orientierung der Rhabdome in den Ocellen der Honigbiene *Apis mellifica carnica* und der Erdwespe *Vespa vulgaris. Zool Jb Physiol 82, 263–271.*—In the ocelli of honeybee workers recent studies revealed a gross fanlike pattern as well as a prevalence of receptors sensitive to vertically polarized light in regions looking at the horizon:

Ribi WA, Warrant EJ, Zeil J (2011): The organization of honeybee ocelli: regional specializations and rhabdom arrangements. *Arthropod Struct Dev 40, 509–520*. Ogawa Y, Ribi WA, Zeil J, Hemmi JM (2017): Regional differences in the preferred e-vector orientation of honeybee ocellar photoreceptors. *J Exp Biol 220, 1701–1708*.—For a revival of interest in such measurements in a variety of insects, see, e.g., Zeil J, Ribi WA, Narendra A (2014): Polarisation vision in ants, bees and wasps. In: Horváth G, ed., Polarized Light and Polarization Vision in Animal Sciences, pp. 41–60. *Berlin, Springer*. Narendra A, Ramirez-Esquivel F, Ribi WA (2016): Compound eye and ocellar structure for walking and flying modes of locomotion in the Australian ant, *Camponotus consobrinus*. *Sci Rep 6, 22331*.

30. For example, Ribi WA, Warrant EJ, Zeil J (2011), *n29*.

31. Link E (1909): Über die Stirnaugen der hemimetabolen Insekten. *Zool Jb Anat Ontog 27, 281–376, quot*. p. 372. Rowell CHF, Reichert H (1986): Three descending interneurons reporting deviation from course in the locust. II. Physiology. *J Comp Physiol A 158, 775–794*. Parsons MM, Krapp HG, Laughlin SB (2010): Sensor fusion in identified visual interneurons. *Curr Biol 20, 624–628*.

32. When moving in uneven and cluttered terrain, the cataglyphs are subject to quite substantial inadvertent pitch movements of their heads. Ardin P, Mangan M, Wystrach A, Webb B (2015): How variation in head pitch could affect image matching algorithms for ant navigation. *J Comp Physiol A 201, 585–597*.

33. For early work in honeybees in this context, see Schricker B (1965): Die Orientierung der Honigbiene in der Dämmerung. Zugleich ein Beitrag zur Frage der Ocellenfunktion bei Bienen. *Z Vergl Physiol 49, 420–458*. Gould JL (1975): Communication of distance information by honeybees. *J Comp Physiol 104, 161–173*.—Nocturnal insects are characterized by especially large lenses and rhabdoms; see, e.g., Berry RP, Wcislo WT, Warrant EJ (2011): Ocellar adaptations for dim light vision in a nocturnal bee. *J Exp Biol 214, 1283–1293*.

34. Note that the presence of ocelli in the worker caste is not special for *Cataglyphis*, but is shared with several formicine ants, e.g., with *Formica* and *Melophorus*, while nearly all myrmicine workers (including *Ocymyrmex*) lack ocelli. As Figure 3.14c shows, some rudiments seem to occur in *Ocymyrmex* ergatoids.

35. For early descriptions of underwater polarization, see Waterman TH (1954): Polarization patterns in submarine illumination. *Science 120, 927–932*. Ivanoff A (1974): Polarization measurements in the sea. In: Jerlov NG, Nielsen ES, eds., Optical Aspects of Oceanography, pp. 151–175. *London, Academic Press*. Jerlov NG (1976): Optical Oceanography. *Amsterdam, Elsevier*.—Contrast enhancement has been treated by Lythgoe JN (1971): Vision. In: Woods JD, Lythgoe JN, eds., Underwater Science, pp. 103–139. *London, Oxford University Press*. Shashar N, Johnsen S, Lerner A, Sabbah S, Chiao CC, Mathger LM, Hanlon RT (2011): Underwater linear polarization: physical limitations to biological functions. *Phil Trans R Soc B 366, 649–654*. Marshall J, Cronin TW (2014): Polarization vision of crustaceans. In: Horváth G, ed., Polarized Light and Polarization Vision in Animal Sciences, pp. 171–216. *Berlin, Springer*.

36. The proto-arthropod fauna has been extensively described and portrayed in Whittington HB (1985): The Burgess Shale. *New Haven CT, Yale University Press*. Hou X, Ramsköld L, Bergström J (1991): Composition and preservation of the Chengjiang fauna—a Lower Cambrian soft-bodied biota. *Zool Scr 20, 395–411*. Conway Morris S (1998): The Crucible of Creation. The Burgess Shale and the Rise of Animals. *Oxford, Oxford University Press*.

37. On trilobite and early crustacean compound eyes, see Clarkson ENK (1997): The eye: morphology, function and evolution. In: Moore RC, Kaesler RL, eds., Treatise on Invertebrate Paleontology, Part O, Arthropoda 1, Trilobita, revised, pp. 114–132. *Boulder CO, Geological Society of America*. Levi-Setti R (2014): The Trilobite Book: A Visual Journey. *Chicago, University of Chicago Press*. Clarkson ENK, Levi-Setti R, Horváth G (2006): The eyes of trilobites: the oldest preserved visual system. *Arthropod Struct Dev 35, 247–259*. Schoenemann B (2013): The eyes of a tiny 'Orsten' crustacean—a compound eye at receptor level? *Vision Res 76, 89–93*.

38. Coulson KL, Dave JV, Sekera Z (1960): Tables Related to Radiation Emerging from a Planetary Atmosphere with Rayleigh Scattering. *Berkeley, University of California Press*. Coemans MAJM, Vos Hzn JJ, Nuboer JFW (1994): The relation between celestial colour gradients and the position

of the sun, with regard to the sun compass. *Vision Res 34, 1461–1470.*

39. Even though the computations performed by the spectral opponent unit in Figure 3.17b are hypothetical, interneurons of the type assumed here have indeed been found in the visual systems of honeybees, bumblebees, and some other insects, e.g., by Chittka L, Beier W, Hertel H, Steinmann E, Menzel R (1992): Opponent colour coding is a universal strategy to evaluate the photoreceptor input in Hymenoptera. *J Comp Physiol A 170, 545–563.* Yang EC, Lin HC, Yung YS (2004): Patterns of chromatic information processing in the lobula of the honeybee. *J Insect Physiol 50, 913–925.* Paulk A, Dacks A, Gronenberg W (2009): Color processing in the medulla of the bumblebee (Apidae: *Bombus impatiens*). *J Comp Neurol 513, 41–456.* Schnaitmann C, Garbers C, Wachtler T, Tanimoto H (2013): Color discrimination with broadband photoreceptors. *Curr Biol 23, 2375–2382.* Kelber A, Henze MJ (2013): Colour vision: parallel pathways intersect in *Drosophila*. *Curr Biol 23, R1043–R1045.*—Let us assume that the azimuth-specific responses of S-neurons are integrated along celestial meridians. Then one would predict that visual interneurons with palisade-like meridional dendritic trees occurred in the dorsal but not ventral half of the visual system; candidates could be types of transmedulla neurons found in honeybees and bumblebees: Pfeiffer K, Kinoshita M (2012): Segregation of visual inputs from different regions of the compound eye in two parallel pathways through the anterior optic tubercle of the bumblebee (*Bombus ignites*). *J Comp Neurol 520, 212–229.* Zeller M, Held M, Bender J, Berz A, Heinloth T, et al., Pfeiffer K (2015): Transmedulla neurons in the sky compass network of the honeybee (*Apis mellifera*) are a possible site of circadian input. *PLoS ONE 10(12), e0143244.*—"Azimuth tuning" for unpolarized light stimuli has been described for medullar neurons in locusts: el Jundi B, Pfeiffer K, Homberg U (2011): A distinct layer of the medulla integrates sky compass signals in the brain of an insect. *PLoS One 6(11), e27855.*—With respect to unpolarized beams of light, honeybees take a long-wavelength stimulus for the sun and expect a short-wavelength stimulus to lie within the antisolar half of the sky. Edrich W, Neumeyer C, von Helversen O (1979): "Anti-sun orientation" of bees with regards to a field of ultraviolet light. *J Comp Physiol 134, 151–157.* Brines ML, Gould JL (1979),

n5. Rossel S, Wehner R (1984): Celestial orientation in bees: the use of spectral cues. *J Comp Physiol A 155, 605–613.*—This behavior directly results from the S-responses as computed in Figure 3.17b.

40. The phototactic response superimposed on the ants' compass bearings can be computationally excluded by performing all experiments in a twin-type way with the sun being either to the left or, by the same angular amount, to the right of the training direction. Wehner R (1997), *n16.* For a systematic study of this phototactic effect under monochromatic conditions, see Lanfranconi B (1982), *n11.*

41. Santschi F (1911): Observations et remarques critiques sur le mécanisme de l'orientation chez les fourmis. *Rev Suisse Zool 19, 305 338.* Santschi's 'mirror experiment' shows that the ants can keep a straight course by using the sun as a visual cue, but it does not show that the ant can set arbitrary directions within an earthbound system of reference centered about the position of the nest.

42. Notice that the spatial sampling interval of the compound eye is given by the interommatidial angle $\Delta\varphi$ (the so-called divergence angle), i.e., by the angle between the optical axes of two adjacent ommatidia. It defines the minimum angular distance by which two positions of a point-light source such as the sun can be resolved. As for the sun compass, $\Delta\varphi$ must be related to the azimuthal distance $\Delta\alpha$ between the two positions of the sun. This depends strongly on the elevation μ_s of the sun (Figure 3.18a): the larger μ_s, the larger $\Delta\alpha$, and hence the lower the precision with which the azimuthal position of the sun can be measured. Moreover, note that in *Cataglyphis* $\Delta\varphi$ increases steadily from 3.0° at the equator to 7.0° at the most dorsal part of the eye (see Figure 1.9b).

43. For details, see Lanfranconi B (1982), *n11.* Müller M, Wehner R (2007): Wind and sky as compass cues in desert ant navigation. *Naturwissenschaften 94, 589–594.*—Even though the ants' precision (the scatter of the data about the mean) varies with the elevation of the sun, the accuracy with which the ants selected their homeward courses (the deviation of the mean value from the true homeward course) does not depend on solar elevation.

44. For the sun and polarized-light compass set in competition and replacing each other, see Lebhardt F, Ronacher B (2014), *n11.* Wystrach A, Schwarz S, Schultheiss P, Baniel A,

Cheng K (2014): Multiple sources of celestial compass information in the Central Australian desert ant *Melophorus bagoti. J Comp Physiol A 200, 591–601.*—Information transfer between the two compass systems has been shown by Wehner R (1997), *n16.* Lebhardt F, Ronacher B (2015): Transfer of directional information between the polarization compass and the sun compass in desert ants. *J Comp Physiol A 201, 599–608.*

45. The first attempts to demonstrate whether an animal's celestial compass would take account of the daily movement of the sun did not yet show time compensation in either ants or bees. Brun R (1914): Die Raumorientierung der Ameisen und das Orientierungsproblem im allgemeinen. *Jena, G. Fischer, pp. 176–184.* Wolf E (1927): Über das Heimfindevermögen der Bienen. II. *Z Vergl Physiol 6, 221–254.* The latter study does not allow for a definite conclusion.

46. The classical accounts, in which time compensation was demonstrated for the first time in bees, ants, crustaceans, spiders, and birds, are von Frisch K (1950): Die Sonne als Kompass im Leben der Bienen. *Experientia 6, 210–222.* Kramer G (1953): Die Sonnenorientierung der Vögel. *Verh Dtsch Zool Ges 52, 72–84.* Papi F (1955): Experiments on the sense of time in *Talitrus saltator* (Crustacea: Amphipoda). *Experientia 11, 201–202.* Renner M (1957): Neue Versuche über den Zeitsinn der Honigbiene. *Z Vergl Physiol 40, 85–118.* Jander R (1957): Die optische Richtungsorientierung der roten Waldameise *(Formica rufa). Z Vergl Physiol 40, 162–238.* Meder E (1958): Über die Einberechnung der Sonnenwanderung bei der Orientierung der Honigbiene. *Z Vergl Physiol 40, 610–641.* Lindauer M (1959): Angeborene und erlernte Komponenten in der Sonnnenorientierung der Bienen. Bemerkungen und Versuche zu einer Mitteilung von Kalmus. *Z Vergl Physiol 42, 43–62.*

47. For honeybees, see Renner M (1959): Über ein weiteres Versetzungsexperiment zur Analyse des Zeitsinns und der Sonnenorientierung der Honigbiene. *Z Vergl Physiol 42, 449–483.* Beier W, Lindauer M (1970): Der Sonnenstand als Zeitgeber für die Biene. *Apidologie 1, 5–28.* Cheeseman JF, Winnebeck EC, Millar CD, Kirkland LS, Sleigh J, et al., Warman GR (2012): General anesthesia alters time perception by phase shifting the circadian clock. *Proc Natl Acad Sci USA 109, 7061–7066.*

48. Brines ML (1980): Dynamic patterns of skylight polarization as clock and compass. *J Theor Biol 86, 507–512.*

49. Gould JL (1980): Sun compensation by bees. *Science 207, 545–547.* Gould JL (1984): Processing of sun-azimuth information by honey bees. *Anim Behav 32, 149–152.*

50. Wehner R, Lanfranconi B (1981): What do the ants know about the rotation of the sky? *Nature 293, 731–733.* Dyer FC (1987): Memory and sun compensation by honey bees. *J Comp Physiol A 160, 621–633.*

51. The skyline as a reference frame for the calibration of the ephemeris function was first proposed by New DAT, New JK (1962): The dances of honeybees at small zenith distances of the sun. *J Exp Biol 39, 271–291.* Dyer FC, Gould JL (1981): Honey bee orientation: a backup system for cloudy days. *Science 214, 1041–1042.* Dyer FC (1987), *n50.*

52. Towne WF, Moscrip H (2008): The connection between landscapes and the solar ephemeris in honeybees. *J Exp Biol 211, 3729–3736.* Towne WF (2008): Honeybees can learn the relationship between the solar ephemeris and a newly-experienced landscape. *J Exp Biol 211, 3737–3743.* Kemfort JR, Towne WF (2013): Honeybees can learn the relationship between the solar ephemeris and a newly experienced landscape: a confirmation. *J Exp Biol 216, 3767–3771.*

53. Dyer FC, Dickinson JA (1994): Development of sun compensation by honeybees: how partially experienced bees estimate the sun's course. *Proc Natl Acad Sci USA 91, 4471–4474.*—For *Cataglyphis,* see Wehner R, Müller M (1993): How do ants acquire their celestial ephemeris function? *Naturwissenschaften 80, 331–333.*

54. Lindauer M (1959), *n46.* Dyer FC, Dickinson JA (1994), *n53.*

55. Dyer FC, Dickinson JA (1996): Sun-compass learning in insects: representation in a simple mind. *Curr Direct Psychol Sci 5, 67–72.*

56. Wehner R, Meier C, Zollikofer CPE (2004): The ontogeny of foraging behaviour in desert ants, *Cataglyphis bicolor. Ecol Entomol 29, 240–250.* Fleischmann PN, Christian M, Müller VL, Rössler W, Wehner R (2016): Ontogeny of learning walks and the acquisition of landmark information in desert ants, *Cataglyphis fortis. J Exp Biol 219, 3137–3145.* Fleischmann PN, Grob R, Wehner R, Rössler W (2017): Species-specific differences in the fine structure of learning walk elements in *Cataglyphis* ants. *J Exp Biol 220, 2426–2435.*

57. Grob R, Fleischmann PN, Grübel K, Wehner R, Rössler W (2017): The role of celestial compass information in *Cataglyphis* ants during learning walks and for neuro-plasticity in the central complex and mushroom bodies. *Front Behav Neurosci 11, 226.* Fleischmann PN, Grob R, Müller VL, Wehner R, Rössler W (2018): The geomagnetic field as a compass cue in *Cataglyphis* ant navigation. *Curr Biol 28, 1440–1444.*

58. Collett TS, Baron J (1994): Biological compasses and the coordinate frame of landmark memories in honey-bees. *Nature 368, 137–140.* Jander R, Jander U (1998): The light and magnetic compass of the weaver ant, *Oecophylla sma-ragdina* (Hymenoptera: Formicidae). *Ethology 104, 743–758.* Riveros AJ, Srygley RB (2008): Do leafcutter ants, *Atta colom-bica,* orient their path-integrated home vector with a magnetic compass? *Anim Behav 75, 1273–1281.* For a review, see Wajn-berg E, Acosta-Avalos D, Alves OC, de Oliveira JF, Srygley RB, Esquivel DMS (2010): Magnetoreception in eusocial insects: an update. *J R Soc Interface 7, S207–S225.*—Moreover, *Cata-glyphis nodus* can be trained to use an artificial disturbance of the geomagnetic field as home-site cue. Bühlmann C, Hansson BS, Knaden M (2012): Desert ants learn vibration and mag-netic landmarks. *PLoS One 7, e33117.*

59. The blind mole rat, *Spalax ehrenbergi,* has been re-ported to use the geomagnetic field for path integration in the dark. Kimchi T, Etienne AS, Terkel J (2004): A subterra-nean mammal uses the magnetic compass for path integra-tion. *Proc Natl Acad Sci USA 101, 1105–1109.*

60. There are some indications that a magneto-mechano-transducer mechanism may be involved in the magnetic sense of at least some insect species. See Lindauer M, Martin H (1968): Die Schwereorientierung der Bienen unter dem Einfluss des Erdmagnetfelds. *Z Vergl Physiol 60, 219–243.* Wehner R, Labhart T (1970): Perception of the geomag-netic field in the fly *Drosophila melanogaster. Experientia 26, 967–968.*—Could the Johnston organ with its bowl-shaped array of mechanoreceptors in the ant's antennae play a role?

61. Wehner R, Duelli P (1971), *n5.* Duelli P (1971): Die Konkurrenz von Polarisationsmuster des Himmels, Mon-dazimut und Windrichtung bei der Orientierung der Wüste-nameise *Cataglyphis bicolor* im Zeitintervall zwischen Son-nenunter- und Sonnenaufgang. *Diploma Thesis, University of Zürich.*—The ants' immediate response to the e-vector pat-tern appearing in the sky at dawn can also be deduced from experiments in which the ants are deprived of any wind in-formation (by testing them with our trolley device). Under these conditions the switch from completely disoriented to fully oriented behavior occurs 30–35 min before sunrise, exactly at the time when in the open the ants switch from wind-based to sky-based orientation.

62. In early experiments Karl Eduard Linsenmair showed that walking tenebrionid and scarabaeoid beetles can keep a constant course relative to the direction of the wind. Linsen-mair KE (1969): Anemotaktische Orientierung bei Tenebri-oniden und Mistkäfern (Insecta, Coleoptera). *Z Vergl Physiol 64, 154–211.* Linsenmair KE (1970): Die Interaktion der paarigen antennalen Sinnesorgane bei der Windorien-tierung laufender Mist- und Schwarzkäfer (Insecta, Coleop-tera). *Z Vergl Physiol 70, 247–277.*

63. Müller M, Wehner R (2007), *n43.*

64. Anatomy: Wehner R, Bernard GD, Geiger E (1975), *n6.* Herrling PL (1976): Regional distribution of three ultra-structural retinula types in the retina of *Cataglyphis bicolor* (Formicidae, Hymenoptera). *Cell Tiss Res 169, 247–266.* Meyer EP (1979): Golgi-EM-study of first and second order neurons in the visual system of *Cataglyphis bicolor* (Hyme-noptera, Formicidae). *Zoomorphologie 92, 115–139.* Räber F (1979): Retinatopographie und Sehfeldtopologie des Kom-plexauges von *Cataglyphis bicolor* (Formicidae, Hymenop-tera). *PhD Thesis, University of Zürich.* Meyer EP (1984): Ret-rograde labelling of photoreceptors in different eye regions of the compound eyes of bees and ants. *J Neurocytol 13, 825–836.* Meyer EP, Nässel DR (1986): Terminations of pho-toreceptor axons from different regions of the compound eye of the desert ant *Cataglyphis bicolor. Proc R Soc B 228, 59–69.* Meyer EP, Domanico V (1999): Microvillar orientation in the photoreceptors of the ant *Cataglyphis bicolor. Cell Tiss Res 295, 355–361.*—Physiology: Mote MI, Wehner R (1980), *n28.* Labhart T (1986): The electrophysiology of photo-ceptors in different eye regions of the desert ant, *Cataglyphis bicolor. J Comp Physiol A 158, 1–7.*—Fan array: Wehner R (1982), *n15.*

65. For early studies on polarization opponency based on anatomical evidence and computational work, see Wehner

R (1982), *n15.*—Physiological evidence: Labhart T (1986), *n64.*—Behavioral evidence: Rossel S, Wehner R (1986), *n22.*—In an elegant functional imaging study the sites of reciprocal interactions between 'crossed analyzer' photoreceptors have been elucidated in individual columns of the *Drosophila* medulla by Weir PT, Henze MJ, Bleul C, Baumann-Klausener F, Labhart T, Dickinson MH (2016): Anatomical reconstruction and functional imaging reveal an ordered array of skylight polarization detectors in *Drosophila. J Neurosci 36, 5397–5404.*—For a review of polarization-opponent neurons in insects, see Heinze S (2014): Polarized light processing in insect brains: recent insights from the desert locust, the monarch butterfly, the cricket, and the fruit fly. In: Horváth G, ed., Polarized Light and Polarization Vision in Animal Sciences, pp. 61–111. *Berlin, Springer.*

66. Theoretical arguments are provided by Nilsson DE, Labhart T, Meyer EP (1987): Photoreceptor design and optical properties affecting polarization sensitivity in ants and crickets. *J Comp Physiol A 161, 645–658.*

67. For *Gryllus:* Labhart T (1988): Polarization opponent interneurons in the insect visual system. *Nature 331, 435–437.* Labhart T, Petzold J (1993): Processing of polarized light information in the visual system of crickets. In: Wiese K, Gribakin FG, Popov AV, Renninger G, eds., Sensory Systems of Arthropods, pp. 158–168. *Basel, Birkhäuser.* Petzold J (2001): Polarisationsempfindliche Neuronen im Sehsystem der Feldgrille, *Gryllus campestris:* Elektrophysiologie, Anatomie und Modellrechnungen. *PhD Thesis, University of Zürich.*—For *Cataglyphis,* see: Labhart T (2000): Polarization-sensitive interneurons in the optic lobes of the desert ant, *Cataglyphis bicolor. Naturwissenschaften 87, 133–136.*

68. In contrast to the POL1 neurons of crickets the medullar POL neurons of locusts respond also to unpolarized light, and most of them do not exhibit polarization opponency. el Jundi B, Homberg U (2010): Evidence for the possible existence of a second polarization-vision pathway in the locust brain. *J Insect Physiol 56, 971–979.* el Jundi B, Pfeiffer K, Homberg U (2011), *n39.*

69. Polarization sensitivities in photoreceptors of the DRA and non-DRA amount to PS>10 and PS<2, respectively: Labhart T (1980): Specialized photoreceptors at the dorsal rim of the honeybee's compound eye: polarizational and angular sensitivity. *J Comp Physiol 141, 19–30.*—The ontogenetic development of the twist is described in Wagner-Boller E (1987): Ontogenese des peripheren visuellen Systems der Honigbiene *(Apis mellifera). PhD Thesis, University of Zürich.*—For rhabdomeric disorders and polarization sensitivity in *Cataglyphis,* see Meyer EP, Domanico V (1999), *n64.* Labhart T (2000), *n67.* Rhabdomeric twist in flies: Smola U, Wunderer H (1981): Fly rhabdomeres twist in vivo. *J Comp Physiol 142, 43–49.*—The functional significance of photoreceptor twist is analyzed in Wehner R, Bernard GD (1993): Photoreceptor twist: a solution to the false-color problem. *Proc Natl Acad Sci USA 90, 4132–4135.* Horváth G, Hegedüs R (2014): Polarization-induced false colours. In: Horváth G, ed., Polarized Light and Polarization Vision in Animal Sciences, pp. 293–302. *Berlin, Springer.*

70. Labhart T, Meyer EP (1999): Detectors for polarized skylight in insects: a survey of ommatidial specializations in the dorsal rim area of the compound eye. *Micr Res Tech 47, 368–379.* Wehner R, Labhart T (2006), *n12.* Heinze S (2014), *n65.*

71. el Jundi B, Foster JJ, Khaldy L, Byrne MJ, Dacke M, Baird E (2016): A snapshot-based mechanism for celestial orientation. *Curr Biol 26, 1456–1462.*—For comparison, see Hölldobler B (1980): Canopy orientation: a new kind of orientation in ants. *Science 210, 86–88.*

72. Schmitt F, Stieb SM, Wehner R, Rössler W (2016): Experience-related reorganization of giant synapses in the lateral complex: potential role in plasticity of the sky-compass pathway in the desert ant *Cataglyphis fortis. Dev Neurobiol 76, 390–404.* Grob R, Fleischmann PN, Grübel K, Wehner R, Rössler W (2017), *n57.* Rössler W (2019): Neuroplasticity in desert ants (Hymenoptera: Formicidae)—importance for the ontogeny of navigation. *Myrm News 29, 1–20.*

73. Wehner R, Müller M (1985), *n16.* In *Cataglyphis,* the two major visual pathways were traced by Schmitt F, Stieb SM, Wehner R, Rössler W (2016), *n72,* and Grob R, Fleischmann PN, Grübel K, Wehner R, Rössler W (2017), *n72.*

74. Heinze S, Homberg U (2007): Map-like representation of celestial E-vector orientations in the brain of an insect. *Science 315, 995–997.*

75. The seminal Heinze and Homberg (2007) paper, *n74,* started amazing research activities on the central polarization pathway in a quite diverse variety of insects, not only in locusts and crickets, but also in monarch butterflies, honeybees, fruit flies, and dung beetles. Important references covering these insect groups are Homberg U, Heinze S, Pfeiffer K, Kinoshita M, el Jundi B (2011): Central neural coding of sky polarization in insects. *Phil Trans R Soc B 366, 680–687.* el Jundi B, Pfeiffer K, Heinze S, Homberg U (2014): Integration of polarization and chromatic cues in the insect sky compass. *J Comp Physiol A 200, 575–589.* Pfeiffer K, Homberg U (2014): Organization and functional roles of the central complex in the insect brain. *Annu Rev Entomol 59, 165–184.* Heinze S (2014), *n65,* a well-designed comprehensive review. Held M, Berz A, Hensgen B, Muenz TS, et al., Pfeiffer K (2016): Microglomerular complexes in the sky-compass network of the honeybee connect parallel pathways from the anterior optic tubercle to the central complex. *Front Behav Neurosci 10: 186.* el Jundi B, Warrant EJ, Pfeiffer K, Dacke M (2018): Neuroarchitecture of the dung beetle central complex. *J Comp Neurol 526, 2612–2630.* For phylogenetic aspects, see Thoen HH, Marshall J, Wolff GH, Strausfeld NJ (2017): Insect-like organization of the stomatopod central complex: functional and phylogenetic implications. *Front Behav Neurosci 11, 12.*

76. Träger U, Wagner R, Bausenwein B, Homberg U (2008): A novel type of microglomerular synaptic complex in the polarization vision pathway of the locust brain. *J Comp Neurol 506, 288–300.* Schmitt F, Stieb SM, Wehner R, Rössler W (2016), *n72.*—For reference to the mammalian auditory pathway, see Borst JGG, Rusu SI (2012): The Calyx of Held synapse. In: Trussell LO, Popper A, Fay R, eds., Synaptic Mechanisms in the Auditory System, pp. 95–134. *New York, Springer.*

77. Strausfeld NJ (2012): Arthropod Brain: Evolution, Functional Elegance, and Historical Significance. *Cambridge MA, Harvard University Press, quot.* p. 310.

78. Right-left bargaining in the central complex: Strauss R (2002): The central complex and the genetic dissection of locomotor behaviour. *Curr Opin Neurobiol 12, 633–638.* The dual compass representation is most likely perpetuated in the organization of the path integrator in the central complex: Stone T, Webb B, Adden A, Wedding NB, Honkanen A, et al., Heinze S (2017): An anatomically constrained model for path integration in the bee brain. *Curr Biol 27, 3069–3085.* Collett M, Collett TS (2017): Path integration: combining optic flow with compass orientation. *Curr Biol 27, R1113–R1116.*

79. The term "compass neurons" has previously been used in neural network studies: Hartmann G, Wehner R (1995): The ant's path integration system: a neural architecture. *Biol Cybern 73, 483–497.*

80. Giraldo YM, Leitch KJ, Ros IG, Warren TL, Weir PT, Dickinson MH (2018): Sun navigation requires compass neurons in *Drosophila. Curr Biol 28, 2845–2852.* Warren TL, Weir PT, Dickinson MH (2018): Flying *Drosophila melanogaster* maintain arbitrary but stable headings relative to the angle of polarized light. *J Exp Biol 221, 1–12.*—In fruit flies the CL1 cells (here called E-PG cells) of the ellipsoid body are described to represent the compass neurons, but also there the TB1 cells of the protocerebral bridge exhibit compass properties. As the two cell types exchange information bidirectionally, the CL1-TB1 system could be considered the inner compass.

81. Seelig JD, Jayaraman V (2015): Neural dynamics for landmark orientation and angular path integration. *Nature 521, 186–191.* Turner-Evans D, Wegener S, Rouault H, Franconville R, Wolff TT, et al., Jayaraman V (2017): Angular velocity integration in a fly heading circuit. *eLife 6, e23496.*—The fly's CL1 neurons are studied by two-photon calcium imaging in head-fixed animals walking within a virtual reality environment. They are equivalent to the head direction cells in the mammalian cortex, also in the sense that they receive information not only about landmarks but also self-generated movements. Taube JS (2007): The head direction signal: origins and sensory-motor integration. *Annu Rev Neurosci 30, 181–207.*

82. Bech M, Homberg U, Pfeiffer K (2014): Receptive fields of locust brain neurons are matched to polarization patterns in the sky. *Curr Biol 24, 2124–2129.*

83. For walking and flying locusts, see Eggers A, Weber T (1993): Behavioural evidence for polarization vision in locusts. *Proc Neurobiol Conf Göttingen 21, 336.* Mappes M, Homberg U (2004): Behavioral analysis of polarization vision in tethered flying locusts. *J Comp Physiol A 190, 61–68.*—Similar behavioral responses have been recorded

in tethered walking or flying crickets, houseflies and fruit flies. The flies align their longitudinal body axis preferably parallel or perpendicularly to the e-vector axis ('polarotaxis'): Brunner D, Labhart T (1987): Behavioural evidence for polarization vision in crickets. *Physiol Entomol 12, 1–10*. von Philipsborn A, Labhart T (1990): A behavioral study of polarization vision in the fly, *Musca domestica. J Comp Physiol A 167, 737–743*. Henze MJ, Labhart T (2007): Haze, clouds and limited sky visibility: polarotactic orientation of crickets under difficult stimulus conditions. *J Exp Biol 210, 3266–3276*. Wernet MF, Velez MM, Clark DA, Baumann-Klausener F, Brown JR, Klovstad M, Labhart T, Clandinin TR (2012): Genetic dissection reveals two separate retinal substrates for polarization vision in *Drosophila. Curr Biol 22, 12–20*. Velez MM, Wernet MF, Clark DA, Clandinin TR (2014): Walking *Drosophila* align with the e-vector of linearly polarized light through directed modulation of angular acceleration. *J Comp Physiol A 200, 603–614*.

84. According to some classical studies large-scale movements of *Schistocerca gregaria* are largely downwind: Rainey, RC (1963): Meteorology and the migration of desert locusts. World Meteorological Organization Technical Note. No. 54, *Geneva*. Schaefer GW (1976): Radar observations of insect flight. *Symp R Ent Soc 7, 157–197*.—There is no satisfactory evidence yet that the sun has any lasting influence on migratory flight orientation (see some mirror experiments): Kennedy JS (1951): The migration of the desert locust *(Schistocerca gregaria). Phil Trans R Soc B 235, 163–290*. Ellis PE, Ashall C (1957): Field studies on diurnal behaviour, movement and aggregation in the desert locust. *Anti Locust Bull 25, 1–94*.—For a short and lucid synopsis of the behavior of solitarious and gregarious locusts, see Homberg U (2015): Sky compass orientation in desert locusts—evidence from field and laboratory studies. *Front Behav Neurosci 9, 346*.

85. Weissmann C (1978): Reverse genetics. *Trends Biochem Sci 3, N109–N111*. Wehner R (2005): Brainless eyes. *Nature 435, 157–158*.—See also Li YF, Costello JC, Holloway AK, Hahn MW (2008): 'Reverse ecology' and the power of population genomics. *Evolution 62, 2984–2994*.

86. Heiligenberg WF (1991): Neural Nets in Electric Fish. *Cambridge MA, MIT Press*.

87. In the study of insect navigation, some recent examples in which 'reverse' and 'forward' approaches have been combined in exciting ways, are Seelig JD, Jayaraman V (2015), *n81*. Martin J, Guo P, Mu L, Harley CM, Ritzmann RE (2015): Central-complex control of movement in the freely walking cockroach. *Curr Biol 25, 2795–2803*. Stone T, Webb B, Adden A, Wedding NB, Honkanen A, et al., Heinze S (2017), *n78*. Giraldo YM, Leitch KJ, Ros IG, Warren TL, Weir PT, Dickinson MH (2018), *n80*.

4. Estimating Distances

1. For the earliest known explicit description of the chip log, see Bourne W (1574): A Regiment for the Sea: Conteyning most profitable Rules, Mathematical experiences, and perfect knowledge of Navigation, for all Coastes and Countreys. *London, Hacket*. References are from Taylor ERG (1963): A Regiment for the Sea and Other Writings on Navigation by William Bourne. *Cambridge, Cambridge University Press*, pp. 126–128.—Early refinements of the log line are described in Hewson JB (1983): A History of the Practice of Navigation. 2nd ed. *Glasgow, Brown, Son and Ferguson*, p. 160.—For the remark on Magellan's voyage in 1521, see Hutchins E (1995): Cognition in the Wild. *Cambridge MA, MIT Press*, p. 104.

2. "Derived from its own movements" is often described as 'idiothetic' behavior. However, when Horst and Marie-Luise Mittelstaedt introduced this term in a study of path integration, they argued that idiothetic information can only be gained by internally generated (e.g., proprioceptive) cues without any reference to external cues. Mittelstaedt H, Mittelstaedt ML (1973): Mechanismen der Orientierung ohne richtende Aussenreize. *Fortschr Zool 21, 46–58*. See also Mittelstaedt H (2000): Triple-loop model of path integration control by head direction and place cells. *Biol Cybern 83, 261–270*.—If external cues are used, the path integration system is usually called 'allothetic.' However, sometimes the use of sensory information obtained by external stimuli only for determining the animal's rate of linear or angular motion, e.g., by exploiting self-induced optic-flow or celestial cues merely to estimate the translational and rotational components of movement, is also subsumed under the term 'idiothetic.' See also Cheung A, Zhang SW, Stricker C, Srinivasan MV (2008): Animal navigation: general properties of directed walks. *Biol Cybern 99, 197–217*. Vickerstaff RJ, Cheung A

(2010): Which coordinate system for modelling path integration? *J Theor Biol 263, 242–261*. Cheung A (2014): Animal path integration: a model of positional uncertainty along tortuous paths. *J Theor Biol 341, 17–33*.—As the terms 'idiothetic' and 'allothetic,' which now flit about the literature on animal navigation, might appear a bit enigmatic, let me hint at their Greek etymological roots: ídios (ιδιος), own, self; állos (αλλος), alien; thetós (θετος), determined.

3. Measuring distance by 'protocounting' within a series of artificial landmarks experimentally set up between the hive and the feeder has been reported for honeybees: in cue-conflict tests, in which the spacing interval was varied, a minority (about 10%–25%) of bees referred to the landmark number; the majority relied on distance flown. Chittka L, Geiger K (1995): Can honey bees count landmarks? *Anim Behav 49, 159–164*. Menzel R, Fuchs J, Nadler L, Weiss B, et al., Greggers U (2010): Dominance of the odometer over serial landmark learning in honeybee navigation. *Naturwissenschaften 97, 763–767*.—For the limited capacity of landmark counting as tested in flight tunnels (upper limit: four landmarks), see Dacke M, Srinivasan MV (2008): Evidence for counting in insects. *Anim Cogn 11, 683–689*.

4. Heran H, Wanke L (1952): Beobachtungen über die Entfernungsmeldungen der Sammelbienen. *Z Vergl Physiol 34, 383–393*. Heran H (1956): Ein Beitrag zur Frage nach der Wahrnehmungsgrundlage der Entfernungsmessung der Biene. *Z Vergl Physiol 38, 168–218*. Scholze E, Pichler H, Heran H (1964): Zur Entfernungsschätzung der Bienen nach dem Kraftaufwand. *Naturwissenschaften 51, 69–70*.—For first doubts on the energy expenditure hypothesis in honeybees, see Goller F, Esch HE (1990): Waggle dances of honey bees: is distance measured through energy expenditure on outward flight? *Naturwissenschaften 77, 594–595*. Esch HE, Burns JE (1996): Distance estimation by foraging honeybees. *J Exp Biol 199, 155–162*.

5. Schäfer M, Wehner R (1993): Loading does not affect measurement of walking distance in desert ants, *Cataglyphis fortis. Verh Dtsch Zool Ges 86, 270*. Pfeffer SE, Wittlinger M (2016): How to find home backwards? Navigation during rearward homing of *Cataglyphis fortis* desert ants. *J Exp Biol 219, 2119–2126*.

6. Medawar PB (1964): Is the scientific paper a fraud? In: Edge D, ed., Experiment: A Series of Scientific Case Histories, pp. 7–13. *London, British Broadcasting Corporation, quot.* p. 7.

7. Heusser D, Wehner R (2002): The visual centering response in desert ants, *Cataglyphis fortis. J Exp Biol 205, 585–590*.—Honeybees flying through the middle of a tunnel have been shown to balance the apparent motions of the images of the walls on the two sides: Srinivasan MV, Lehrer M, Kirchner WH, Zhang SW (1991): Range perception through apparent image speed in freely-flying honeybees. *Vis Neurosci 6, 519–535*.

8. Wittlinger M, Wehner R, Wolf H (2006): The ant odometer: stepping on stilts and stumps. *Science 312, 1965–1967*. Wittlinger M, Wehner R, Wolf H (2007): The desert ant odometer: a stride integrator that accounts for stride length and walking speed. *J Exp Biol 210, 198–207*.

9. For the necessary normalization of stride length with respect to body size and running speed, see Wittlinger M, Wehner R, Wolf H (2007), n8.—For further considerations, see Collett M, Collett TS, Srinivasan MV (2006): Insect navigation: measuring travel distance across ground and though air. *Curr Biol R887–R890*. Walls ML, Layne JE (2009): Direct evidence for distance measurement via flexible stride integration in the fiddler crab. *Curr Biol 19, 25–29*.

10. Piéron H (1904): Du rôle du sens musculaire dans l'orientation des fourmis. *Bull Inst Gén Psychol 4, 168–186*.—Henri Piéron coined the term 'pedometer.'

11. Wittlinger M, Wolf H, Wehner R (2007): Hair plate mechanoreceptors associated with body segments are not necessary for three-dimensional path integration in desert ants, *Cataglyphis fortis. J Exp Biol 210, 375–382*. Steck K, Wittlinger M, Wolf H (2009): Estimation of homing distance in desert ants, *Cataglyphis fortis,* remains unaffected by disturbances of walking behaviour. *J Exp Biol 212, 2893–2901*. Wittlinger M, Wolf H (2013): Homing distance in desert ants, *Cataglyphis fortis,* remains unaffected by disturbance of walking behaviour and visual input. *J Physiol 107, 130–136*.—For arguments that stride length is most likely derived from proprioceptive inputs from the legs, see also Collett M, Collett TS, Srinivasan MV (2006), n9.

12. Pfeffer SE, Wittlinger M (2016), n5.

13. Similar to desert ants, fiddler crabs track the location of their burrow by path integration. Michael Walls and John Layne at the University of Cincinnati, Ohio, let the crabs home on a slippery acetate sheet, and by this showed that the animals measure distance by integrating strides rather than optic flow. Walls ML, Layne JE (2009), *n9*.

14. Ronacher B, Wehner R (1995): Desert ants, *Cataglyphis fortis,* use self-induced optic flow to measure distance travelled. *J Comp Physiol A 177, 21–27.*—The ants' overshooting and undershooting is not influenced by the pattern wavelength of the black-and-white gratings and thus shows that the estimation of distance depends on self-induced image speed rather than temporal contrast frequency.

15. Ronacher B, Gallizzi K, Wohlgemuth S, Wehner R (2000): Lateral optic flow does not influence distance estimation in the desert ant *Cataglyphis fortis. J Comp Physiol A 203, 1113–1121.*

16. Hingston RWG (1923): A Naturalist in Hindustan. *London, H. F. and G. Witherby, quot.* p. 281.—In *Cataglyphis,* carrying behavior has been described, e.g., in Délye G (1967): Physiologie et comportement de quelques fourmis (Hym. Formicidae) du Sahara en rapport avec les principaux facteurs du climat. *Insect Soc 14, 323–338.* Harkness RD (1977): The carrying of ants *(Cataglyphis bicolor)* by others of the same nest. *J Zool 183, 419–430.* Cerdá X, Retana J (1992): A behavioural study of transporter workers in *Cataglyphis iberica* ant colonies (Hymenoptera, Formicidae). *Ethol Ecol Evol 4, 359–374.* Dahbi A, Retana J, Lenoir A, Cerdá X (2008): Nest-moving by the polydomous ant *Cataglyphis iberica. J Ethol 26, 119–126.* Amor F, Ortega P, Jowers MJ, Cerdá X, et al., Boulay RR (2011): The evolution of worker-queen polymorphism in *Cataglyphis* ants: interplay between individual- and colony-level selections. *Behav Ecol Sociobiol 65, 1473–1482.*—In ants, the mode of adult transport is a largely taxon-specific trait; see Hölldobler B, Wilson EO (1990): The Ants. *Cambridge MA, Harvard University Press,* pp. 279–285. Plate 71 in Hölldobler B, Wilson EO (2009): The Superorganism. *New York, Norton.*

17. Duelli P (1973): Astrotaktisches Heimfindevermögen tragender und getragener Ameisen (*Cataglyphis bicolor,* Hymenoptera, Formicidae). *Rev Suisse Zool 80, 712–719.* Duelli P (1976): Distanzdressuren von getragenen Ameisen. *Rev Suisse Zool 83, 413–418.* Fourcassié V, Dahbi A, Cerdá X (2000): Orientation and navigation during adult transport between nests in the ant *Cataglyphis iberica. Naturwissenschaften 87, 355–359.*

18. Pfeffer SE, Wittlinger M (2016): Optic flow odometry operates independently of stride integration in carried ants. *Science 353, 1155–1157.*

19. Wohlgemuth S, Ronacher B, Wehner R (2002): Distance estimation in the third dimension in desert ants. *J Comp Physiol A 188, 273–281.* Thiélin-Bescond M, Beugnon G (2005): Vision-independent odometry in the ant *Cataglyphis cursor. Naturwissenschaften 92, 193–197.*

20. Seidl T, Knaden M, Wehner R (2006): Desert ants: is active locomotion a prerequisite for path integration? *J Comp Physiol A 192, 1125–1131.*

21. Wolf H, Wittlinger M, Pfeffer SE (2018): Two distance memories in desert ants—modes of interaction. *PLoS ONE, e0204664.*

22. Arnoldi KV (1932): Biologische Beobachtungen an der neuen paläarktischen Sklavenhalterameise *Rossomyrmex proformicarum,* nebst einigen Bemerkungen über die Beförderungsweise der Ameisen. *Z Morph Ökol Tiere 24, 319–326.* Marikovsky PY (1974): The biology of the ant *Rossomyrmex proformicarum. Insect Soc 21, 301–308.* Tinaut A, Ruano F (1998): Implications phylogénétiques du mechanisme de recrutement chez *Rossomyrmex minuchae* (Hymenoptera, Formicidae). *Act Coll Insect Soc 11, 125–132.* Ruano F, Sanllorente O, Lenoir A, Tinaut A (2013): *Rossomyrmex,* the slave-maker ants from the arid steppe environments. *Psyche 2013, Article ID 541804.*

23. Ruano F, Tinaut A (1999): Raid process, activity pattern and influence of abiotic conditions in the slave-making ant *Rossomyrmex minuchae* (Hymenoptera, Formicidae). *Insect Soc 46, 341–347.* Ruano F, Tinaut A (2004): The assault process of the slave-making ant *Rossomyrmex minuchae* (Hymenoptera, Formicidae). *Sociobiology 43, 201–209.*

24. The quotations are from Ward PS (2014): The phylogeny and evolution of ants. *Annu Rev Ecol Evol Syst 45, 23–43;* Danforth BN (2013): Social insects: are ants just wingless bees? *Curr Biol 23, R1011–R1012.*—For phylogenies, see especially Johnson BR, Borowiec ML, Chiu JC, Lee EK, Atallah J, Ward PS (2013): Phylogenomics resolve

evolutionary relationships among ants, bees, and wasps. *Curr Biol 23, 2058–2062*.

25. Srinivasan MV, Zhang SW, Lehrer M, Collett TS (1996): Honeybee navigation en route to the goal: visual flight control and odometry. *J Exp Biol 199, 237–243*. Srinivasan MV, Zhang SW, Bidell NJ (1997): Visually mediated odometry in honeybees. *J Exp Biol 200, 2513–2522*. Srinivasan MV, Zhang SW, Altwein M, Tautz J (2000): Honeybee navigation: nature and calibration of the odometer. *Science 287, 851–853*. Esch HE, Zhang SW, Srinivasan MV, Tautz J (2001): Honeybee dances communicate distances measured by optic flow. *Nature 411, 581–583*.—Optic-flow odometry has also been reported for stingless bees in the Amazon rain forest: Hrncir M, Jarav S, Zucchi R, Barth FG (2003): A stingless bee *(Melipona seminigra)* uses optic flow to estimate flight distances. *J Comp Physiol A 189, 761–768*.—These bees use optic flow also for locating food sources in the vertical plane. Eckles MA, Roubik DW, Nieh JC (2012): A stingless bee can use visual odometry to estimate both height and distance. *J Exp Biol 215, 3150–3160*.—Vertical optic flow is also exploited by honeybees: Dacke M, Srinivasan MV (2007): Honeybee navigation: distance estimation in the third dimension. *J Exp Biol 210, 845–853*.

26. Esch HE, Burns JE (1995): Honeybees use optic flow to measure the distance of a food source. *Naturwissenschaften 82, 38–40*. Esch HE, Burns JE (1996), *n4*.

27. Von Frisch K (1965): Tanzsprache und Orientierung der Bienen. *Berlin, Springer*. Von Frisch K (1973): Decoding the Language of the Bee. *Nobel Lecture, December 12, 1973*.

28. Esch HE, Zhang SW, Srinivasan MV, Tautz J (2001), *n25*.—For the distance calibration function of the waggle dance, see von Frisch K, Kratky O (1962): Über die Beziehung zwischen Flugweite und Tanztempo bei der Entfernungsmeldung der Bienen. *Naturwissenschaften 49, 409–417*.—The different distance calibration functions reported for different races of honeybees might well result from structural differences in the bees' foraging environments rather than from congenital racial differences.

29. Wittlinger M, Wolf H (2013), *n11*.—It is not only the flow meter but also the pedometer that certainly dates back to the ants' flying ancestors, and most likely even further back. Chittka L, Williams NM, Rasmussen H, Thomson JD (1999):

Navigation without vision: bumblebee orientation in complete darkness. *Proc R Soc B 266, 45–50*.

30. Ibbotson MR (2001): Evidence for velocity-tuned motion-sensitive descending neurons in the honeybee. *Proc R Soc B 268, 2195–2201*.—The bee's visual odometer receives its input from the long-wavelength type of receptor: Chittka L, Tautz J (2003): The spectral input to honeybee visual odometry. *J Exp Biol 206, 2393–2397*.

31. Stone T, Webb B, Adden A, Wedding NB, Honkanen A, et al., Heinze S (2017): An anatomically constrained model for path integration in the bee brain. *Curr Biol 27, 3069–3085*.

32. Wohlgemuth S, Ronacher B, Wehner R (2001): Ant odometry in the third dimension. *Nature 411, 795–798*. Wohlgemuth S, Ronacher B, Wehner R (2002), *n19*.

33. Maronde K, Wohlgemuth S, Ronacher B, Wehner R (2003): Ground instead of walking distances determine the direction of the home vector in 3-D path integration of desert ants. *Proc Neurobiol Conf Göttingen 29, 548–549*.—For further 3-D experiments, see Grah G, Wehner R, Ronacher B (2005): Path integration in a three-dimensional maze: ground distance estimation keeps desert ants *(Cataglyphis fortis)* on course. *J Exp Biol 208, 4005–4011*.—The computation of 3-D path integration vectors has been discussed but not systematically explored in jumping spiders and rodents. Hill DE (1979): Orientation by jumping spiders of the genus *Phidippus* (Araneae: Salticidae) during the pursuit of prey. *Behav Ecol Sociobiol 5, 301–322*. Bardunias PM, Jander R (2000): Three dimensional path integration in the house mouse *(Mus domestica)*. *Naturwissenschaften 87, 532–534*.—The rodent's map is considered to be metrically organized only in the horizontal plane, and thus 'metrically flat,' whereas the vertical dimension is encoded in some nonmetric way; see Jeffery KJ, Jovalekic A, Verriotis M, Hayman R (2013): Navigating in a three-dimensional world. *Behav Brain Sci 36, 523–587*.

34. Walls ML, Layne JE (2009): Fiddler crabs accurately measure two-dimensional distance over three-dimensional terrain. *J Exp Biol 212, 3236–3240*.—In contrast, when honeybees are trained to fly in 3-D tunnel systems, the distances signaled in the waggle dances depends on the total rather than the ground distances traveled. Moreover, the elevation of a food source is not communicated. Von Frisch K (1965),

n27, pp. 165–172. Esch HE, Burns JE (1995), *n26*. Dacke M, Srinivasan MV (2007), *n25*.

35. In intricate behavioral experiments *C. fortis* discriminated slope differences as small as at least 12.5°, but this might not be the lower limit. Wintergerst S, Ronacher B (2012): Discrimination of inclined path segments by the desert ant *Cataglyphis fortis. J Comp Physiol A 198, 363–373.*—Wood ants, honeybees, and ground beetles detect 3°–5° slope deviations from the horizontal. Bückmann D (1955): Zur Leistung des Schweresinnes bei Insekten. *Naturwissenschaften 42, 78–79.* Markl H (1962): Borstenfelder an den Gelenken als Schweresinnesorgane bei Ameisen und anderen Hymenopteren. *Z Vergl Physiol 45, 475–569.* Von Frisch K (1965), *n27*, p. 148.

36. Markl H (1971): Proprioceptive gravity perception in Hymenoptera. In: Gordon S, Cohen M, eds., Gravity and the Organism, pp. 185–194. *Chicago, University of Chicago Press.*

37. Wohlgemuth S, Ronacher B, Wehner R (2002), *n19*. Wittlinger M, Wolf H, Wehner R (2007), *n11*.

38. Reviews of insect cuticular mechanoreceptors and chordotonal organs are given in Keil TA (1997): Functional morphology of insect mechanoreceptors. *Microsc Res Techn 39, 506–531.* Field LH, Matheson T (1998): Chordotonal organs in insects. *Adv Insect Physiol 27, 1–228.*—The involvement of such receptors in any kind of odometry is hypothesized in Seidl T, Wehner R (2008): Walking on inclines: how do desert ants monitor slope and step length? *Front Zool 5, 8.*—For mechanoreceptors that detect forces as cuticular strains in the insect exoskeleton, see e.g., Zill SN, Chaudhry S, Büschges A, Schmitz J (2013): Directional specificity and encoding of muscle forces and loads by stick insect tibial campaniform sensilla, including receptors with round cuticular caps. *Arthropod Struct Dev 42, 455–467.* Wöhrl T, Reinhardt L, Blickhan R (2017): Propulsion in hexapod locomotion: how do desert ants traverse slopes? *J Exp Biol 220, 1618–1625.*

39. Seidl T, Wehner R (2008), *n38*. Hess D, Koch J, Ronacher B (2009): Desert ants do not rely on sky compass information for the perception of inclined paths. *J Exp Biol 212, 1528–1534.* Weihmann T, Blickhan R (2009): Comparing inclined locomotion in a ground-living and a climbing ant species: sagittal plane kinematics. *J Comp Physiol A 195, 1011–1020.*

40. Wohlgemuth S, Ronacher B, Wehner R (2002), *n19*. Hess D, Koch J, Ronacher B (2009), *n39*.

41. Grah G, Wehner R, Ronacher B (2007): Desert ants do not acquire and use a three-dimensional global vector. *Front Zool 4, 12.* Grah G, Ronacher B (2008): Three-dimensional orientation in desert ants: context-independent memorisation and recall of sloped path segments. *J Comp Physiol A 194, 517–522.*

42. Cornetz V (1909): Le sentiment topographique chez les fourmis. *Rev Idées 6, 452–458,* "pour la fourmi le monde est plan," *quot.* p. 458.—Cornetz referred to a remark made by the philosopher Remy de Gourmont. De Gourmont R (1909): Promenades philosophiques. Chapter 8: Le sens topographique chez les fourmi, pp. 153–158. *Paris, Mercure de France, quot.* p. 153.

43. Wehner R (2016): Early ant trajectories: spatial behaviour before behaviourism. *J Comp Physiol A 202, 247–266.*

5. Integrating Paths

1. Homer, Odyssey 8, 557f.

2. Mittelstaedt H, Mittelstaedt ML (1973): Mechanismen der Orientierung ohne richtende Aussenreize. *Fortschr Zool 21, 46–58.*

3. The term 'dead reckoning' was first used and characterized by William Barlow, the later archdeacon of Salisbury, who called it "an uncertaine ghesse": Barlow W (1597): The Navigator's Supply. *London, Bishop, Newbery and Barker, quot.* p. H3 [published in facsimile by *Theatrum Orbis Terrarum, Amsterdam, 1972*].—It has often been claimed that etymologically 'dead reckoning' derives from 'deduced reckoning,' but this is still a matter of debate. The term could also have the meaning of 'dead aim' or 'dead right,' denoting precision: Cotter CH (1978): Early dead reckoning navigation. *J Navig 31, 20–27.* Huth JE (2013): The Lost Art of Finding Our Way. *Cambridge MA, Harvard University Press,* p. 23.—A short account on the history of dead reckoning is given in Taylor EGR (1950): Five centuries of dead reckoning. *J Navig 3, 280–285.*—Charles Darwin used the term when he referred to observations made by Ferdinand von Wrangel during extensive expeditions to northern Siberia. There "the natives kept a true course towards a particular spot whilst passing . . . through hummocky ice with incessant changes of direction."

Darwin C (1873): Origin of certain instincts. *Nature 7, 417–418;* comment by Murphy JJ (1873): Instinct: a mechanical analogy. *Nature 7, 483.*

4. Wehner R (1982): Himmelsnavigation bei Insekten. Neuphysiologie und Verhalten. *Neujahrsbl Naturf Ges Zürich 184, 1–132.* Wehner R (1983): Celestial and terrestrial navigation: human strategies—insect strategies. In: Huber F, Markl H, eds., Neuroethology and Behavioral Physiology, pp. 366–381. *Berlin, Springer.* Collett M, Collett TS (2000): How do insects use path integration for their navigation? *Biol Cybern 83, 245–259.*—The term 'vector navigation' is also used, though in a different context and meaning, in the study of bird migration. Schmidt-König K (1973): Über die Navigation der Vögel. *Naturwissenschaften 60, 88–94.*—In this general account the term describes an inherited spatiotemporal vector program, some kind of clock-and-compass strategy, which enables even first-year migrants to steer their courses to their unfamiliar winter quarters. Mouritsen H, Mouritsen O (2000): A mathematical expectation model for bird navigation based on the clock-and-compass strategy. *J Theor Biol 207, 283–291.*

5. Ministry of Defence Staff (1960): Admiralty Manual of Navigation. Vol. 1. *London, Stationery Office, quot.* p. 72.

6. Morison SE (1944): Admiral of the Ocean Sea: A Life of Christopher Columbus. *Boston, Little, Brown.* Fuson RH (1987): The Log of Christopher Columbus. *Camden ME, International Marine.* Peck DT (2011): The empirical reconstruction of Columbus' navigational log and track of his 1492–1493 discovery voyage. *J Navigation 64, 193–205.*

7. Schmidt L, Collett TS, Dillier FX, Wehner R (1992): How desert ants cope with enforced detours on their way home. *J Comp Physiol A 171, 285–288.* Wehner R (2003): Desert ant navigation: how miniature brains solve complex tasks. *J Comp Physiol A 189, 579–588.*

8. Besides ants in particular and insects in general, several other groups of animals have also become subjects of experimental studies in path integration. Some principal accounts:

Crabs: Zeil J (1998): Homing in fiddler crabs (*Uca lactea* and *Uca vomeris:* Ocypodidae). *J Comp Physiol A 183, 367–377.* Layne JE, Barnes WJP, Duncan LMJ (2003): Mechanisms of homing in the fiddler crab *Uca rapax.* 2. Information sources and frame of reference for a path integration system. *J Exp Biol 206, 4425–4442.* Walls ML, Layne JE (2009): Direct evidence for distance measurement via flexible stride integration in the fiddler crab. *Curr Biol 19, 25–29.*

Spiders: Görner P (1958): Die optische und kinästhetische Orientierung der Trichterspinne *Agelena labyrinthica. Z Vergl Physiol 41, 111–153.* Görner P, Claas B (1985): Homing behavior and orientation in the funnel-web spider, *Agelena labyrinthica.* In: Barth FG, ed., Neurobiology of Arachnids, pp. 275–297. *Berlin, Springer.* Seyfarth EA, Hergenröder R, Ebbes H, Barth FG (1982): Idiothetic orientation of a wandering spider: compensation of detours and estimates of goal distances. *Behav Ecol Sociobiol 11, 139–148.* Moller P, Görner P (1994): Homing by path integration in the spider *Agelena labyrinthica. J Comp Physiol A 174, 221–229.* Nørgaard T, Henschel JR, Wehner R (2006): The night-time temporal window of locomotor activity in the Namib Desert long-distance wandering spider, *Leucorchestris arenicola. J Comp Physiol A 192, 365–372.* Ortega-Escobar J (2006): Role of anterior lateral eyes of the wolf spider *Lycosa tarantula* (Araneae, Lycosidae) during path integration. *J Arachn 34, 51–61.*

Birds: von Saint-Paul U (1982): Do geese use path integration for walking home? In: Papi F, Wallraff HG, eds., Avian Navigation, pp. 298–307. *Berlin, Springer.*

Mammals: Mittelstaedt ML, Mittelstaedt H (1980): Homing by path integration in a mammal. *Naturwissenschaften 67, 566–567.* Séguinot V, Cattet J, Benhamou S (1998): Path integration in dogs. *Anim Behav 55, 787–797.* Etienne AS, Jeffery KJ (2004): Path integration in mammals. *Hippocampus 14, 180–192.* Kimchi T, Etienne AS, Terkel J (2004): A subterranean mammal uses the magnetic compass for path integration. *Proc Natl Acad Sci USA 101, 1105–1109.*

Humans: Klatzky RL, Loomis JM, Golledge RG, Cicinelli JG, Doherty S, Pellegrino JW (1990): Acquisition of route and survey knowledge in the absence of vision. *J Mot Behav 22, 19–43.* Mittelstaedt ML, Glasauer S (1991): Idiothetic navigation in gerbils and humans. *Zool Jb Allg Zool Physiol Tiere 95, 427–435.* Loomis JM, Klatzky RL, Golledge RG, Philbeck JW (1999): Human navigation by path integration. In: Golledge RG, ed., Wayfinding Behavior, pp. 125–151. *Baltimore, Johns Hopkins University Press.* Wang RF, Spelke ES (2003): Comparative approaches to human navigation. In: Jeffery KJ, ed., The Neurobiology of Spatial Behaviour, pp. 119–143. *Oxford, Oxford University Press.* Foo P, Warren WH, Duchon A, Tarr MJ (2005): Do humans integrate routes into a cognitive map? Map- versus landmark-based navigation of novel shortcuts. *J Exp Psychol Learn Mem Cogn 31, 195–215.* Petzschner FH, Glasauer S (2011): Iterative Bayesian estimation as an explanation for range and regression effects: a study on human path integration. *J Neurosci 31, 17220–17229.*

9. The two standard coordinate systems sketched out in Figure 5.3 are subsets of extended classes of systems, in which a larger number of reference directions might render the systems neurobiologically more plausible. For a full treatment, see Vickerstaff RJ, Cheung A (2010): Which coordinate system for modelling path integration? *J Theor Biol 263, 242–261* and references therein. Cheung A, Vickerstaff RJ (2010): Finding the way with a noisy brain. *PLoS Comput Biol 6(11), e1000992.*—Two examples for applying geocentric systems: Müller M, Wehner R (1988): Path integration in desert ants, *Cataglyphis fortis. Proc Natl Acad Sci USA 85, 5287–5290.* Hartmann G, Wehner R (1995): The ant's path integration system: a neural architecture. *Biol Cybern 73, 483–497.*—Two examples for considering egocentric systems: Benhamou S, Sauvé JP, Bovet P (1990): Spatial memory in large scale movements: efficiency and limitations of the egocentric coding process. *J Theor Biol 145, 1–12.* Merkle T, Rost M, Alt W (2006): Egocentric path integration models and their application to desert arthropods. *J Theor Biol 240, 385–399.*

10. Mittelstaedt H (1962): Control systems of orientation in insects. *Annu Rev Entomol 7, 177–198.* Mittelstaedt H (1963): Bikomponenten-Theorie der Orientierung. *Ergeb Biol 26, 253–258.*—As sketched out in Figure 5.4, Mittelstaedt's model considers path integration during the outbound journey plus homing: the sine and cosine components modulated by odometric data and integrated independently are cross-correlated with the sine and cosine input components (i.e., the current compass direction) to generate the proper steering command.

11. Vickerstaff RJ, Cheung A (2010), *n9.* Cheung A, Vickerstaff RJ (2010), *n9.* Cheung A (2014): Animal path integration: a model of positional uncertainty along tortuous paths. *J Theor Biol 341, 17–33.* Cheung A, Vickerstaff RJ (2015): Sensory and update errors which can affect path integration. *J Theor Biol 372, 217–221.* As theoretically out-

lined by Cheung and Vickerstaff (2010), the animal's movements can be projected onto multiple axes resulting in multi-component systems rather than only a 'bi-component' one, see also *n9.*

12. For definitions of the terms 'idiothetic' and 'allothetic,' and especially for the ambiguity in defining idiothetic path integration, see Chapter 4, *n2.*

13. Murphy JJ (1873), *n3.* Barlow JS (1964): Inertial navigation as a basis for animal navigation. *J Theor Biol 6, 76–117.* Mayne R (1974): A system concept of the vestibular organs. In: Kornhuber AA, ed., Handbook of Sensory Physiology, vol. 6/2, pp. 493–580. *Berlin, Springer.* Wiener W, Berthoz A (1993): Forebrain structures mediating the vestibular contribution during navigation. In: Berthoz A, ed., Multisensory Control of Movement, pp. 427–456. *Oxford, Oxford University Press.*

14. Barlow JS (1964), *n13.*—In long-distance homing, pigeons completely disregard inertial information. Neither anesthesia during transport nor surgical operations in the semicircular system impaired their homing abilities. Keeton WT (1974): The orientational and navigational basis of homing in birds. *Adv Study Behav 5, 47–132.*

15. Evidence of short-range idiothetic path integration in small mammals, which seem to employ labyrinthine information, is provided by Mittelstaedt ML, Mittelstaedt H (1980), *n8.* Etienne AS (1980): The orientation of the golden-hamster to its nest-site after the elimination of various sensory cues. *Experientia 36, 1048–1050.* Mittelstaedt H, Mittelstaedt ML (1982): Homing by path integration. In: Papi F, Wallraff HG, eds., Avian Navigation, pp. 290–297. *Berlin, Springer.* Mittelstaedt ML, Glasauer S (1991), *n8.* McNaughton BL, Knierim JL, Wilson MA (1994): Vector encoding and the vestibular foundation of spatial cognition. In: Gazzaniga M, ed., The Cognitive Neurosciences, pp. 585–595. *Cambridge MA, MIT Press.* Etienne AS, Maurer R, Séguinot V (1996): Path integration in mammals and its interaction with visual landmarks. *J Exp Biol 199, 201–209.* Bardunias PM, Jander R (2000): Three dimensional path integration in the house mouse *(Mus domestica). Naturwissenschaften 87, 532–534.*

16. Seyfarth EA, Barth FG (1972): Compound slit sense organs on the spider legs: mechanoreceptors involved in kinesthetic orientation. *J Comp Physiol 78, 176–191.* Seyfarth

EA, Hergenröder R, Ebbes H, Barth FG (1982), *n8.* Durier V, Rivault C (1999): Path integration in cockroach larvae, *Blatella germanica* (Insecta: Dictyoptera): direction and distance estimation. *Anim Learn Behav 27, 108–118.*—For small-scale idiothetic path integration in funnel-web spiders and fiddler crabs, see Görner P, Claas B (1985), *n8.* Layne JE, Barnes WJP, Duncan LMJ (2003), *n8.*

17. Cheung A, Zhang SW, Stricker C, Srinivasan MV (2007): Animal navigation: the difficulty of moving in a straight line. *Biol Cybern 97, 47–61.* Cheung A, Zhang SW, Stricker C, Srinivasan MV (2008): Animal navigation: general properties of directed walks. *Biol Cybern 99, 197–217.*

18. Souman JL, Frissen I, Sreenivasa MN, Ernst MO (2009): Walking straight into circles. *Curr Biol 19, 1538–1542.*—Without a compass, and with directional errors being biased, a directed walk is expected to approximate a logarithmic spiral: Cheung A, Zhang SW, Stricker C, Srinivasan MV (2008), *n17.*

19. Wehner R, Gallizzi K, Frei C, Vesely M (2002): Calibration processes in desert ant navigation: vector courses and systematic search. *J Comp Physiol A 188, 683–693.*—When the open-jaw experiment is performed with a forced-detour channel device rather than in the open fields, the resulting recalibration of the ants' vector navigation system is very similar to the one depicted in Figure 5.6: Collett M, Collett TS, Wehner R (1999): Calibration of vector navigation in desert ants. *Curr Biol 9, 1031–1034.* Collett M, Collett TS (2000), *n4.*

20. A recent study performed in *Melophorus* ants within a landmark-rich (grass-tussock and scattered-tree) environment starts to dwell on the set of questions raised in the text. Freas CA, Cheng K (2018): Limits of vector calibration in the Australian desert ant, *Melophorus bagoti. Insect Soc 65, 141–152.*

21. In the main text, recalibration is considered only with respect to the directional vector component. Experimentally, however, recalibration of the distance component has been tested as well. Such open-jaw experiments were performed by using straight channels, in which the foraging ants experienced different outbound and inbound travel distances. Cheng K, Wehner R (2002): Navigating desert ants *(Cataglyphis fortis)* learn to alter their search patterns on their homebound journey. *Physiol Entomol 27, 285–290.*

Bolek S, Wittlinger M, Wolf H (2012): Establishing food site vectors in desert ants. *J Exp Biol 215, 653–656.*—Main result: the lengths of the ants' recalibrated PI vectors are intermediate between the outbound and inbound distances.—In honeybees, open-jaw experiments have been performed by Friedrich Otto, then a PhD student of Karl von Frisch. In this study, the waggle dances of the bees indicated the bisector of the outward and inward direction. Otto F (1959): Die Bedeutung des Rückfluges für die Richtungs- und Entfernungsangabe der Bienen. *Z Vergl Physiol 42, 303–333.* See also Lindauer M (1963): Kompassorientierung. *Ergeb Biol 26, 158–181.* Von Frisch K (1965): Tanzsprache und Orientierung der Bienen. *Berlin, Springer,* pp. 116–120, 172–183.—However, in honeybees no data are available yet to assess the $\delta_{in} / \delta_{out}$ relationship. As to the distance component, in tunnel experiments the bees registered only the outbound distance. Srinivasan MV, Zhang SW, Bidwell NJ (1997): Visually mediated odometry in honeybees. *J Exp Biol 200, 2513–2522.*—In contrast, stingless bees referred to the return distance. Hrncir M, Jarav S, Zucchi R, Barth FG (2003): A stingless bee *(Melipona seminigra)* uses optic flow to estimate flight distances. *J Comp Physiol A 189, 761–768.*—Systematically varying subtle differences in experimental design are needed to resolve these discrepancies and arrive at a full picture of PI calibration in bees.

22. The various ways of how path integration could be used for navigation draw upon Collett M, Collett TS, Wehner R (1999), *n19.* Collett M, Collett TS (2000), *n4.* Wehner R, Cheng K, Cruse H (2014): Visual navigation strategies in insects: lessons from desert ants. In: Werner J, Chalupa L, eds., The New Visual Neurosciences, pp. 1153–1163. *Cambridge MA, MIT Press.* I am grateful for comments by Stefan Sommer and Thierry Hoinville.—For considerations of geo- and egocentric systems of reference, see the fundamental accounts by Allen Cheung and Robert Vickerstaff, *n11.*

23. As in the channel device the ants had only partial views of the sky, their polarized-light compass induces systematic errors (see Figure 3.13a). If these errors are taken into account, the position of B as indicated by the ant's path integrator shifts slightly from the real position of B (at the 4 m mark) toward site A.

24. Knaden M, Wehner R (2005): Nest mark orientation in desert ants *Cataglyphis:* what does it do to the path inte-

grator? *Anim Behav 70, 1349–1354.* Knaden M, Wehner R (2006): Ant navigation: resetting the path integrator. *J Exp Biol 209, 26–31.*

25. In deciding between the modes of path integration sketched out in Figure 5.7, behavioral experiments can be informative. For example, the 'discontinuous' and the 'push-pull' mode might not easily satisfy the outcome of the shortcut experiment (Figure 5.8).

26. Until recently, there have been only few experimental attempts to investigate how precision and accuracy of the ant's path integration system (if unaffected by other navigational routines) depend on path length and beeline distance: Sommer S, Wehner R (2004): The ant's estimation of distance travelled: experiments with desert ants, *Cataglyphis fortis. J Comp Physiol A 190, 1–6.* Merkle T, Knaden M, Wehner R (2006): Uncertainty about nest position influences systematic search strategies in desert ants. *J Exp Biol 209, 3545–3549.* Merkle T, Wehner R (2010): Desert ants use foraging distance to adopt the nest search to the uncertainty of the path integrator. *Behav Ecol 21, 349–355.*—Note that in Figure 5.11 the ants frequent a familiar feeder, so that path length and beeline distance are almost identical. For the general foraging case, in which the two measures do not coincide, see the theoretical treatment by Cheung A (2014), *n11.*

27. Sommer S, Wehner R (2004), *n26.*—The inbound / outbound relation of travel distances is shown in Figure 7.8b*. For a leaky-integrator simulation, see figure 12 in Vickerstaff RJ, Di Paolo EA (2005): Evolving neural models of path integration. *J Exp Biol 208, 3349–3366.*—In cluttered environments the breaking off of the straight homeward runs may occur after various fractions of the home vector have been run off; see, e.g., Narendra A (2007): Homing strategies of the Australian desert ant *Melophorus bagoti.* I. Proportional path-integration takes the ant half-way home. *J Exp Biol 210, 1798–1803.* Cheung A, Hiby L, Narendra A (2013): Ant navigation: fractional use of the home vector. *PLoS ONE 7, e50451.* Schwarz S, Wystrach A, Cheng K (2017): Ants' navigation in an unfamiliar environment is influenced by their experience of a familiar route. *Sci Rep 7, 14161.*—As in all these situations the break-off point depends on the varied influence of landmark information on the path integration routine, it will be considered only later (in Chapter 7).

28. For *Cataglyphis* and *Ocymyrmex* species, see, e.g., Wehner R, Räber F (1979): Visual spatial memory in desert ants, *Cataglyphis bicolor* (Hymenoptera: Formicidae). *Experientia 35, 1569–1571.* Åkesson S, Wehner R (2002): Visual navigation in desert ants, *Cataglyphis fortis:* are snapshots coupled to a celestial system of reference? *J Exp Biol 205, 1971–1978.* Wehner R, Müller M (2010): Piloting in desert ants: pinpointing the goal by discrete landmarks. *J Exp Biol 213, 4174–4179.*

29. Seidl T, Wehner R (2006): Visual and tactile learning of ground structures in desert ants. *J Exp Biol 209, 3336–3344.* Merkle T (2009): Surface structure helps desert ants return to known feeding sites. *Commun Integr Biol 2, 27–28.*

30. Steck K, Hansson BS, Knaden M (2009): Smells like home: desert ants, *Cataglyphis fortis,* use olfactory landmarks to pinpoint the nest. *Front Zool 6, 5.* Steck K, Hansson BS, Knaden M (2011): Desert ants benefit from combining visual and olfactory landmarks. *J Exp Biol 214, 1307–1312.* Bühlmann C, Hansson BS, Knaden M (2013): Flexible weighing of olfactory and vector information in the desert ant *Cataglyphis fortis. Biol Lett 9, 20130070.*

31. Note that an arithmetic spiral samples the surface homogeneously, i.e., assumes a flat target probability distribution. In this respect, a logarithmic spiral, which spends more time at the center, would seem more appropriate.

32. Wehner R, Srinivasan MV (1981): Searching behaviour of desert ants, genus *Cataglyphis* (Formicidae, Hymenoptera). *J Comp Physiol 142, 325–338.* Wehner R, Wehner S (1986): Path integration in desert ants: approaching a long-standing puzzle in insect navigation. *Monit Zool Ital NS 20, 309–331.* Müller M, Wehner R (1994): The hidden spiral: systematic search and path integration in desert ants, *Cataglyphis fortis. J Comp Physiol A 175, 525–530.* Merkle T, Wehner R (2009): How flexible is the systematic search behaviour of desert ants? *Anim Behav 77, 1051–1056.* Vickerstaff RJ, Merkle T (2012): Path integration mediated systematic search: a Bayesian model. *J Theor Biol 307, 1–19.* See also Schultheiss P, Cheng K (2011): Finding the nest: inbound searching behaviour in the Australian desert ant, *Melophorus bagoti. Anim Behav 81, 1031–1038.* Schultheiss P, Cheng K, Reynolds AM (2015): Searching behavior in social Hymenoptera. *Learn Motiv 50, 59–67.*

33. For data, see Wehner R (1992): Arthropods. In: Papi F, ed., Animal Homing, pp. 45–144. *London, Chapman and Hall.* Merkle T, Knaden M, Wehner R (2006), *n26.* Merkle T, Wehner R (2010), *n26.*—For theoretical treatment, see Heinze S, Narendra A, Cheung A (2018): Principles of insect path integration. *Curr Biol 28, R1043–R1058.* If one assumes Gaussian target uncertainty (in a geocentric representation with an allothetic compass) and an exponential detection law, the search should follow a parabolic distribution. Then the theory of optimal search predicts that both the imprecision of path integration and the search width should increase with path length, but the latter less quickly than the former. This is in accord with the data in Merkle and Wehner (2010), *n26,* but more extensive and detailed experimental work is necessary to substantiate the prediction.

34. Turner CH (1907): The homing of ants: an experimental study of ant behaviour. *J Comp Neurol Psychol 17, 367–434.*—For further historical details, see Wehner R (2016): Early ant trajectories: spatial behaviour before behaviourism. *J Comp Physiol A 202, 247–266.*—Dethier VG (1976): The Hungry Fly. *Cambridge MA, Harvard University Press,* pp. 27–30. Kim IS, Dickinson MH (2017): Idiothetic path integration in the fruit fly *Drosophila melanogaster. Curr Biol 27, 2227–2238.*

35. Fourcassié V, Traniello JFA (1994): Food search behaviour in the ant *Formica schaufussi* (Hymenoptera, Formicidae): response of naive foragers to protein and carbohydrate food. *Anim Behav 48, 69–79.* Bolek S, Wittlinger M, Wolf H (2012): What counts for ants? How return behaviour and food search of *Cataglyphis* ants are modified by variations in food quality and experience. *J Exp Biol 215, 3218–3222.* Schultheiss P, Cheng K (2013): Finding food: outbound searching behavior in the Australian desert ant *Melophorus bagoti. Behav Ecol 24, 128–135.* Pfeffer S, Bolek S, Wolf H, Wittlinger M (2015): Nest and food search behaviour in desert ants, *Cataglyphis:* a critical comparison. *Anim Cogn 18, 885–894.*—The ants' search routines have often been discussed in terms of modified random walk strategies; see also Hoffmann G (1983a, b), *n36.* In particular, they have been interpreted as a composite Brownian walk strategy, a mixture of two random walks, in which two search distributions (one with shorter and one with longer mean segment lengths) are overlaid: Narendra A, Cheng K, Sulikowski D,

Wehner R (2008): Search strategies of ants in landmark-rich habitats. *J Comp Physiol A 194, 929–938.* Schultheiss P, Cheng K (2011), *n32.* Reynolds AM, Schultheiss P, Cheng K (2014): Does the Australian desert ant *Melophorus bagoti* approximate a Lévy search by an intrinsic bi-modal walk? *J Theor Biol 340, 17–22.*

36. Hoffmann G (1983a): The random elements in the systematic search behavior of the desert isopod *Hemilepistus reaumuri. Behav Ecol Sociobiol 13, 81–92.* Hoffmann G (1983b): The search behavior of the desert isopod *Hemilepistus reaumuri* as compared with a systematic search. *Behav Ecol Sociobiol 13, 93–106.* Hoffmann G (1985): The influence of landmarks on the systematic search behaviour of the desert isopod *Hemilepistus reaumuri.* I. Role of the landmark made by the animal. II. Problems with similar landmarks and their solution. *Behav Ecol Sociobiol 17, 325–334, 335–348.* Alt W (1995): Elements of a systematic search in animal behavior and model simulations. *Biosystems 34, 11–26.*

37. Warburg MR, Linsenmair KE, Bercovitz K (1984): The effect of climate on the distribution and abundance of isopods. *Symp Zool Soc Lond 53, 339–367.*—The life cycles, especially the extended brood care systems of these xerophilous isopods and the suite of involved physiological, behavioral, and ecological adaptations have been thoroughly studied by Karl Eduard Linsenmair; see, e.g., Linsenmair KE (2007): Sociobiology of terrestrial isopods. In: Duffy JE, Thiel M, eds., Evolutionary Ecology of Social and Sexual Systems, pp. 339–364. *Oxford, Oxford University Press.*

38. Alcock J (1976): The behaviour of the seed-collecting larvae of a carabid beetle (Coleoptera). *J Nat Hist 10, 367–375.* Durier V, Rivault C (1999), *n16.*

39. A survey of the search behavior of lost people is given in Syrotuck W (2000): Analysis of Lost Person Behavior: An Aid to Search Planning. *Mechanicsburg PA, Barkleigh.* Huth JE (2013), *n3,* pp. 32–38.—For a mythological account, see Hyltén-Cavallius GO (1863): Wärend och Wirdarne. *Stockholm, Nordstedt and Söner.*

40. The first formulation of optimal search strategies and the quotation are from Koopman BO (1946): The Summary Reports Group of the Columbia University Division of War Research. *OEG Report No. 56, quot.* p. 113.—For extensive treatises, see Stone LD (1975): Theory of Optimal Search.

New York, Academic Press. Koopman BO (1979): Search and Screening. *New York, Pergamon.*

41. Dahmen H, Wahl VL, Pfeffer SE, Mallot HA, Wittlinger M (2017): Naturalistic path integration of *Cataglyphis* desert ants on an air-cushioned lightweight spherical treadmill. *J Exp Biol 220, 634–644.* Bühlmann C, Fernandes ASD, Graham P (2018): The interaction of path integration and terrestrial visual cues in desert ants: what can we learn from path characteristics? *J Exp Biol 221, jeb167304.*

42. Kim D, Hallam JCT (2000): Neural network approach in path integration for homing navigation. In: Meyer JA, Berthoz A, Floreano D, Roitblat HL, Wilson SW, eds., From Animals to Animats 6, pp. 228–235. *Cambridge MA, MIT Press.* Vickerstaff RJ, Di Paolo EA (2005), *n27.* Consult again Vickerstaff RJ, Cheung A (2010), *n9.*—The ants can adjust the spatial details of the search pattern in various ways. For example, the pattern may blend from its radially symmetric form into an asymmetric one: Wehner R, Gallizzi K, Frei C, Vesely M (2002), *n19.* Familiar landmarks may further affect the shape of the search: Narendra A (2007): Homing strategies of the Australian desert ant *Melophorus bagoti.* II. Interaction of the path integrator with visual cue information. *J Exp Biol 210, 1804–1812.*

43. Brun R (1914): Die Raumorientierung der Ameisen und das Orientierungsproblem im allgemeinen. *Jena, G. Fischer, quot.* p. 192 (translation mine). For a historical treatise, see Wehner (2016): Early ant trajectories: spatial behaviour before behaviourism. *J Comp Physiol A 202, 247–266.*

44. Müller M, Wehner R (1988), *n9; especially* Müller M (1989): Mechanismus der Wegintegration bei *Cataglyphis fortis* (Hymenoptera: Formicidae). *PhD Thesis, University of Zürich.*—Systematic 'inbound errors' have also been observed in a number of other studies on desert ants, e.g., Wehner R, Boyer M, Loertscher F, Sommer S, Menzi U (2006): Desert ant navigation: one-way routes rather than maps. *Curr Biol 16, 75–79.* Collett M, Collett TS (2009): The learning and maintenance of local vectors in desert ant navigation. *J Exp Biol 212, 895–990.*

45. For experimental data and the original computational model, see Müller M, Wehner R (1988), *n9.*—For modeling the data on the basis of leaky integration, see figure 11 in Vickerstaff RJ, Di Paolo EA (2005), *n27.*—Note, however, that the leakage rate explaining over-turns in the detour experiment is an order of magnitude larger than the one fitting underestimated distance data.—For neural network models, see Hartmann G, Wehner R (1995), *n9.* Wittmann T, Schwegler H (1995): Path integration—a network model. *Biol Cybern 73, 569–575.*

46. Bisetzky AR (1957): Die Tänze der Bienen nach einem Fussweg zum Futterplatz, unter besonderer Berücksichtigung von Umwegversuchen. *Z Vergl Physiol 40, 264–288.* Görner P (1958), *n8.* Lindauer M (1963), *n21.* Seyfarth EA, Hergenröder R, Ebbes H, Barth FG (1982), *n8.* Chapuis N, Varlet C (1987): Short cuts by dogs in natural surroundings. *Q J Exp Psychol 35B, 213–219.* Sauvé JP (1989): L'orientation spatiale: Formalisation d'un modèle de mémorisation égocentrée et expérimentation chez l'homme. *PhD Thesis, University of Aix-Marseille.* Séguinot V, Maurer R. Etienne AS (1993): Dead reckoning in a small mammal—the evaluation of distance. *J Comp Physiol A 173, 103–113.*

47. Sommer S, Wehner R (2005): Vector navigation in desert ants, *Cataglyphis fortis:* celestial compass cues are essential for the proper use of distance information. *Naturwissenschaften 92, 468–471.* Ronacher B, Westwig E, Wehner R (2006): Integrating two-dimensional paths: do desert ants process distance information in the absence of celestial compass cues? *J Exp Biol 209, 3301–3308.*—For linkages between the memories of direction and distance of traveled vector routes, see also Fernandes ASD, Philippides A, Collett TS, Niven JE (2015): Acquisition and expression of memories of distance and direction in navigating wood ants. *J Exp Biol 218, 3580–3588.*

48. Cataglyphs are able to monitor travel distances completely in the dark. Thiélin-Bescond M, Beugnon G (2005): Vision-independent odometry in the ant *Cataglyphis cursor. Naturwissenschaften 92, 193–197.*

49. For a first experiment of that kind in bees, see Bisetzky AR (1957), *n46.*—In bees, the odometer used in path integration is gated by information from the celestial compass, but the odometer that signals the total distance traveled is not: Dacke M, Srinivasan MV (2008): Two odometers in honeybees? *J Exp Biol 211, 3281–3286.*—See also Dacke M, Srinivasan MV (2007): Honeybee navigation: distance estimation in the third dimension. *J Exp Biol 210, 845–853.* Evangelista C, Kraft P, Dacke M, Labhart T, Srinivasan MV

(2014): Honeybee navigation: critically examining the role of the polarization compass. *Phil Trans R Soc B 369, 20130037.*

50. Hartmann G, Wehner R (1995), *n9.* Wittmann T, Schwegler H (1995), *n45.*

51. Müller M, Homberg U, Kühn A (1997): Neuroarchitecture of the lower division of the central body in the brain of the locust *(Schistocerca gregaria). Cell Tiss Res 288, 159–176.*

52. Stone T, Webb B, Adden A, Wedding NB, Honkanen A, et al., Heinze S (2017): An anatomically constrained model for path integration in the bee brain. *Curr Biol 27, 3069–3085.*—This publication is a state-of-the-art neurocomputational account that sets the stage for unraveling the neural underpinnings of highly advanced behavioral routines. For overviews, opinions, and considerations based on the neuroarchitecture presented in that paper, see Heinze S (2017): Unraveling the neural basis of insect navigation. *Curr Opin Ins Sci 24, 58–67.* Collett M, Collett TS (2017): Path integration: combining optic flow with compass orientation. *Curr Biol 27, R1113–R1116.* Heinze S, Narendra A, Cheung A (2018), *n33.* Beetz MJ, el Jundi B (2018): Insect orientation: stay on course with the sun. *Curr Biol 28, R933–R936.*—Similar topics are treated for *Drosophila* flies in Seelig JD, Jayaraman V (2015): Neural dynamics for landmark orientation and angular path integration. *Nature 521, 186–191.* Kim SS, Rouault H, Druckmann S, Jayaraman V (2017): Ring attractor dynamics in the *Drosophila* central brain. *Science 356, 849–853.*—As fruit flies have recently been shown also to be able to perform path integration (Kim IS, Dickinson MH (2017), *n34*), they will offer a great potential for neurogenetically dissecting the path integrator.

53. As it is beyond the scope of this book to present Stone et al.'s model in any detail, I refer the interested reader to the references in *n52.* In the model, elaborate as it is, several aspects are still hypothetical. For example, at the time of this writing, there are no physiological data available yet for the integrating CPU4 cells, the core of path integration. It is completely on the basis of their anatomical connectivity patterns that they are assumed to encode the continually updated home vector.

54. Hartmann G, Wehner R (1995), *n9.* Wittmann T, Schwegler H (1995), *n45.* Vickerstaff RJ, Di Paolo EA (2005), *n27.* Haferlach T, Wessnitzer J, Mangan M, Webb B (2007):

Evolving a neural model of insect path integration. *Adapt Behav 15, 273–287.* Goldschmidt D, Manoonpong P, Dasgupta S (2017): A neurocomputational model of goal-directed navigation in insect-inspired artificial agents. *Front Neurorobot 11, 20.*

6. Using Landmarks

1. Bowditch N (2002): The American Practical Navigator: An Epitome of Navigation. *Bethesda MD, National Imagery and Mapping Agency.*—For a similar use of the term 'pilotage' in avian and mammalian navigation, see Bingman VP (1998): Spatial representation and homing pigeon navigation. In: Healy S, ed., Spatial Representation in Animals, pp. 69–85. *Oxford, Oxford University Press.* Wallace DG, Hines DJ, Gorny JH, Whishaw IQ (2003): A role for the hippocampus in dead reckoning: an ethological analysis using natural exploratory and food-carrying tasks. In: Jeffery KJ, ed., The Neurobiology of Spatial Behaviour, pp. 31–47. *Oxford, Oxford University Press.*—In bird navigation 'piloting' has often been defined in a wider and sometimes ambiguous way: Griffin DR (1952): Bird navigation. *Biol Rev Cambridge Phil Soc 27, 359–400.* Able KP (2000): The concept and terminology of bird navigation. *J Avian Biol 31, 174–183.*—For a telling example of the effect of piloting in *Ocymyrmex* ants, see Wehner R, Müller M (2010): Piloting in desert ants: pinpointing the goal by discrete landmarks. *J Exp Biol 213, 4174–4179.*

2. Bouvier EL (1901): Les habitudes des *Bembex* (Monographie biologique). *Ann Psychol 7, 1–68.* Tinbergen N (1932): Über die Orientierung des Bienenwolfes *(Philanthus triangulum). Z Vergl Physiol 16, 305–335.* Tinbergen N, Kruyt W (1938): Über die Orientierung des Bienenwolfes. III. Die Bevorzugung bestimmter Wegmarken. *Z Vergl Physiol 25, 292–334.* Van Beusekom G (1948): Some experiments on the orientation in *Philanthus triangulum. Behaviour 1, 195–225.* Van Iersel JJA (1975): The extension of the orientation system of *Bembix rostrata* as used in the vicinity of the nest. In: Baerends GP, Beer C, Manning A, eds., Function and Evolution in Behaviour, pp. 142–168. *Oxford, Clarendon Press.*—For the sand wasp *Ammophila,* see Baerends GP (1941): Fortpflanzungsverhalten und Orientierung der Grabwespe *Ammophila campestris. Tijdschr Entomol 84, 68–275.*

3. 'Anticipatory foraging behavior': Baerends GP (1941), *n2*. Rosenheim JA (1987): Host location and exploitation by the cleptoparasitic wasp *Argochrysis armilla*: the role of learning (Hymenoptera: Chrysididae). *Behav Ecol Sociobiol 21, 401–406*. Van Nouhuys S, Kaartinen R (2008): A parasitoid wasp uses landmarks while monitoring potential resources. *Proc R Soc Lond B 275, 377–385*.

4. Seidl T, Wehner R (2006): Visual and tactile learning of ground structures in desert ants. *J Exp Biol 209, 3336–3344*. Bühlmann C, Hansson BS, Knaden M (2012): Desert ants learn vibration and magnetic landmarks. *PLoS ONE 7, e33117*.

5. Wehner R, Räber F (1979): Visual spatial memory in desert ants, *Cataglyphis bicolor* (Hymenoptera: Formicidae). *Experientia 35, 1569–1571*. Wehner R, Michel B, Antonsen P (1996): Visual navigation in insects: coupling of egocentric and geocentric information. *J Exp Biol 199, 129–140*. Cheng K, Narendra A, Sommer S, Wehner R (2009): Traveling in clutter: navigation in the central Australian desert ant *Melophorus bagoti*. *Behav Processes 80, 261–268*.

6. For similar experiments in wood ants, see Graham P, Durier V, Collett TS (2004): The binding and recall of snapshot memories in wood ants *(Formica rufa)*. *J Exp Biol 207, 393–398*.

7. In wood ants the importance of the frontal retina for viewing discrete parts of the visual scene is emphasized in Graham P, Collett TS (2002): View-based navigation in insects: how wood ants *(Formica rufa)* look at and are guided by extended landmarks. *J Exp Biol 205, 2499–2509*. Durier V, Graham P, Collett TS (2003): Snapshot memories and landmark guidance in wood ants. *Curr Biol 13, 1614–1618*.

8. Several early displacement experiments in honeybees already indicate that memorizing visual scenes is important for the insect's homing abilities. Romanes G (1885): Homing faculty of Hymenoptera. *Nature 32, 630*. Wolf E (1926, 1927): Über das Heimfindevermögen der Bienen. I; II. *Z Vergl Physiol 3, 615–691; 6, 221–254*. Uchida T, Kuwabara M (1951): The homing instinct of the honey bee, *Apis mellifica*. *J Fac Sci Hokkaido Univ Ser VI Zool 10, 87–96*. Becker L (1958): Untersuchungen über das Heimfindevermögen der Bienen. *Z Vergl Physiol 41, 1–25*.—It has often been observed that when the entrance hole of a honeybee colony is displaced by only some tens of centimeters, the returning for-

agers first fly to the old site of the hole and hover there for some time in a dense cloud about an imagery point in space, until they finally locate the new site. Upon return from their subsequent foraging trips they continue to approach first the old site and fly from there to the new site, for hours and sometimes even days; e.g., Butler CG, Fletcher DJC, Watler D (1970): Hive entrance finding by honeybee *(Apis mellifera)* foragers. *Anim Behav 18, 78–91*.—For more recent studies on food site rather than nest-site localization, see Lehrer M (1980): Die Lokalisation eines Raumpunktes bei der optische Nahorientierung der Honigbiene *(Apis mellifera)*. *PhD Thesis, University of Zürich*. Cartwright BA, Collett TS (1983): Landmark learning in bees: experiments and models. *J Comp Physiol A 151, 521–543*.

9. See the classic studies of the behavior of sphecid wasps upon leaving and returning to the goal cited in *n2* and van Iersel JJA, van der Assem J (1964): Aspects of orientation in the digger wasp *Bembix rostrata*. *Anim Behav Suppl 1, 145–162*. Zeil J (1993): Orientation flights of solitary wasps *(Cerceris*; Sphecidae; Hymenoptera). I. Description of flight. II. Similarities between orientation and return flights and the use of motion parallax. *J Comp Physiol A 172, 189–205, 207–222*.

10. Collett TS, Land MF (1975): Visual spatial memory in a hoverfly. *J Comp Physiol 100, 59–84*.—In a similar vein, guard bees of the meliponine bee *Trigona angustula* hover stably in midair in front of the entrance of the nest for many minutes. They persistently return to their hovering position after they have briefly left it, e.g., for chasing off a robber-bee intruder. They even maintain this position when the nest entrance is slowly moved sideways, upward or downward: Zeil J, Wittmann D (1989): Visually controlled station-keeping by hovering guard bees of *Trigona (Tetragonisca) angustula* (Apidae, Meliponinae). *J Comp Physiol A 165, 711–718*. Kelber A, Zeil J (1990): A robust procedure of visual stabilization of hovering flight position in guard bees of *Trigona (Tetragonisca) angustula* (Apidae, Meliponinae). *J Comp Physiol A 167, 569–577*.

11. Ofstad TA, Zuker CS, Reiser MB (2011): Visual place learning in *Drosophila melanogaster*. *Nature 474, 204–207*.—The thermal-visual arena used by the authors was inspired by the famous Morris water maze frequently used in spatial orientation tasks in rodents, and a heat maze, in which

cockroaches and crickets had previously been tested: Morris RGM (1984): Development of a water-maze procedure for studying spatial-learning in the rat. *J Neurosci Methods 11, 47–60.* Mizunami M, Weibrecht JM, Strausfeld NJ (1998): Mushroom bodies of the cockroach: their participation in place memory. *J Comp Neurol 402, 520–537.* Wessnitzer J, Mangan M, Webb B (2008): Place memory in crickets. *Proc R Soc B 275, 915–921.*—For a review of the work done in *Drosophila,* see Ostrowski D, Zars T (2014): Place learning. In: Dubnau J, ed., Behavioral Genetics of the Fly *(Drosophila melanogaster),* pp. 125–134. *Cambridge, Cambridge University Press.*

12. Lent DD, Graham P, Collett TS (2009): A motor component to the memories of habitual foraging routes in wood ants? *Curr Biol 19, 115–121.* Lent DD, Graham P, Collett TS (2010): Image-matching during ant navigation occurs through saccade-like body turns controlled by learned visual features. *Proc Natl Acad Sci USA 107, 16348–16353.* Lent DD, Graham P, Collett TS (2013): Phase-dependent visual control of the zigzag paths of navigating wood ants. *Curr Biol 23, 1–7.* Collett TS, Lent DD, Graham P (2014): Scene perception and visual control of travel direction in navigating wood ants. *Proc R Soc B 369, 20130035.*

13. After a fixation episode the ants are even able to continue their trajectories for substantial distances, though less precisely, when the goal landmark has been made to disappear. For short-term memory of orientation direction in *Drosophila,* see Strauss R, Pichler J (1998): Persistence of orientation toward a temporarily invisible landmark in *Drosophila melanogaster. J Comp Physio A 182, 411–423.* Neuser K, Triphan T, Mronz M, Poeck B, Strauss R (2008): Analysis of spatial orientation memory in *Drosophila. Nature 453, 1244–1247.*—Head direction cells in the central complex of *Drosophila* retain a short-term orientation for up to 30 s when landmarks are temporarily out of sight: Seelig JD, Jayaraman V (2015): Neural dynamics for landmark orientation and angular path integration. *Nature 521, 186–191.*

14. Judd SPD, Collett TS (1998): Multiple stored views and landmark guidance in ants. *Nature 392, 710–714.* Nicholson DJ, Judd SPD, Cartwright BA, Collett TS (1999): Learning walks and landmark guidance in wood ants, *Formica rufa. J Exp Biol 202, 1831–1838.*

15. For the importance of global features (center of mass, fractional partition of mass), see Harris RA, Graham P, Collett TS (2007): Visual cues for the retrieval of landmark memories by navigating wood ants. *Curr Biol 17, 93–102.* Lent DD, Graham P, Collett TS (2013), *n12.* Woodgate JL, Bühlmann C, Collett TS (2016): When navigating wood ants use the centre of mass of a shape to extract directional information from a panoramic skyline. *J Exp Biol 219, 1689–1696, quot.* p. 1696. Bühlmann C, Woodgate JL, Collett TS (2016): On the encoding of panoramic visual scenes in navigating wood ants. *Curr Biol 26, 2022–2027.*—As to feature extraction, neural correlates have been found in some insect species, e.g., O'Carroll D (1993): Feature-detecting neurons in dragonflies. *Nature 362, 541–543.* Maddess T, Yang E (1997): Orientation-sensitive neurons in the brain of the honey bee *(Apis mellifera). J Insect Physiol 43, 329–336.* Seelig JD, Jayaraman V (2013): Feature detection and orientation in the *Drosophila* central complex. *Nature 503, 262–266.*

16. The first TM model was developed by Cartwright BA, Collett TS (1983), *n8.*—For extensions and further modeling approaches, see Möller R, Lambrinos D, Pfeifer R, Labhart T, Wehner R (1998): Modeling ant navigation with an autonomous agent. In: Pfeifer R, Blumberg B, Meyer JA, Wilson SW, eds., From Animals to Animats 5, pp. 185–194. *Cambridge MA, MIT Press.* Möller R, Lambrinos D, Roggendorf T, Pfeifer R, Wehner R (2001): Insect strategies of visual homing in mobile robots. In: Webb B, Consi TR, eds., Biorobotics: Methods and Applications, pp. 37–66. *Menlo Park CA, AAAI Press; Cambridge MA, MIT Press.* Möller R (2001): Do insects use templates or panoramas for landmark information? *J Theor Biol 210, 33–45.* Möller R, Vardy A (2006): Local visual homing by matched-filter descent in image distances. *Biol Cybern 95, 413–430.*

17. For a better-fitting 'partial TM model,' in which matching occurs sequentially with individual landmarks, see Möller R, Lambrinos D, Pfeifer R, Wehner R (1999): Do desert ants use partial image matching for landmark navigation? *Proc Neurobiol Conf Göttingen 27, 430.* Nicholson DJ, Judd SPD, Cartwright BA, Collett TS (1999), *n14.*—Some further snapshot-based accounts: Möller R (2001), *n16.* Graham P, Durier V, Collett TS (2004), *n6.* Narendra A, Si A, Sulikowski D, Cheng K (2007): Learning, retention and coding of nest-associated cues by the Aus-

tralian desert ant, *Melophorus bagoti. Behav Ecol Sociobiol 61, 1543–1553.*

18. Wehner R, Müller M (1985): Does interocular transfer occur in visual navigation by ants? *Nature 315, 228–229.* Antonsen P, Wehner R (1995): Visual field topology of the desert ant's snapshot. *Proc Neurobiol Conf Göttingen 23, 42.* Wehner R, Michel B, Antonsen P (1996), *n5.*

19. For frontal fixation of terrestrial cues, see especially Judd SPD, Collett TS (1998) and Nicholson DJ, Judd SPD, Cartwright BA, Collett TS (1999), *n14.* Laboratory experiments in wood ants have also shown that snapshots may be extensive and include widely separated objects. Durier V, Graham P, Collett TS (2003), *n7.*—For the use of whole panoramic scenes, see Fourcassié V (1991): Landmark orientation in natural situations in the red wood ant *Formica lugubris* (Hymenoptera, Formicidae). *Ethol Ecol Evol 3, 89–99.* Pastergue-Ruiz I, Beugnon G, Lachaud JP (1995): Can the ant *Cataglyphis cursor* (Hymenoptera: Formicidae) encode global landmark-landmark relationships in addition to isolated landmark-goal relationships? *J Insect Behav 8, 115–132.* Wehner R, Michel B, Antonsen P (1996), *n5.* Graham P, Cheng K (2009): Which portion of the natural panorama is used for view-based navigation in the Australian desert ant? *J Comp Physiol A 195, 681–689.* Reid SF, Narendra A, Hemmi JM, Zeil J (2011): Polarized skylight and the landmark panorama provide night-active bull ants with compass information during route following. *J Exp Biol 214, 363–370.* —When approaching a goal surrounded by a set of artificial landmarks, *Ocymyrmex robustior* does not seem to fixate the individual landmarks with its frontal field of view, but runs at a rather steady pace toward the goal. Wehner R, Müller M (2010), *n1.*

20. Åkesson S, Wehner R (2002): Visual navigation in desert ants, *Cataglyphis fortis:* are snapshots coupled to a celestial system of reference? *J Exp Biol 205, 1971–1978.*—For a study in honeybees, see Dickinson JA (1994): Bees link local landmarks with celestial compass cues. *Naturwissenschaften 81, 465–467.*

21. The Sahabot projects were performed in cooperation with Rolf Pfeifer from the Informatics Institute in Zürich; Dimitrios Lambrinos and Ralf Möller, then graduate students in informatics; and Thomas Labhart in our neuroethology group: Lambrinos D, Kobayashi H, Pfeifer R, Maris M, Labhart T, Wehner R (1997): An autonomous agent navigating with polarized light. *Adapt Behav 6, 131–161.* Lambrinos D, Möller R, Labhart T, Pfeifer R, Wehner R (2000): A mobile robot employing insect strategies. *Robot Autonom Syst 30, 39–64.* Möller R, Lambrinos D, Roggendorf T, Pfeifer R, Wehner R (2001), *n16.*

22. For example, see two early position papers by Barbara Webb: Webb B (2000): What does robotics offer animal behavior? *Anim Behav 60, 545–558.* Webb B (2001): Can robots make good models of biological behavior? *Behav Brain Sci 24, 1033–1050.*—Moreover: Dean J (1998): Animats and what they can tell us. *Trends Cogn Sci 2, 60–67.*

23. Lambrinos D, Möller R, Labhart T, Pfeifer R, Wehner R (2000), *n21.* Möller R (2000): Insect visual homing strategies in a robot with analog processing. *Biol Cybern 83, 231–243.* Hafner V (2001): Adaptive homing—robotic exploration tours. *Adapt Behav 9, 131–141.*—In the original AVL model distance information is ignored. For extensions, see Yu SE, Kim DE (2011): Landmark vectors with quantified distance information for homing navigation. *Adapt Behav 19, 121–141.* Lee C, Yu SE, Kim DE (2017): Landmark-based homing navigation using omnidirectional depth information. *Sensors 17, 1928.*—Ralf Möller and Andrew Vardy have provided a systematic review and a classificatory scheme of various methods of local visual homing. Möller R, Vardy A (2006), *n16.* Möller R (2012): A model of ant navigation based on visual prediction. *J Theor Biol 305, 118–130.*—In trying to mimic visual place memories of crickets, Michael Mangan and Barbara Webb have compared different models including versions of the ones mentioned in the present account. Mangan M, Webb B (2009): Modelling place memory in crickets. *Biol Cybern 101, 307–323.*

24. Zeil J, Hofmann MI, Chahl JS (2003): Catchment areas of panoramic snapshots in outdoor scenes. *J Opt Soc Am A 20, 450–469.* Stürzl W, Zeil J (2007): Depth, contrast and view-based matching homing in outdoor scenes. *Biol Cybern 96, 519–531.*—See also Möller R, Vardy A, Kreft S, Ruwitsch S (2007): Visual homing in environments with anisotropic landmark distribution. *Autonom Robots 23, 231–245.*

25. Franz MO, Schölkopf B, Mallot HA, Bülthoff HH (1998): Where did I take that snapshot? Scene-based homing by image motion. *Biol Cybern 79, 191–202.* Möller R, Krzykawski M, Gerstmayr L (2010): Three 2D-warping

schemes for visual robot navigation. *Autonom Robot 29, 253–291.* Möller R (2012), *n23.* Zhu Q, Liu C, Cai C (2015): A novel robot visual homing method based on SIFT features. *Sensors 15, 26063–26084.*

26. Wystrach A, Beugnon G, Cheng K (2011): Landmarks or panoramas: what do navigating ants attend to for guidance? *Front Zool 8, 21.* Reid SF, Narendra A, Hemmi JM, Zeil J (2011), *n19.*—For a discussion of the use of local and global features for guidance in *Formica* ants, see Collett TS, Lent DD, Graham P (2014), *n12.*—Note also the insightful early conclusion drawn in Vowles DM (1958): The perceptual world of ants. *Anim Behav 6, 115–116.*

27. Wystrach A, Dewar ADM, Philippides A, Graham P (2016): How do field of view and resolution affect the information content of panoramic scenes for visual navigation? A computational investigation. *J Comp Physiol A 202, 87–95.*— For the information content of spatially low-pass filtered natural scenes, see Vardy A, Möller R (2005): Biologically plausible visual homing methods based on optic flow techniques. *Connect Sci 17, 47–89.* Stürzl W, Mallot HA (2006): Efficient visual homing based on Fourier transformed panoramic images. *Robot Autonom Syst 54, 300–313.*—The benefits that large-field vision with low spatial and high temporal resolution provide for fast panoramic motion perception have led to the design of biomimetic compound eyes; see, e.g., Floreano D, Pericet-Camara R, Viollet S, Ruffier F, Brückner A, et al., Franceschini N (2013): Miniature curved artificial compound eyes. *Proc Natl Acad Sci USA 110, 9267–9272.* Song YM, Xie Y, Malyarchuk V, Xiao J, Jung I, et al., Rogers JA (2013): Digital cameras with designs inspired by the arthropod eye. *Nature 497, 95–99.*

28. Gronenberg W, Hölldobler B (1999): Morphologic representation of visual and antennal information in the ant brain. *J Comp Neurol 412, 229–240.*

29. Wehner R (1982): Himmelsnavigation bei Insekten. Neurophysiologie und Verhalten. *Neujahrsbl Naturf Ges Zürich 184, 1–132.*

30. Möller R (2002): Insects could exploit UV-green contrast for landmark navigation. *J Theor Biol 214, 619–631.* Kollmeier T, Röben F, Schenck W, Möller R (2007): Spectral contrast for landmark navigation. *J Opt Soc Am A 24, 1–10.* Stone T, Mangan M, Ardin B, Webb B (2014): Sky segmentation with ultraviolet images can be used for navigation. In:

Proceedings of Robotics: Science and Systems. Berkeley CA. Differt D, Möller R (2015): Insect models of illumination-invariant skyline extraction from UV and green channels. *J Theor Biol 380, 444–462.* Differt D, Möller R (2016): Spectral skyline separation: extended landmark databases and panoramic imaging. *Sensors 16: s16101614.*

31. Schultheiss P, Wystrach A, Schwarz S, Tack A, Delor J, Nooten SS, et al., Cheng K (2016): Crucial role of ultraviolet light for desert ants in determining direction from the terrestrial panorama. *Anim Behav 115, 19–28.*

32. Menzel R, Blakers M (1975): Functional organization of an insect ommatidium with fused rhabdom. *Cytobiol 11, 279–298.* Mote MI, Wehner R (1980): Functional characteristics of photoreceptors in the compound eye and ocellus of the desert ant, *Cataglyphis bicolor. J Comp Physiol 137, 63–71.* Labhart T (1986): The electrophysiology of photoreceptors in different eye regions of the desert ant, *Cataglyphis bicolor. J Comp Physiol A 158, 1–7.*

33. Briscoe A, Chittka L (2001): The evolution of colour vision in insects. *Annu Rev Entomol 46, 471–510.* Ogawa Y, Falkowski M, Narendra A, Zeil J, Hemmi JM (2015): Three spectrally distinct photoreceptors in diurnal and nocturnal Australian ants. *Proc R Soc B 282, 20150673.*—Blue-sensitive opsin gene expression has been found in some *Acromyrmex, Harpegnathos,* and *Camponotus* species: Yilmaz A, Lindenberg A, Albert S, Grübel K, Spaethe J, Rössler W, Groh C (2016): Age-related and light-induced plasticity in opsin gene expression and in primary and secondary visual centers of the nectar-feeding ant *Camponotus rufipes. Dev Neurobiol 76, 1041–1057.*

34. Graham P, Cheng K (2009), *n19.* Graham P, Cheng K (2009): Ants use the panoramic skyline as a visual cue during navigation. *Curr Biol 19, R935–R937.*—Simulations in virtual desert ant environments show that the skyline provides sufficient information for visually based route guidance: Basten K, Mallot HA (2010): Simulated visual homing in desert ant natural environments: efficiency of skyline cues. *Biol Cybern 102, 413–425.* Schwarz S, Julle-Daniere E, Morin L, Schultheiss P, Wystrach A, Ives J, Cheng K (2014): Desert ants *(Melophorus bagoti)* navigating with robustness to distortions of the natural panorama. *Insect Soc 61, 371–383.*

35. Hölldobler B, Wilson EO (1990): The Ants. *Cambridge MA, Harvard University Press*, pp. 265–286. Hölldobler B, Wilson EO (2009): The Superorganism. *New York, Norton*, quot. p. 61. Dussutour A, Nicolis SC, Shephard G, Beekman M, Sumpter DJT (2009): The role of multiple pheromones in food recruitment by ants. *J Exp Biol 212, 2337–2348*. Czaczkes TJ, Grüter C, Ratnieks FLW (2015): Trail pheromones: an integrative view of their role in social insect colony organization. *Annu Rev Entomol 60, 581–599*.—For details of 'antennating' in trail sampling, see Draft RW, McGill MR, Kapoor V, Murthy VN (2018): Carpenter ants use diverse antennae sampling strategies to track odor trails. *J Exp Biol 221, jeb185124*.

36. For a review on ants in general, see Bos N, D'Ettore P (2012): Recognition of social identity in ants. *Front Psychol 3:83*.—Studies on mate recognition in *Cataglyphis* species are referenced in Chapter 1, *n16*.

37. Steck K, Hansson BS, Knaden M (2009): Smells like home: desert ants, *Cataglyphis fortis,* use olfactory landmarks to pinpoint the nest. *Front Zool 6, 5*. Steck K, Hansson BS, Knaden M (2010): Do desert ants smell the scenery in stereo? *Anim Behav 79, 939–945*. Steck K, Hansson BS, Knaden M (2011): Desert ants benefit from combining visual and olfactory landmarks. *J Exp Biol 214, 1307–1312*. Bühlmann C, Hansson BS, Knaden M (2012): Path integration controls nest-plume following in desert ants. *Curr Biol 22, 645–649*. Bühlmann C, Graham P, Hansson BS, Knaden M (2014): Desert ants locate food by combining high sensitivity to food odors with extensive crosswind runs. *Curr Biol 24, 960–964*. Bühlmann C, Graham P, Hansson BS, Knaden M (2015): Desert ants use olfactory scenes for navigation. *Anim Behav 106, 99–105*. Huber R, Knaden M (2018): Desert ants possess distinct memories for food and nest odors. *Proc Natl Acad Sci USA 115, 10470–10474*.

38. The cataglyphs' behavior in small-scale odor arrays is reminiscent of Auguste Forel's somewhat vague concept of a 'topochemical sense.' Forel A (1902): Die psychischen Fähigkeiten der Ameisen und einiger anderer Insekten, mit einem Anhang über die Eigentümlichkeiten des Geruchssinnes bei jenen Tieren. *Munich, Reinhardt*.—For a critical consideration of Forel's concept, see Brun R (1916): Weitere Untersuchungen über die Fernorientierung der Ameisen. *Biol Centralbl 36, 261–303*.

39. Collett M, Collett TS, Bisch S, Wehner R (1998): Local and global vectors in desert ant navigation. *Nature 394, 269–272*. Bisch S, Wehner R (1998): Visual navigation in ants: evidence for site-based vectors. *Proc Neurobiol Conf Göttingen 26, 417*.—For honeybees, see Collett TS, Baron J, Sellen K (1996): On the encoding of movement vectors by honeybees. Are distance and direction represented independently? *J Comp Physiol A 179, 395–406*. Srinivasan MV, Zhang SW, Bidwell NJ (1997): Visually mediated odometry in honeybees. *J Exp Biol 200, 2513–2522*. Collett M, Harland D, Collett TS (2002): The use of landmarks and panoramic context in the performance of local vectors by navigating honeybees. *J Exp Biol 205, 807–814*.

40. Collett TS, Collett M, Wehner R (2001): The guidance of desert ants by extended landmarks. *J Exp Biol 204, 1635–1639*. Bisch-Knaden S, Wehner R (2001): Egocentric information helps desert ants to navigate around familiar obstacles. *J Exp Biol 204, 4177–4184*. Modifying the discussion in the latter study, the landmark cues provided by the barrier have been made rather (but not fully) inconspicuous. Hence, visual cues as well as motor memories may have contributed to the ants' behavior. Further note that when the ants have reached the end of the barrier, they are provided with a path integration vector pointing back to the point of release.

41. Collett M, Collett TS, Bisch S, Wehner R (1998), *n39*.

42. Knaden M, Lange C, Wehner R (2006): The importance of procedural knowledge in desert-ant navigation. *Curr Biol 16, 916–917*. Finally, see also a sophisticated study showing that depending on their motivational state (fed or unfed) zero-vector ants exhibit homeward or foodward local vectors, respectively: Fernandes ASD, Philippides A, Collett TS, Niven JE (2015): Acquisition and expression of memories of distance and direction in navigating wood ants. *J Exp Biol 218, 3580–3588*.

43. In Figure 6.13b the learned motor component could mean: turn to the right when having reached the end of the channel. As to the general role of learned motor components, at present there are too few data at hand about how ants learn to negotiate T- or Y-mazes in the dark. The most amazing results in this respect have been reported for neotropical *Gigantiops* ants by Macquart D, Latil G, Beugnon G

(2008): Sensorimotor sequence learning in the ant *Gigantiops destructor. Anim Behav 75, 1693–1701;* but see the remark on *Lasius niger* in Jones S (2013): Chemical based communication and its role in decision making within the social insects. *PhD Thesis, University of Sussex,* p. 60.—For experiments in which walking ants experienced visual stimuli that jumped or disappeared on an LED screen, see Lent DD, Graham P, Collett TS (2009), *n12.*—Honeybees can recall learned sequences of motor commands in the absence of visual cues: Collett TS, Fry SN, Wehner R (1993): Sequence learning by honeybees. *J Comp Physiol A 172, 693–706.*—For route-segment odometry in honeybees, see Srinivasan MV, Zhang SW, Bidwell NJ (1997), *n39.* Collett M, Harland D, Collett TS (2002), *n39.*

44. See also Collett M, Collett TS (2009): The learning and maintenance of local vectors in desert ant navigation. *J Exp Biol 212, 895–900.* Collett TS, Collett M (2015): Route-segment odometry and its interaction with global path-integration. *J Comp Physiol A 201, 617–630.*

45. For impressive spatial representations of trunk trail systems in various ant species, see, e.g., Hölldobler B (1976): Recruitment behavior, home range orientation and territoriality in harvester ants, *Pogonomyrmex. Behav Ecol Sociobiol 1, 3–44.* Neumeyer R (1994): Strategie der Nahrungsbeschaffung syntoper Arten der Ernteameisengattung *Messor* im mitteltunesischen Steppengebiet. *PhD Thesis, University of Zürich.* Wirth R, Herz H, Ryel RJ, Beyschlag W, Hölldobler B (2003): Herbivory of Leaf-Cutting Ants: A Case Study on *Atta colombica* in the Tropical Rainforest of Panama. *Berlin, Springer.* Jackson DE, Martin SJ, Holcombe M, Ratnieks FLW (2006): Longevity and detection of persistent foraging trails in Pharaoh's ants, *Monomorium pharaonis. Anim Behav 71, 351–359.*

46. Route guidance by means of visual spatial memories has been reported in a number of early papers for several *Formica, Camponotus,* and *Pogonomyrmex* species as well as some ponerine ants (species of *Paltothyreus, Pachycondyla, Odontomachus, Paraponera,* and *Dinoponera*). A few exemplary accounts are Rosengren R (1977): Foraging strategies of wood ants (*Formica rufa* group). I. Age polyethism and topographic traditions. *Acta Zool Fenn 149, 1–30.* Fresneau D (1985): Individual foraging and path fidelity in a ponerine ant. *Insect Soc 32, 109–116.* Oliveira PS, Hölldobler B

(1989): Orientation and communication in the neotropical ant *Odontomachus bauri* Emery (Hymenoptera, Formicidae, Ponerinae). *Ethology 83, 154–166.* Fourcassié V, Henriques A, Fontella C (1999): Route fidelity and spatial orientation in the ant *Dinoponera gigantea* (Hymenoptera, Formicidae) in a primary forest: a preliminary study. *Sociobiology 34, 505–524.*

47. In route-following experiments, in which the significance of visual versus chemical cues has been assessed, the visual cues have usually been the dominant ones; see, e.g., David CT, Wood DL (1980): Orientation to trails by a carpenter ant, *Camponotus modoc* (Hymenoptera: Formicidae), in a giant sequoia forest. *Canad Entomol 112, 993–1000.* Breed MD, Fewell JH, Moore AJ, Williams KR (1987): Graded recruitment in a ponerine ant. *Insect Soc 34, 222–226.* Beugnon G, Fourcassié V (1988): How do red wood ants orient during diurnal and nocturnal foraging in a three dimensional system? II. Field experiments. *Insect Soc 35, 106–124.* Aron S, Deneubourg JL, Pasteels JM (1988): Visual cues and trail-following idiosyncrasies in *Leptothorax unifasciatus:* an orientation process during foraging. *Insect Soc 35, 355–366.* Harrison JF, Fewell JH, Stiller TM, Breed MD (1989): Effects of experience on use of orientation cues in the giant tropical ant. *Anim Behav 37, 869–871.* Card A, McDermott C, Narendra A (2016): Multiple orientation cues in the Australian trunk-trail-forming ant, *Iridomyrmex purpureus. Austr J Zool 64, 227–232.*—For detailed studies on this subject in the mass-recruiting black garden ant, see Grüter C, Czaczkes TJ, Ratnieks FLW (2011): Decision making in ant foragers *(Lasius niger)* facing conflicting private and social information. *Behav Ecol Sociobiol 65, 141–148.* Jones S (2013), *n43,* pp. 59–78.

48. Rosengren R (1971): Route fidelity, visual memory and recruitment behaviour in foraging wood ants of the genus *Formica* (Hymenoptera, Formicidae). *Acta Zool Fenn 133, 1–106.* Rosengren R, Pamilo P (1978): Effect of winter timber felling on behaviour of foraging wood ants (*Formica rufa* group) in early spring. *Mem Zool 29, 143–155.* Ebbers BC, Barrows EM (1980): Individual ants specialize on particular aphid herds (Hymenoptera: Formicidae; Homoptera: Aphid2dae). *Proc Ent Soc Wash 82, 405–407.* Rosengren R, Fortelius W (1986): Ortstreue in foraging ants of the *Formica rufa* group—hierarchy of orientation cues and

long-term memory. *Insect Soc 33, 306–337.* Fourcassié V (1991), *n19.*

49. Habitual foraging and homing routes have been recorded in several *Cataglyphis, Ocymyrmex,* and *Melophorus* species: Wehner R, Michel B, Antonsen P (1996), *n5.*—This is the first study of idiosyncratic route traveling by zero-vector ants.—Kohler M, Wehner R (2005): Idiosyncratic route-based memories in desert ants, *Melophorus bagoti:* how do they interact with path-integration vectors? *Neurobiol Learn Mem 83, 1–12.* Muser B, Sommer S, Wolf H, Wehner R (2005): Foraging ecology of the thermophilic Australian desert ant, *Melophorus bagoti. Austr J Zool 53, 301–311.* Sommer S, von Beeren C, Wehner R (2008): Multiroute memories in desert ants. *Proc Natl Acad Sci USA 105, 317–322.* Mangan M, Webb B (2012): Spontaneous formation of multiple routes in individual desert ants *(Cataglyphis velox). Behav Ecol 23, 944–954.* Sommer S, Weibel D, Blaser N, Furrer A, Wenzler NE, Rössler W, Wehner R (2013): Group recruitment in a thermophilic desert ant, *Ocymyrmex robustior. J Comp Physiol A 199, 711–722.*

50. Wehner R, Boyer M, Loertscher F, Sommer S, Menzi U (2006): Desert ant navigation: one-way routes rather than maps. *Curr Biol 16, 75–79.*

51. Field experiments in *Cataglyphis fortis* also show that landmark-based local vectors are used only in the navigational context in which they have been learned. Bisch-Knaden S, Wehner R (2003): Local vectors in desert ants: context-dependent learning during outbound and inbound runs. *J Comp Physiol A 189, 181–187.*—When in laboratory experiments wood ants experienced a single black landmark along a short outbound route, they could acquire some landmark information later used during their inbound run, and vice versa. Obviously, this learning occurs when the ants briefly reverse direction and retrace their steps for a short distance. Graham P, Collett TS (2006): Bi-directional route learning in wood ants. *J Exp Biol 209, 3677–3684.*

52. See also Harris RA, Hempel de Ibarra N, Graham P, Collett TS (2005): Ant navigation: Priming of visual route memories. *Nature 438, 302.*

53. Sommer S, von Beeren C, Wehner R (2008), *n49.*—As shown for *Cataglyphis velox,* desert ants can acquire multiple inward and outward routes also spontaneously: Mangan M, Webb B (2012), *n49.*—Landmark memories of nest sites are long-lasting and retained without reinforcement for periods

that surpass the foragers' lifespan in the field: at least for 8 days in *M. bagoti,* 10 days in *C. bicolor,* and 20 days in *C. fortis;* longer times not tested: Narendra A, Si A, Sulikowski D, Cheng K (2007), *n17.* Wehner R (1981): Spatial vision in arthropods. In: Autrum H, ed., Handbook of Sensory Physiology, vol. 7/6C, pp. 287–616. *Berlin, Springer.* Ziegler PE, Wehner R (1997): Time-courses of memory decay in vector-based and landmark-based systems of navigation in desert ants, *Cataglyphis fortis. J Comp Physiol A 181, 13–20.*—See also Freas CA, Whyte C, Cheng K (2017): Skyline retention and retroactive interference in the navigating Australian desert ant, *Melophorus bagoti. J Comp Physiol A 203, 353–367.*

54. For manipulations of route landmarks in field studies on desert ants, see Collett TS, Dillmann E, Giger A, Wehner R (1992): Visual landmarks and route following in desert ants. *J Comp Physiol A 170, 435–442.* Wehner R, Michel B, Antonsen P (1996), *n5.* Bregy P, Sommer S, Wehner R (2008): Nest-mark orientation versus vector navigation in desert ants. *J Exp Biol 211, 1868–1873.* Collett M (2010): How desert ants use a visual landmark for guidance along a habitual route. *Proc Natl Acad Sci USA 107, 11638–11643.* Wystrach A, Schwarz S, Schultheiss P, Beugnon G, Cheng K (2011): Views, landmarks, and routes: how do desert ants negotiate an obstacle course? *J Comp Physiol A 197, 167–179.* Wystrach A, Beugnon G, Cheng K (2012): Ants might use different view-matching strategies on and off the route. *J Exp Biol 215, 44–55.*—In several laboratory studies performed in wood ants and rain forest ants artificial landmarks have been used to investigate route learning in considerable detail: e.g., in Graham P, Collett TS (2002), *n7.* Graham P, Fauria K, Collett TS (2003): The influence of beacon aiming on the routes of wood ants. *J Exp Biol 206, 535–541.* Macquart D, Gamier L, Combe M, Beugnon G (2006): Ant navigation en route to the goal: signature routes facilitate way-finding of *Gigantiops destructor. J Comp Physiol A 192, 221–234.*

55. Macquart D, Gamier L, Combe M, Beugnon G (2006), *n54.*—For the number of ommatidia per compound eye of *Gigantiops destructor,* see Gronenberg W, Hölldobler B (1999), *n28.*

56. Kohler M, Wehner R (2005), *n49.*

57. For considerations of how insects may partition their routes into segments defined by distinct landmark

cues, and how they may recall visual route memories by sequentially retrieving sets of visual landmark memories, see Collett TS, Fry SN, Wehner R (1993), *n43*. Zhang S, Mizutani A, Srinivasan MV (2000): Maze navigation by honeybees: learning path regularity. *Learn Mem 7, 363–374*. Collett TS, Collett M (2002): Memory use in insect navigation. *Nat Rev Neurosci 3, 542–552*. Collett TS, Graham P, Harris RA, Hempel de Ibarra N (2006): Navigational memories in ants and bees: memory retrieval when selecting and following routes. *Adv Stud Behav 36, 123–172*.

58. Riabinina O, Hempel de Ibarra N, Howard L, Collett TS (2011): Do wood ants learn sequences of visual stimuli? *J Exp Biol 214, 2739–2748*.—A theoretical approach of linking together local navigation methods (implemented on a gantry robot platform and tested in a simple visual environment) revealed limitations of such local linking strategies: Smith L, Philippides A, Graham P, Baddeley B, Husbands P (2007): Linked local navigation for visual route guidance. *Adapt Behav 15, 257–271*.

59. Scanning movements of route-traveling ants were first video-recorded in *Melophorus*: Wystrach A, Philippides A, Aurejac A, Cheng K, Graham P (2014): Visual scanning behaviours and their role in the navigation of the Australian desert ant *Melophorus bagoti*. *J Comp Physiol A 200, 615–626*.—For saccadic turning movements recorded in the laboratory in wood ants, see Lent DD, Graham P, Collett TS (2010) and (2013), *n12*.—The term 'alignment image matching' was introduced in Collett M, Chittka L, Collett TS (2013): Spatial memory in insect navigation. *Curr Biol 23, R789–R800*.

60. Kodzhabashev A, Mangan M (2015): Route following without scanning. In: Wilson SP, Verschure PFMJ, Mura A, Prescott TJ, eds., Biomimetic and Biohybrid Systems, pp. 199–210. *Barcelona, Springer*.—For klinokinesis, see Gomez-Marin A, Louis M (2012): Active sensation during orientation behavior in the *Drosophila* larva: more sense than luck. *Curr Opin Neurobiol 22, 208–215*.

61. By applying the learning algorithm 'Infomax,' Baddeley and his colleagues trained simulated ants along artificial routes and adapted the synaptic weightings in a two-layer network by the input information derived from the images that the ants successively experienced along the route. This way, they have completely dispensed with the assumption that routes are divided into segments, which are demarcated by landmarks. Lulham A, Bogacz R, Vogt S, Brown MW (2011): An Infomax algorithm can perform both familiarity discrimination and feature extraction in a single network. *Neural Comput 23, 909–926*. Baddeley B, Graham P, Philippides A, Husbands P (2011): Holistic visual encoding of ant-like routes: navigation without waypoints. *Adapt Behav 19, 3–15*. Baddeley B, Graham P, Husbands P, Philippides A (2012): A model of route navigation driven by scene familiarity. *PLoS Comput Biol 8, e1002336*. Philippides A, Graham P, Baddeley B, Husbands P (2015): Using neural networks to understand the information that guides behavior: a case study in visual navigation. *Methods Mol Biol 1260, 227–244*.

62. Wystrach A, Mangan M, Philippides A, Graham P (2013): Snapshots in ants? New interpretations of paradigmatic experiments. *J Exp Biol 216, 1766–1770*.

63. Duelli P (1975): A fovea for e-vector orientation in the eye of *Cataglyphis bicolor* (Formicidae, Hymenoptera). *J Comp Physiol 102, 43–56*. Pfeffer SE, Wittlinger M (2016): Optic flow odometry operates independently of stride integration in carried ants. *Science 353, 1155–1157*.

64. Ardin P, Mangan M, Wystrach A, Webb B (2015): How variation in head pitch could affect image matching algorithms for ant navigation. *J Comp Physiol A 201, 585–597*.

65. Raderschall CA, Narendra A, Zeil J (2016): Head role stabilisation in the nocturnal bull ant *Myrmecia pyriformis*: implications for visual navigation. *J Exp Biol 219, 1449–1457*. Freas CA, Wystrach A, Narendra A, Cheng K (2018): The view from the trees: nocturnal bull ants, *Myrmecia midas*, use the surrounding panorama while descending from trees. *Front Psychol 9, 16*.

66. Bowditch N (2002), *n1, quot.* p. 105.

67. 'Novices' are defined as all unmarked ants that appear at the nest opening after all foragers have been marked by a day-specific color code during the preceding at least three days. For rationale, see Wehner R, Müller M (1993): How do ants acquire their celestial ephemeris function? *Naturwissenschaften 80, 331–333*.

68. Wehner R, Meier C, Zollikofer CPE (2004): The ontogeny of foraging behaviour in desert ants, *Cataglyphis bicolor. Ecol Entomol 29, 240–250*. Stieb SM, Hellwig A, Wehner R, Rössler W (2012): Visual experience affects both behav-

ioral and neural aspects in the individual life history of the desert ant *Cataglyphis fortis*. *Dev Neurobiol 72, 729–742*. Fleischmann PN, Christian M, Müller VL, Rössler W, Wehner R (2016): Ontogeny of learning walks and the acquisition of landmark information in desert ants, *Cataglyphis fortis*. *J Exp Biol 219, 3137–3145*.

69. For relearning walks, see Müller M (1984): Interokularer Transfer bei der Wüstenameise *Cataglyphis fortis* (Formicidae, Hymenoptera). *Diploma Thesis, University of Zürich*. Müller M, Wehner R (2010): Path integration provides a scaffold for landmark learning in desert ants. *Curr Biol 20, 1368–1371*.

70. The fine structure of the pirouettes and other types of motor routines in the learning walks of desert ants have been analyzed by Müller M, Wehner R (2010), *n69*. Fleischmann PN, Grob R, Wehner R, Rössler W (2017): Species-specific learning walk choreographies in *Cataglyphis* desert ants. *J Exp Biol 219, 3137–3145*. Jayatilaka P, Murray T, Narendra A, Zeil J (2018): The choreography of learning walks in the Australian jack jumper ant *Myrmecia croslandi*. *J Exp Biol 221, jeb185306*.

71. In 1829 the French entomologist Gaspard Auguste Brullé participated in a Morea (Peloponnese) expedition, in which he described *Cataglyphis nodus* [*Formica nodus*] on the basis of worker specimens, but did not provide a differential diagnosis. Brullé GA (1832): Expédition scientifique de Morée. Section de sciences physiques, vol. 3, pt. 1, Zoologie, sect. 2, Des animaux articulés, *Strasbourg, F. G. Levrault*, pp. 326–327, plate 48 / 1.

72. Graham P, Philippides A, Baddeley B (2010): Animal cognition: multimodal interactions in ant learning. *Curr Biol 20, R639–R640*.—For more extensive modeling, see Baddeley B, Graham P, Husbands P, Philippides A (2012), *n61*. Dewar ADM, Philippides A, Graham P (2014): What is the relationship between visual environment and the form of the ant learning walks? An *in silico* investigation of insect navigation. *Adapt Behav 22, 163–179*.—Furthermore, Narendra A, Gourmaud S, Zeil J (2013): Mapping the navigational knowledge of individually foraging ants, *Myrmecia croslandi*. *Proc R Soc B 280, 20130683*. Zeil J, Narendra A, Stürzl W (2014): Looking and homing: how displaced ants decide where to go. *Phil Trans R Soc B 369, 20130034*. Stürzl W, Grixa I, Mair E,

Narendra A, Zeil J (2015): Three-dimensional models of natural environments and the mapping of navigational information. *J Comp Physiol A 201, 563–584*.

73. The use of absolute distance in localizing a goal has been demonstrated in honeybees and halictid bees. Lehrer M, Collett TS (1994): Approaching and departing bees learn different cues to the distance of a landmark. *J Comp Physiol A 175, 171–177*. Brünnert U, Kelber A, Zeil J (1994): Ground-nesting bees determine the location of their nest relative to a landmark by other than angular size cues. *J Comp Physiol A 175, 363–369*. It may be worth mentioning that in arena experiments, in which cylindrical landmarks are covered by the same random dot patterns as the surroundings, honeybees can locate a goal by motion cues alone. Dittmar L, Stürzl W, Baird E, Boeddeker N, Egelhaaf M (2010), *n31*. Dittmar L (2011): Static and dynamic snapshots for goal localization in insects? *Comm Integr Biol 4, 17–20*.

74. Von Berlepsch A (1860): Die Biene und die Bienenzucht. *Mühlhausen, F. Heinrichshofen*, pp. 36, 176. Von Buttel-Reepen H (1900): Sind die Bienen Reflexmaschinen? Viertes Stück. *Biol Zbl 20, 209–224*. Bates HW (1864): The Naturalist on the River Amazons. 2nd ed. *London, J. Murray*, pp. 221–222. Belt T (1874): The Naturalist in Nicaragua. *London, J. Murray*, p. 136. Peckham GW, Peckham EG (1898): On the instincts and habits of solitary wasps. *Bull Wisc Geol Nat Hist Surv 2, 1–245*, pp. 215–218. Wagner W (1907): Psychobiologische Untersuchungen an Hummeln. *Zoologica 19, 1–239*, pp. 51–77.

75. The first and already paradigmatic video study on learning flights in hymenopterans is Zeil J (1993), *n9*.—For detailed recent studies, see Philippides A, Hempel de Ibarra N, Riabinina O, Collett TS (2013): Bumblebee calligraphy: the design and control of flight motifs in the learning and return flights of *Bombus terrestris*. *J Exp Biol 216, 1093–1104*. Riabinina O, Hempel de Ibarra N, Philippides A, Collett TS (2014): Head movements and the optic flow generated during the learning flights of bumblebees. *J Exp Biol 217, 2633–2642*. Stürzl W, Zeil J, Boeddeker N, Hemmi JM (2016): How wasps acquire and use views for homing. *Curr Biol 26, 470–482*.—For learning flights of vespid wasps on departure from a newly discovered feeder, see Collett TS, Lehrer M (1993): Looking and learning: a spatial pattern in

the orientation flight of the wasp *Vespula vulgaris*. *Proc R Soc Lond B 252, 129–134.*

76. After the observations made by August von Berlepsch and Hugo von Buttel-Reepen in the nineteenth century, *n72*, the role of the honeybees' orientation flights in learning large-scale landscape features was first systematically studied by Lore Becker at the University of München, Germany, followed and extended by Elizabeth Capaldi and her associates in Michigan and Illinois. In these studies Capaldi was also the first to apply harmonic radar techniques for recording orientation flights (only four years after this method had been introduced in tracking honeybees). Becker L (1958), *n8.* Capaldi EA, Dyer EC (1999): The role of orientation flights on homing performance in honeybees. *J Exp Biol 202, 1655–1666.* Riley JR, Smith AD, Reynolds DR, Edwards AS, Osborne JL, et al., Poppy GM (1996): Tracking bees with harmonic radar. *Nature 379, 29–30.* Capaldi EA, Smith AD, Osborne JL, Fahrbach SE, Farris SM, et al., Riley JR (2000): Ontogeny of orientation flight in the honeybee revealed by harmonic radar. *Nature 403, 537–540.*—For an extensive recent radar study, see Degen J, Kirbach A, Reiter L, Lehmann K, Norton P, et al., Menzel R (2015): Exploratory behaviour in honeybees during orientation flights. *Anim Behav 102, 45–57.*

77. In general, relearning walks or flights are elicited, when the returning foragers experience an altered visual environment around the goal. For references, see *n9.* Relearning walks in the cataglyphs: Müller M (1984), *n69.* Fleischmann PN, Christian M, Müller VL, Rössler W, Wehner R (2016), *n68.*—Even though the former relearning study precedes the learning study by more than 30 years, the data are directly comparable, as in both cases identical experimental procedures and arrays of landmarks have been used.—Novice honeybees leaving their hive for the first time exhibit a straight-way-out and straight-way-in flight pattern. Later learning (orientation) flights are longer and lead in different directions. When subsequently displaced (as zero-vector bees) to various release sites around the hive, the bees return faster and along straighter flights from areas explored during their orientation flights than from unexplored areas: Degen J, Kirbach A, Reiter L, Lehmann K, Norton P, et al., Menzel R (2016): Honeybees learn landscape features during exploratory orientation flights. *Curr Biol 26, 2800–2804.*

78. Fleischmann PN, Rössler W, Wehner R (2018): Early foraging life: spatial and temporal aspects of landmark learning in the ant *Cataglyphis nodus*. *J. Comp. Physiol. A 204, 579–592.* The quotation is adopted from Feynman RP (1960): There's plenty of room at the bottom. *Caltech Engin Sci 23, 22–36.* For the use of water-filled moats in studying *Drosophila* behavior, see Grover D, Katsuki T, Grenspan RJ (2016): Flyception: imaging brain activity in freely walking fruit flies. *Nature Methods 13, 569–572.*

79. Freas CA, Cheng K (2018): Landmark learning, cue conflict, and outbound view sequence in navigating desert ants. *J Exp Psychol Anim Learn Cogn, 44 409–421.*—For turn-backs during the outbound path, see Zeil J, Narendra A, Stürzl W (2014), *n72*, and (in wood ants) upon leaving a food patch, see Nicholson DJ, Judd SPD, Cartwright BA, Collett TS (1999), *n14.*

80. Wystrach A, Philippides A, Aurejac A, Cheng K, Graham P (2014), *n59.* Zeil J, Narendra A, Stürzl W (2014), *n72.*—The data given in the text refer to the latter study performed in *Myrmecia* ants.

81. Dujardin F (1850): Mémoire sur le système nerveux des insectes. *Ann Sci Nat Sér 3 (Zool Biol Anim) 14, 195–206.* Kenyon FC (1896): The meaning and structure of the so-called "mushroom bodies" of the hexapod brain. *Am Nat 30, 643–650.*—For illuminating reviews of mushroom body function, see, e.g., Heisenberg M (1998): What do the mushroom bodies do for the insect brain? An introduction. *Learn Mem 5, 1–10.* Strausfeld NJ, Hansen L, Li Y, Gomez RS, Ito K (1998): Evolution, discovery, and interpretations of arthropod mushroom bodies. *Learn Mem 5, 11–37.* Fahrbach SE (2006): Structure of the mushroom bodies of the insect brain. *Annu Rev Entomol 51, 209–232.* Strausfeld NJ, Sinakevitch I, Brown SM, Farris SM (2009): Ground plan of the insect mushroom body: functional and evolutionary implications. *J Comp Neurol 513, 265–291.* See especially the extensive treatment in Nicholas Strausfeld's magnificent opus magnum: Strausfeld NJ (2012): Arthropod Brains: Evolution, Functional Elegance, and Historical Significance. *Cambridge MA, Harvard University Press,* pp. 242–309, 450–451, 548. Farris SM, van Dyke JW (2015): Evolution and function of the insect mushroom bodies: contributions from comparative and model systems studies. *Curr Opin Ins Sci 12, 19–25.*—From the early 1980s onward, studies in honeybees and fruit flies have re-

vealed the mushroom bodies as focal brain areas for olfactory associative learning and memory. For review and reflections, see Zars T (2000): Behavioral function of the insect mushroom body. *Curr Opin Neurobiol 10, 790–795.* Heisenberg M (2003): Mushroom body memoir: from maps to models. *Nature Neurosci 4, 266–275.*—Recently, by applying genetic manipulation technologies, a host of research schools has raised *Drosophila* to *the* model organism for unraveling the cellular details of the associative learning circuits housed in the mushroom bodies: e.g., Aso Y, Hattori D, Yu Y, Johnston RM, Iyer NA, et al., Rubin GM (2015): The neuronal architecture of the mushroom body provides a logic for associative learning. *eLife 3, e04577.*

82. Vowles DM (1955): The structure and connections of the corpora pedunculata in bees and ants. *Quart J Microsc Sci 96, 239–255, quot.* p. 252.

83. Gronenberg W, Hölldobler B (1999), *n28.*

84. For numbers of Kenyon cells per mushroom body in *Drosophila* (about 2,000 neurons), see Technau GM, Heisenberg M (1982): Neural reorganization during metamorphosis of the corpora pedunculata in *Drosophila melanogaster. Nature 295, 405–407.* Heisenberg M (2003), *n78.* Aso Y, Grübel K, Busch S, Friedrich AB, Siwanowicz I, Tanimoto H (2009): The mushroom body of adult *Drosophila* characterized by GAL4 drivers. *J Neurogen 23, 156–172.*—For *Apis mellifera* workers (about 180,000 neurons), see Witthöft W (1967): Absolute Anzahl und Verteilung der Zellen im Hirn der Honigbiene. *Z Morph Tiere 61, 160–184.* Strausfeld NJ (2002): Organization of the honey bee mushroom body: representation of the calyx within the vertical and gamma lobes. *J Comp Neurol 450, 4–33.*—For *Camponotus* worker ants (at least 130,000 neurons), see Ehmer B, Gronenberg W (2004): Mushroom body volumes and visual interneurons in ants: comparison between sexes and castes. *J Comp Neurol 469, 198–213.*

85. Darwin C (1871): The Descent of Man, and Selection in Relation to Sex. *London, J. Murray.* Reprint by *Princeton University Press,* 1981, *quot.* p. 145.—For in-depth accounts on the early anatomical work on mushroom bodies especially in bees, flies, and cockroaches, see Strausfeld NJ (2012), *n81.*

86. Pietschker H (1911): Das Gehirn der Ameise. *Jena Z Med Naturwiss 47, 43–114.* Thompson A (1913): A comparative study of the brains of three genera of ants, with special reference to the mushroom bodies. *J Comp Neurol 23, 515–572.* Pandazis G (1930): Über die relative Ausbildung der Gehirnzentren bei biologisch verschiedenen Ameisenarten. *Z Morph Ökol Tiere 18, 114–169.*

87. Ramón y Cajal S, Sánchez y Sánchez D (1915): Contribución al conocimiento de los centros nerviosos de los insectos. Part 1. Retina y centros opticós. *Trab Lab Invest Biol Univ Madrid 13, 1–168.*

88. The estimated number of microglomeruli per collar is based on confocal imaging studies of Stieb SM, Münz TS, Wehner R, Rössler W (2010): Visual experience and age affected synaptic organization in the mushroom bodies of the desert ant *Cataglyphis fortis. Dev Neurobiol 70, 408–423.* Grob R, Fleischmann PN, Grübel K, Wehner R, Rössler W (2017): The role of celestial compass information in *Cataglyphis* ants during learning walks and for neuroplasticity in the central complex and mushroom bodies. *Front Behav Neurosci 11, 226.* Rössler W (2019): Neuroplasticity in desert ants (Hymenoptera: Formicidae)—importance for the ontogeny of navigation. *Myrm News 29, 1–20.*

89. Steiger U (1967): Über die Funktion des Neuropils im Corpus pedunculatum der Waldameise. *Z Zellforsch 81, 511–53.* Although microglomeruli had briefly been described already five years earlier, this previous study did not allow a decision on the polarity of the synapses and thus an understanding of the functional significance of these synaptic complexes: Trujillo-Cenóz O, Melamed J (1962): Electron microscope observations on the calyces of the insect brain. *J Ultrastr Res 7, 389–398.*—First studies in bees and flies: Ganeshina O, Menzel R (2001): GABA-immunoreactive neurons in the mushroom bodies of the honeybee: an electron microscopic study. *J Comp Neurol 437, 335–349.* Yasuyama K, Meinertzhagen IA, Schürmann FW (2002): Synaptic organization of the mushroom body calyx in *Drosophila melanogaster. J Comp Neurol 445, 211–226.*

90. On mushroom body architecture: Strausfeld NJ, Hansen L, Li Y, Gomez RS, Ito K (1998): Evolution, discovery, and interpretation of arthropod mushroom bodies. *Learn Mem 5, 11–37.* Strausfeld NJ (2012), *n81.*—For the structure of mushroom bodies in ants, see Gronenberg W (1999): Modality-specific segregation of input to ant mushroom bodies. *Brain Behav Evol 54, 85–95.* Gronenberg W,

Hölldobler B (1999), *n28*. Gronenberg W (2001): Subdivisions of hymenopteran mushroom body calyces by their afferent supply. *J Comp Neurol 436, 474–489*. Ehmer B, Gronenberg W (2004), *n84*. Gronenberg W, Lopez-Riquelme GO (2004): Multisensory convergence in the mushroom bodies of ants and bees. *Acta Biol Hung 55, 31–37*. Gronenberg W (2008): Structure and function of ant (Hymenoptera: Formicidae) brains: strength in numbers. *Myrm News 11, 25–36*.

91. Painstaking neurometric studies on the divergence-convergence ratios are currently performed by Wolfgang Rössler and his collaborators in *Cataglyphis* and other ants. In honeybees the convergence of Kenyon cells to mushroom body output cells might exceed even 500:1. For data, see Mobbs P (1982): The brain of the honeybee *Apis mellifera*. I. The connections and spatial organization of the mushroom bodies. *Phil Trans R Soc B 298, 309–354*. Rybak J, Menzel R (1993): Anatomy of the mushroom bodies in the honey bee brain: the neuronal connections of the alpha-lobe. *J Comp Neurol 334, 444–465*. Strausfeld NJ (2002), *n84*. For sparsening, see Jortner RA, Farivar SS, Laurent G (2007): A simple connectivity scheme for sparse coding in an olfactory system. *J Neurosci 27, 1659–1669*. Honegger KS, Campbell RA, Turner GC (2011): Cellular-resolution population imaging reveals robust sparse coding in the *Drosophila* mushroom body. *J Neurosci 31, 11772–11785*. See also Rössler W, Groh C (2012): Plasticity of synaptic microcircuits in the mushroom-body calyx of the honeybee. In: Galizia CG, Eisenhardt D, Giurfa M, eds., Honeybee Neurobiology and Behavior. *Heidelberg, Springer*, pp. 141–153.

92. Vowles DM (1955), *n82*. Vowles DM (1964): Olfactory learning and brain lesions in the wood ant *(Formica rufa)*. *J Comp Physiol Psychol 58, 105–111*. Vowles DM (1967): Interocular transfer, brain lesions, and maze learning in the wood ant, *Formica rufa*. In: Corning WC, Ratner SC, eds., Chemistry of Learning: Invertebrate Research, pp. 425–447. *New York, Springer*.—For the discovery of the input synapses to the lobes, see Schürmann FW (1972): Über die Struktur der Pilzkörper des Insektenhirns. II. Synaptische Schaltungen im Alpha-Lobus des Heimchens *Acheta domesticus*. *Z Zellforsch 127, 240–257*.—Besides his work on the mushroom bodies, Vowles performed quite a number of experiments on the orientation behavior in ants: Vowles DM (1950): Sensitivity of ants to polarized light. *Nature 165,*

282–283. Vowles DM (1955): The foraging of ants. *Brit J Anim Behav 3, 1–13*. Vowles DM (1958), *n26*. Vowles DM (1965): Maze learning and visual discrimination in the wood ant *(Formica rufa)*. *Brit J Psychol 56, 15–31*.

93. Goll W (1967): Strukturuntersuchungen am Gehirn von *Formica*. *Z Morph Ökol Tiere 59, 143–210*.—The lamination of the pedunculus and the lobes had already caught the eye of early neuroanatomists including Frederick Kenyon, but Goll's degeneration experiments revealed a correspondence between the calycal rings (concentric organization) and the layers in the pedunculus and lobes (slablike organization; Goll erroneously concluded that the concentric organization was maintained). The types of Kenyon cells portrayed by Goll for ants correspond well to those described 15 years later by Peter Mobbs for honeybees. Mobbs PG (1982), *n91*.—For detailed recent analyses, see, e.g., Strausfeld NJ (2002), *n84*. Strausfeld NJ, Sinakevitch I, Brown SM, Farris SM (2009), *n81*.

94. Gronenberg W, Heeren S, Hölldobler B (1996): Age-dependent and task-related morphological changes in the brain and the mushroom bodies of the ant *Camponotus floridanus*. *J Exp Biol 199, 2011–2019*. Gronenberg W, Hölldobler B (1999), *n28*. Hoyer SC, Liebig J, Rössler W (2005): Biogenic amines in the ponerine ant *Harpegnathos saltator*: serotonin and dopamine immunoreactivity in the brain. *Arthropod Struct Develop 34, 429–440*. Kühn-Bühlmann S, Wehner R (2006): Age-dependent and task-related volume changes in the mushroom bodies of visually guided desert ants, *Cataglyphis bicolor*. *J Neurobiol 66, 511–521*. Stieb SM, Münz TS, Wehner R, Rössler W (2010), *n88*.

95. The 'social brain hypothesis' first developed for primates—Dunbar RIM (1998): The social brain hypothesis. *Evol Anthropol 6, 178–190*—is not supported for insects: Gronenberg W, Riveros AJ (2009): Social brains and behavior: past and present. In: Gadau J, Fewell J, eds., Organization of Insect Societies, pp. 377–401. *Cambridge MA, Harvard University Press*. Farris SM, Schulmeister S (2011): Parasitoidism, not sociality, is associated with the evolution of elaborate mushroom bodies in the brains of hymenopteran insects. *Proc R Soc B 278, 940–951*. Farris SM (2013): Evolution of complex higher brain centers and behaviors: Behavioral correlates of mushroom body elaboration in insects. *Brain Behav Evol 82, 9–18*.—For the behavioral challenges to

be met by parasitoid wasps, see *n2* and *n3* as well as Hoedjes KM, Kruidhof HM, Huigens ME, Dicke M, Vet LEM, Smid HM (2011): Natural variation in learning rate and memory dynamics in parasitoid wasps: opportunities for converging ecology and neuroscience. *Proc R Soc B 278, 889–897.*—For non-hymenopterans, see Farris SM, Roberts NS (2005): Co-evolution of generalist feeding ecologies and gyrencephalic mushroom bodies in insects. *Proc Natl Acad Sci USA 102, 17394–17399.* Kinoshita M, Shimohigasshi M, Tominaga Y, Arikawa K, Homberg U (2015): Topographically distinct visual and olfactory inputs to the mushroom body in the Swallowtail butterfly, *Papilio xuthus. J Comp Neurol 523, 162–182.*

96. Gronenberg W (2001), *n90.* Gronenberg W, Riveros AJ (2009), *n95.* Groh R, Kelber C, Grübel K, Rössler W (2014): Density of mushroom-body synaptic complexes limits intraspecies brain miniaturization in highly polymorphic leafcutting ants. *Proc R Soc B 281, 20140432.* Vogt K, Aso Y, Hige T, Knapek S, Ichinose T, et al., Tanimoto H (2016): Direct neural pathways convey distinct visual information to *Drosophila* mushroom bodies. *eLife 5, e14009.* Grob R, Fleischmann PN, Grübel K, Wehner R, Rössler W (2017), *n88.* Rössler W (2019), *n88.*

97. Wessnitzer J, Young YM, Armstrong JD, Webb B (2012): A model of non-elemental olfactory learning in *Drosophila. J Comp Neurosci 32, 197–212.* Ardin P, Peng F, Mangan M, Lagogiannis K, Webb B (2016): Using an insect mushroom body circuit to encode route memory in complex natural environments. *PLoS Comput Biol 12, e1004683.* This is a brilliant account on how mushroom body circuitries can be used to model the ants' route following behavior. In addition: Webb B, Wystrach A (2016): Neural mechanisms of insect navigation. *Curr Opin Ins Sci 15, 27–39.* See also Peng F, Chittka L (2017): A simple computational model of the bee mushroom body can explain seemingly complex forms of olfactory learning and memory. *Curr Biol 27, 224–230.*

98. Mizunami M, Weibrecht JM, Strausfeld NJ (1998), *n11.* See also Lent DD, Pintér M, Strausfeld NJ (2007): Learning with half a brain. *Dev Neurobiol 67, 740–751.*—Lutz CC; Robinson GE (2013): Activity-dependent gene expression in honey bee mushroom bodies in response to orientation flight. *J Exp Biol 216, 2031–2038.*

99. The first accounts on morphological changes occurring in the mushroom bodies of honeybees, social wasps, and ants during the indoor-outdoor activity transition are the following: Withers GS, Fahrbach SE, Robinson GE (1993): Selective neuroanatomical plasticity and division of labour in the honeybee. *Nature 364, 238–240.* Gronenberg W, Heeren S, Hölldobler B (1996), *n94.* O'Donnell S, Donlan NA, Jones TA (2004): Mushroom body structural change is associated with division of labor in eusocial wasp workers (*Polybia aequatorialis,* Hymenoptera: Vespidae). *Neurosci Lett 356, 159–162.*—For the cataglyphs, see Kühn-Bühlmann S, Wehner R (2006): Age-dependent and task-related volume changes in the mushroom bodies of visually guided desert ants, *Cataglyphis bicolor. J Neurobiol 66, 511–521.*

100. Fahrbach SE, Strande JL, Robinson GE (1995): Neurogenesis is absent in the brains of adult honey bees and does not explain behavioral neuroplasticity. *Neurosci Lett 197, 145–148.* Gronenberg W, Heeren S, Hölldobler B (1996), *n94.* Ishii Y, Kubota K, Hara K (2005): Postembryonic development of the mushroom bodies in the ant, *Camponotus japonicus. Zool Sci 22, 743–753.*—Contrasting the situation in bees and ants, adult neurogenesis has been detected in crickets, some moths, and beetles. For example: Bieber M, Fuldner D (1979): Brain growth during the adult stage of a holometabolous insect. *Naturwissenschaften 66, 426.* Dufour MC, Gadenne C (2006): Adult neurogenesis in a moth brain. *J Comp Neurol 495, 635–664.* Cayre M, Scotto-Lomassese S, Malaterre J, Strambi C, Strambi A (2007): Understanding the regulation and function of adult neurogenesis: contribution from an insect model, the house cricket. *Chem Senses 32, 385–395.* Farris SM, Pettrey C, Daly KC (2011): A subpopulation of mushroom body intrinsic neurons is generated by protocerebral neuroblasts in the tobacco hornworm moth, *Manduca sexta* (Sphingidae, Lepidoptera). *Arthropod Struct Dev 40, 395–408.*

101. Synaptic plasticity within the microglomeruli became apparent only after immunohistochemical methods for double-labeling pre- and postsynaptic sites had become routine practice The two pioneering studies are Frambach I, Rössler W, Winkler M, Schürmann FW (2004): F-actin at identified synapses in the mushroom body neuropil of the insect brain. *J Comp Neurol 475, 303–314.* Groh C, Tautz J, Rössler W (2004): Synaptic organization in the adult honey bee brain is influenced by brood-temperature control during pupal development. *Proc Natl Acad Sci USA 101,*

4268–4273.—For a recent review, see Fahrbach SE, van Nest BN (2016): Synapsin-based approaches to brain plasticity in adult social insects. *Curr Opin Ins Sci 18, 27–34.*

102. For *Cataglyphis,* see Seid MA, Wehner R (2009): Delayed axonal pruning in the ant brain: a study of developmental trajectories. *Dev Neurobiol 69, 350–364.* Stieb SM, Münz TS, Wehner R, Rössler W (2010), *n88.* Stieb SM, Hellwig A, Wehner R, Rössler W (2012), *n68.*—For a first comprehensive review on neural plasticity of the lateral complex and the mushroom bodies in the *Cataglyphis* brain as related to behavioral activities, see Rössler W (2019), *n88.*—For honeybees, see Groh C, Lu Z, Meinertzhagen IA, Rössler W (2012): Age-related plasticity in the synaptic ultrastructure of neurons in the mushroom body calyx of the adult honeybee *Apis mellifera. J Comp Neurol 520, 3509–3527.* Scholl C, Wang Y, Krischke M, Müller MJ, Amdam GV, Rössler W (2014): Light exposure leads to reorganization of microglomeruli in the mushroom bodies and influences juvenile hormone levels in the honeybee. *Dev Neurobiol 74, 1141–1153.*

103. For increased density of energy consumption in densely packed synaptic complexes, and miniaturized nervous systems in general, see Laughlin SB, de Ruyter van Steveninck RR, Anderson JC (1998): The metabolic cost of neural information. *Nat Neurosci 1, 36–41.* Attwell D, Laughlin SB (2001): An energy budget for signaling in the grey matter of the brain. *J Cereb Blood Flow Metab 21, 1133–1145.*—For brains as 'expensive tissues,' see also Niven JE, Laughlin SB (2008): Energy limitation as a selective pressure on the evolution of sensory systems. *J Exp Biol 211, 1792–1804.*

104. Grob R, Fleischmann PN, Grübel K, Wehner R, Rössler W (2017), *n88.*—For similar structural changes in olfactory learning of bees and ants, see Hourcade B, Münz TS, Sandoz JC, Rössler W, Devaud JM (2010): Long-term memory leads to synaptic reorganization in the mushroom bodies: a memory trace in the insect brain? *J Neurosci 30, 6461–6465.* Falibene A, Roces F, Rössler W (2015): Long-term avoidance memory formation is associated with a transient increase in mushroom body synaptic complexes in leaf-cutting ants. *Front Behav Neurosci 9, 84.*

105. Dean P, Porrill J, Ekerot CF, Jörntell H (2009): The cerebellar microcircuit as an adaptive filter: experimental and computational evidence. *Nature Rev Neurosci 11, 30–43.*

106. Kenyon FC (1896), *n81,* p. 647. Schürmann FW (1974): Bemerkungen zur Funktion der Corpora pedunculata im Gehirn der Insekten aus morphologischer Sicht. *Exp Brain Res 19, 406–432.* Li YS, Strausfeld NJ (1997): Morphology and sensory modality of mushroom body extrinsic neurons in the brain of the cockroach, *Periplaneta americana. J Comp Neurol 387, 631–650.* Li YS, Strausfeld NJ (1999): Multimodal efferent and recurrent neurons in the medial lobes of the of cockroach mushroom bodies. *J Comp Neurol 409, 647–663.* Farris SM (2011): Are mushroom bodies cerebellum-like structures? *Arthropod Struct Dev 40, 368–379.*

107. For the classic formulation of the reafference principle, see von Holst E, Mittelstaedt H (1950): Das Reafferenzprinzip. *Naturwissenschaften 37, 464–476,* and the concomitant account Sperry RW (1950): Neural basis if the spontaneous optokinetic response produced by visual inversion. *J Comp Physiol Psychol 43, 482–489.* In these papers, the authors introduce the concept of the 'efference copy' or 'corollary discharge,' respectively. Far-reaching implications and extensions of the reafference principle have been provided by Webb B (2004): Neural mechanisms for prediction: Do insects have forward models? *Trends Neurosci 27, 278–282.*

7. Organizing the Journey

1. "Le vermisseau n'est qu'un vermisseau. C'est-à-dire que la petitesse qui vous dérobe son organisation lui ôte son merveilleux." Asséat J, Tourneux M, eds. (1875–1877): Œuvres complètes de Diderot. Tome 2. *Paris, Garnier, quot.* p. 133.

2. Bregy P, Sommer S, Wehner R (2008): Nest-mark orientation versus vector navigation in desert ants. *J Exp Biol 211, 1868–1873.*—For tests of similar cue-conflict situations, see, e.g., Collett TS, Dillmann E, Giger A, Wehner R (1992): Visual landmarks and route following in desert ants. *J Comp Physiol A 170, 435–442.* Collett TS, Collett M, Wehner R (2001): The guidance of desert ants by extended landmarks. *J Exp Biol 204, 1635–1639.* Legge EL, Wystrach A, Spetch ML, Cheng K (2014): Combining sky and earth: desert ants (*Me-*

lophorus bagoti) show weighted integration of celestial and terrestrial cues. *J Exp Biol 217, 4159–4166.*—Integration of multiple cues in *Myrmecia pyriformis* and *M. croslandi*: Reid SF, Narendra A, Hemmi JM, Zeil J (2011): Polarized skylight and the landmark panorama provide night-active bull ants with compass information during route following. *J Exp Biol 214, 363–370.* Narendra A, Gourmaud S, Zeil J (2013): Mapping the navigational knowledge of individually foraging ants, *Myrmecia croslandi. Proc R Soc B 280, 20130683.*—For a review, see Wehner R, Hoinville T, Cruse H, Cheng K (2016): Steering intermediate courses: desert ants combine information from various navigational routines. *J Comp Physiol A 202, 459–472.*

3. Collett M (2012): How navigational guidance systems are combined in a desert ant. *Curr Biol 22, 927–932.*

4. Wystrach A, Mangan M, Webb B (2015): Optimal cue integration in ants. *Proc R Soc B 282, 20151484.*

5. For Bayesian integration—multiple sources of information are combined optimally when they are weighted in inverse proportion to their variances—as treated in humans and several vertebrate species, see, e.g., Ernst MO, Banks MS (2002): Humans integrate visual and haptic information in a statistically optimal fashion. *Nature 415, 429–433.* Cheng K, Shettleworth SJ, Huttenlocher J, Rieser JJ (2007): Bayesian integration of spatial information. *Psychol Bull 133, 625–637.* Ma WJ, Jazayeri M (2014): Neural coding of uncertainty and probability. *Annu Rev Neurosci 37, 205–220.*

6. Bühlmann C, Cheng K, Wehner R (2011): Vector-based and landmark-guided navigation in desert ants inhabiting landmark-free and landmark-rich environments. *J Exp Biol 214, 2845–2853.*

7. Interspecific and intraspecific differences in the ants' propensity to rely more on landmark-guidance or path-integration systems have been described and discussed especially in Bühlmann C, Cheng K, Wehner R (2011), *n6*. Cheng K, Middleton EJT, Wehner R (2012): Vector-based and landmark-guided navigation in desert ants of the same species inhabiting landmark-free and landmark-rich environments. *J Exp Biol 215, 3169–3174.*—For comparison, see also Narendra A (2007a): Homing strategies of the Australian desert ant *Melophorus bagoti*. I. Proportional path-integration takes the ant half-way home. *J Exp Biol 210, 1798–1803.* Narendra A (2007b): Homing strategies of the Australian desert ant *Melophorus bagoti*. II. Interaction of the path integrator with visual cue information. *J Exp Biol 210, 1804–1812.*

8. Fukushi T (2001): Homing in wood ants, *Formica japonica*: use of the skyline panorama. *J Exp Biol 204, 2063–2072.* Beugnon G, Lachaud JP, Chagné P (2005): Use of long-term stored vector information in the neotropical ant *Gigantiops destructor. J Insect Behav 18, 415–432.*—See also the behavior of a bull ant species that starts foraging during the evening twilight period: Freas CA, Narendra A, Cheng K (2017): Compass cues used by a nocturnal bull ant, *Myrmecia midas. J Exp Biol 220, 1578–1585.*

9. Schwarz S, Wystrach A, Cheng K (2017): Ants' navigation in an unfamiliar environment is influenced by their experience of a familiar route. *Sci Rep 7, 14161.* Freas CA, Cheng K (2017): Learning and time-dependent cue choice in the desert ant, *Melophorus bagoti. Ethology 123, 503–515.*—See also Narendra A (2007a, b), *n7*, and Cheung A, Hiby L, Narendra A (2012): Ant navigation: fractional use of the home vector. *PLoS ONE 7, e50451.*

10. Andel D, Wehner R (2004): Path integration in desert ants, *Cataglyphis*: how to make a homing ant run away from home. *Proc R Soc B 271, 1485–1489.* Wystrach A, Schwarz S, Graham P, Cheng K (2019): Running paths to nowhere: repetition of routes shows how navigating ants modulate online the weights accorded to cues. *Anim Cogn 22, 213–222.*

11. Knaden M, Wehner R (2003): Nest defense and conspecific enemy recognition in the desert ant *Cataglyphis fortis. J Insect Behav 16, 717–730.* Knaden M, Wehner R (2004): Path integration in desert ants controls aggressiveness. *Science 305, 60.* Bühlmann C, Hansson BS, Knaden M (2012): Path integration controls nest-plume following in desert ants. *Curr Biol 22, 645–649.*

12. There is an extensive literature on the traditional Micronesian and Polynesian navigation strategies. For general considerations, see Aveni AF (1981): Tropical archaeoastronomy. *Science 213,161–171.* Finney BR (1991): Myth, experiment, and the reinvention of Polynesian voyaging. *Amer Anthropol 93, 383–404.*—The original target expansion strategy is described in Gladwin T (1970): East Is a Big Bird: Navigation and Logic on Puluwat Atoll. *Cambridge MA, Harvard University Press*, pp. 195–200. Lewis D (1994): We, the Navigators. 2nd ed. *Honolulu, University of Hawaii Press*, pp. 195–294.

13. Hoffmann G (1985): The influence of landmarks on the systematic search behaviour of the desert isopod *Hemilepistus reaumuri*. I. Role of the landmark made by the animal. II. Problems with similar landmarks and their solution. *Behav Ecol Sociobiol 17, 325–334, 335–348.*

14. Wolf H, Wehner R (2000): Pinpointing food sources: olfactory and anemotactic orientation in desert ants, *Cataglyphis fortis. J Exp Biol 203, 857–868.*—See also food detection by crosswind walks: Bühlmann C, Graham P, Hansson BS, Knaden M (2014): Desert ants locate food by combining high sensitivity to food odors with extensive crosswind runs. *Curr Biol 24, 960–964.*—The ants' behavior within a food odor plume resembles that of flying male moths within pheromone odor plumes. For a review, see Cardé RT (2016): Moth navigation along pheromone plumes. In: Allison JD, Cardé RT, eds., Pheromone Communication in Moths: Evolution, Behavior and Application, pp. 173–189. *Berkeley, University of California Press.*

15. Post W, Gatty H (1931): Around the World in Eight Days: The Flight of the Winnie Mae. *New York, Chicago: Rand McNally;* 1989 ed., *New York: Orion Books, quot.* pp. 123–124.

16. Gatty H (1958): Nature Is Your Guide: How to Find Your Way on Land and Sea. *London, Collins, quot.* pp. 70–71.

17. Wolf H, Wehner R (2005): Desert ants, *Cataglyphis fortis,* compensate navigation uncertainty. *J Exp Biol 208, 4223–4230.* Wolf H (2008): Desert ants adjust their approach to a foraging site according to experience. *Behav Ecol Sociobiol 62, 415–425.*—The concept of the 'error compensation strategy' was introduced to us by Robert Biegler from the Norwegian University of Science and Technology at Trondheim. Biegler R (2000): Possible use of path integration in animal navigation. *Anim Learn Behav 28, 257–277.*

18. Straight paths: Sommer S, Wehner R (2004): The ant's estimation of distance travelled: experiments with desert ants, *Cataglyphis fortis. J Comp Physiol A 190, 1–6.*—Detour paths: Müller M, Wehner R (1988): Path integration in desert ants, *Cataglyphis fortis. Proc Natl Acad Sci USA 85, 5287–5290.* Wehner R, Wehner S (1990): Insect navigation: use of maps or Adriane's thread? *Ethol Ecol Evol 2, 27–48.*—For a theoretical treatment, see also Vickerstaff RJ, Di Paolo EA (2005): Evolving neural models of path integration. *J Exp Biol 208, 3349–3366.*

19. The ant's heuristic of compensating for passive displacements by jets of air has been elegantly examined by Antoine Wystrach and Sebastian Schwarz in *Melophorus bagoti:* Wystrach A, Schwarz S (2013): Ants use a predictive mechanism to compensate for passive displacement by wind. *Curr Biol 23, R1083–R1085.*

20. Wystrach A, Schwarz S, Baniel A, Cheng K (2013): Backtracking behaviour in lost ants: an additional strategy in their navigational toolkit. *Proc R Soc B 280, 20131677.*—For the importance of recent visual experience (temporal information), see also Collett M (2014): A desert ant's memory of recent visual experience and the control of route guidance. *Proc R Soc B 281, 1787.*—Witty comments are given in Graham P, Mangan M (2015): Insect navigation: do ants live in the now? *J Exp Biol 218, 819–823.*

21. Pfeffer SE, Wahl VL, Wittlinger M (2016): How to find home backwards? Locomotion and inter-leg coordination during rearward walking of *Cataglyphis fortis* desert ants. *J Exp Biol 219, 2110–2118.* Pfeffer SE, Wittlinger M (2016): How to find home backwards? Navigation during rearward homing of *Cataglyphis fortis* desert ants. *J Exp Biol 219, 2119–2126.*

22. Schwarz S, Mangan M, Zeil J, Webb B, Wystrach A (2017): How ants use vision when homing backward. *Curr Biol 27, 401–407.* Ardin P, Mangan M, Webb B (2016): Ant homing ability is not diminished when traveling backwards. *Front Behav Neurosci 10, 69.* See also Collett M, Collett TS (2018): How does the insect central complex use mushroom body output for steering? *Curr Biol 28, R733–R734.*

23. Nachtigall W (1979): Schiebeflug bei der Schmeissfliege *Calliphora erythrocephala* (Diptera: Calliphoridae). *Entom Gen 5, 255–265.* Collett M, Graham P, Collett TS (2017): Insect navigation: what backward walking reveals about the control of movement. *Curr Biol 27, R141–R144.*

24. The principle of a decentralized architecture describing the insect's navigational tool set and arguing against a unique 'navigation module' or global spatial representation was qualitatively outlined in Wehner R (2009): The architecture of the desert ant's navigational toolkit (Hymenoptera: Formicidae). *Myrm News 12, 85–96.*—It was followed by the design of an artificial neural network (navigation network, 'Navinet'): Cruse H, Wehner R (2011): No need for a cogni-

tive map: decentralized memory for insect navigation. *PLoS Comput Biol 7, 3.*—Rather than implementing a decentralized model based on dominance hierarchy, as done in the former account, the Navinet model discussed in the text assumes that guidance systems continuously cooperate and optimally integrate their information: Hoinville T, Wehner R (2018): Optimal multiguidance integration in insect navigation. *Proc Natl Acad Sci USA 115, 2824–2829.*

25. Seelig JD, Jayaraman V (2015): Neural dynamics for landmark orientation and angular path integration. *Nature 521, 186–191.* Heinze S (2015): Neuroethology: unweaving the senses of direction. *Curr Biol 25, R1034–R1037.* Stone T, Webb B, Adden A, Wedding NB, Honkanen A, et al., Heinze S (2017): An anatomically constrained model for path integration in the bee brain. *Curr Biol 27, 3069–3085.*

26. Ardin P, Peng F, Mangan M, Webb B (2016): Using an insect mushroom body circuit to encode route memory in complex natural environments. *PLoS Comput Biol 12, e1004683.*

27. Strausfeld NJ (1999): A brain region in insects that supervises walking. *Progr Brain Res 123, 273–284.*—Much direct evidence is derived from the work of Roland Strauss on locomotor disorders in *Drosophila* mutants with defined genetic defects in the central complex, e.g., Strauss R (2002): The central complex and the genetic dissection of locomotor behaviour. *Curr Opin Neurobiol 12, 633–638.*—In cockroaches directed movements could be elicited by electrically stimulating central complex neurons: Martin JP, Guo P, Mu L, Harley CM, Ritzmann RE (2015): Central-complex control of movement in the freely walking cockroach. *Curr Biol 25, 2795–2803.*

28. The quotations are from Cheeseman JF, Millar CD, Greggers U, Lehmann K, Pawley MDM, et al., Menzel R (2014a): Way-finding in displaced clock-shifted bees proves bees use a cognitive map. *Proc Natl Acad Sci USA 111, 8949–8954, quot.* p. 8953; Menzel R, Brembs B, Giurfa M (2007): Cognition in Invertebrates. In: Kaas JH, ed., Evolution of Nervous Systems, vol. 2, pp. 403–442. *Amsterdam, Academic Press, quot.* p. 429.—The map interpretation of the honeybee data has already been criticized in a multiauthored rebuttal to the Cheeseman et al. (2014a) paper: Cheung A, Collett M, Collett TS, Dewar ADM, Dyer FC, et al., Zeil J (2014): Still

no convincing evidence for cognitive map use by honeybees. *Proc Natl Acad Sci USA 111, E4396–E4397.* See also the original authors' reply: Cheeseman JF, Millar CD, Greggers U, Lehmann K, Pawley MDM, et al., Menzel R (2014b): Reply to Cheung et al.: The cognitive map hypothesis remains the best interpretation of the data in honeybee navigation. *Proc Natl Acad Sci USA 111, E4398.*

29. Menzel R, Greggers U (2015): The memory structure of navigation in honeybees. *J Comp Physiol A 201, 547–561, quot.* p. 548.—For constructing a metric map by combining egocentric landmark-based information with geocentric path integration information, see Gallistel CR, Cramer AE (1996): Computations on metric maps in mammals: getting oriented and choosing a multi-destination route. *J Exp Biol 199, 211–217,* and figure 1 therein.

30. Foraging ranges of *C. albicans, C. fortis,* and *C. bicolor* (see Figure 2.7) as well as *C. nodus* have been documented in Wehner R, Harkness RD, Schmid-Hempel P (1983): Foraging Strategies in Individually Searching Ants, *Cataglyphis bicolor* (Hymenoptera: Formicidae). *Mainz, G. Fischer.* Wehner R (1987): Spatial organization of foraging behaviour in individually searching desert ants, *Cataglyphis* (Sahara Desert) and *Ocymyrmex* (Namib Desert). *Experientia Suppl 54, 15–42.* Bühlmann C, Graham P, Hansson BS, Knaden M (2014), *n14.*

31. For studies on the probability and consistency of returning to previously visited food sites, see Schmid-Hempel P (1987): Foraging characteristics of the desert ant *Cataglyphis. Experientia Suppl 54, 43–61.* Bolek S, Wittlinger M, Wolf H (2012): What counts for ants? How return behaviour and food search of *Cataglyphis* ants are modified by variations in food quality and experience. *J Exp Biol 215, 3218–3222.* Pfeffer S, Bolek S, Wolf H, Wittlinger H (2015): Nest and food search behaviour in desert ants, *Cataglyphis:* a critical comparison. *Anim Cogn 18, 885–894.*

32. Wehner R, Meier C, Zollikofer CPE (2004): The ontogeny of foraging behaviour in desert ants, *Cataglyphis bicolor. Ecol Entomol 29, 240–250.*—For enhancing foraging success by sector fidelity in heterogeneous resource distributions, see Buchkremer EM, Reinhold K (2008): Sector fidelity: an advantageous foraging behaviour resulting from a heuristic search strategy. *Behav Ecol 19, 984–989.*

33. Lehrer M (1991): Bees which turn back and look. *Naturwissenschaften 78, 274–276*. Collett TS, Lehrer M (1993): Looking and learning: a spatial pattern in the orientation flight of the wasp *Vespula vulgaris*. *Proc R Soc Lond B 252, 129–134*. Lehrer M (1993): Why do bees turn back and look? *J Comp Physiol A 173, 23–32*.

34. Compare the effect of relearning walks in figures 30–33 in Müller M (1984): Interokularer Transfer bei der Wüstenameise *Cataglyphis fortis* (Formicidae, Hymenoptera). *Diploma Thesis, University of Zürich*, with that of learning walks in figures 3A–C in Fleischmann PN, Christian M, Müller VL, Rössler W, Wehner R (2016): Ontogeny of learning walks and the acquisition of landmark information in desert ants, *Cataglyphis fortis. J Exp Biol 219, 3137–3145*.

35. Wehner R (1981): Spatial vision in arthropods. In: Autrum H, ed., Handbook of Sensory Physiology, vol. 7/6C, pp. 287–616, figure 64 therein. *Berlin, Springer*. Ziegler PE, Wehner R (1997): Time-courses of memory decay in vector-based and landmark-based systems of navigation in desert ants, *Cataglyphis fortis. J Comp Physiol A 181, 13–20*. Cheng K, Narendra A, Wehner R (2006): Behavioral ecology of odometric memories in desert ants: acquisition, retention and integration. *Behav Ecol 17, 227–235*. Sommer S, von Beeren C, Wehner R (2008): Multiroute memories in desert ants. *Proc Natl Acad Sci USA 105, 317–322*.

36. Bumblebees may engage in circuitous 'exploration flights' interspersed among straight exploitation flights to familiar food patches, but these flights might be part of searching for new food sources. Woodgate JL, Makinson JC, Lim KS, Reynolds AM, Chittka L (2016): Life-long radar tracking of bumblebees. *PLoS ONE 11(8), e0160333*.

37. For the functional significance of linking path integration coordinates to places, see some intriguing arena experiments performed in hamsters: Etienne AS, Maurer R, Boulens V, Levy A, Rowe T (2004): Resetting the path integrator: a basic condition for route-based navigation. *J Exp Biol 207, 1491–1508*. Etienne AS, Jeffery KJ (2004): Path integration in mammals. *Hippocampus 14, 180–192*.—For comments, see Collett TS, Graham P (2004): Animal navigation: path integration, visual landmarks and cognitive maps. *Curr Biol 14, R475–R477*.

38. Gould JL (1986): The locale map of honey bees: do insects have cognitive maps? *Science 232, 861–863*. Gould JL, Towne WF (1987): Evolution of the dance language. *Am Nat 130, 317–338*.—The quotation is from Gould JL, Gould CG (2012): Nature's Compass: The Mystery of Animal Navigation. *Princeton NJ, Princeton University Press*, p. 35.—For valid alternative explanations to invoking a locale map in bees, see Dyer FC, Seeley TD (1989): On the evolution of the dance language. *Am Nat 133, 580–590*. In their rebuttal, Gould and Towne (1989: *Am Nat 134, 156–159*) still claim that their evidence strongly indicates a cognitive map in bees. Carefully designed critical experiments later performed by Fred Dyer clearly show that this claim is unsubstantiated: Dyer FC (1991): Bees acquire route-based memories but not cognitive maps in a familiar landscape. *Anim Behav 41, 239–246*. Dyer FC, Berry NA, Richard AS (1993): Honey bee spatial memory: use of route-based memories after displacement. *Anim Behav 45, 1028–1030*.—See also the summary of results of several similar studies in Wehner R (1992): Arthropods. In: Papi F, ed., Animal Homing, pp. 45–144, figure 3.43 therein. *London, Chapman and Hall*.

39. Gould JL, Gould CG (1982): The insect mind: physics or metaphysics? In: Griffin DR, ed., Animal Mind—Human Mind, pp. 269–298. *Berlin, Springer*.—For further experiments and discussion, see Tautz J, Zhang SW, Spaethe J, Brockmann A, Si A, Srinivasan M (2004): Honeybee odometry: performance in varying natural terrain. *PLoS Biol 2, 915–923*.

40. A masterpiece of sophisticated experimentation and thoughtful consideration is Wray MK, Klein BA, Mattila HR, Seeley TD (2008): Honeybees do not reject dances for 'implausible' locations: reconsidering the evidence for cognitive maps in insects. *Anim Behav 76, 261–269*.

41. Menzel R, Greggers U, Smith A, Berger S, Brandt R, et al., Watzl S (2005): Honey bees navigate according to a map-like spatial memory. *Proc Natl Acad Sci USA 102, 3040–3045*. Menzel R, Kirbach A, Haass WD, Fischer B, Fuchs J, et al., Greggers U (2011): A common frame of reference for learned and communicated vectors in honeybee navigation. *Curr Biol 21, 645–650*. Cheeseman JF, Millar CD, Greggers U, Lehmann K, Pawley MDM, et al., Menzel R (2014a), n28.—See rebuttal: Cheung A, Collett M, Collett TS, Dewar

ADM, Dyer FC, et al., Zeil J (2014), *n28,* and the original authors' reply: Cheeseman JF, Millar CD, Greggers U, Lehmann K, Pawley MDM, et al., Menzel R (2014b), *n28.* —Menzel R, Greggers U (2015), *n29.*

42. For the Navinet simulation of the experiments described in Cheeseman JF, Millar CD, Greggers U, Lehmann K, Pawley MDM, et al., Menzel R (2014a), *n28,* see Hoinville T, Wehner R (2018), *n24.*—The concept of combining a 'bearing map' ("a coarse map of metric relations between locations") with multiple 'sketch maps' as proposed in Menzel R, Greggers U (2015), *n29, quot.* p. 559, is more descriptive than operational.

43. In a classical study tropical orchid bees, *Euglossa surinamensis,* have been observed to repeatedly visit the same plants along feeding routes extending up to 23 km from the nest, and do so in the order in which the plants have originally been incorporated into the route: Janzen DH (1971): Euglossine bees are long-distance pollinators of tropical plants. *Science 171, 203–205.*—See also Wikelski M, Moxley J, Eaton-Mordas A, López-Uribe MM, Holland R, et al., Kays R (2010): Large-range movements of neotropical orchid bees observed via radio telemetry. *PLoS ONE 5, e10738.*

44. In the study of how insects may develop spatial representations of trapline routes, the following accounts are exemplary in conceptual and experimental design: Lihoreau M, Raine NE, Reynolds AM, Stelzer RJ, Lim KS, et al., Chittka L (2012): Radar tracking and motion-sensitive cameras on flowers reveal the development of pollinator multi-destination routes over large spatial scales. *PLoS Biol 10, e1001392.* Reynolds AM, Lihoreau M, Chittka L (2013): A simple iterative model accurately captures complex trapline formation by bumblebees across spatial scales and flower arrangements. *PLoS Comput Biol 9, e1002938.* Lihoreau M, Raine NE, Reynolds AM, Stelzer RJ, Kim KS, et al., Chittka L (2013): Unravelling the mechanisms of traplining foraging in bees. *Comm Integr Biol 6, 1–4.*

45. The question of whether views of familiar landmark sceneries can activate long-term memories of previous path integration states has been treated especially in Sassi S, Wehner R (1997): Dead reckoning in desert ants, *Cataglyphis fortis:* can homeward-bound vectors be reactivated by familiar landmark configurations? *Proc Neurobiol Conf Göt-*

tingen 25, 484. Collett M, Collett TS, Chameron S, Wehner R (2003): Do familiar landmarks reset the global path integration system of desert ants? *J Exp Biol 206, 877–882.*—Resetting of the path integrator occurs only inside the nest: Knaden M, Wehner R (2005): Nest mark orientation in desert ants *Cataglyphis:* what does it do to the path integrator? *Anim Behav 70, 1349–1354.* Knaden M, Wehner R (2006): Ant navigation: resetting the path integrator. *J Exp Biol 209, 26–31.* Certainly, the potential combination of landmark-based and path-integration information needs further analysis.

46. Müller M, Wehner R (2010): Path integration provides a scaffold for landmark learning in desert ants. *Curr Biol 20, 1368–1371.* Fleischmann PN, Christian M, Müller VL, Rössler W, Wehner R (2016), *n34.* Freas CA, Cheng K (2017), *n9.*

47. Compare also the ants' different behaviors in Wystrach A, Beugnon G, Cheng K (2012): Ants might use different view-matching strategies on and off the route. *J Exp Biol 215, 44–55,* and Freas CA, Cheng K (2017), *n9.*

48. The example is taken from Narendra A (2007b), *n7.*—For a similar example, see Wehner R, Michel B, Antonsen P (1996): Visual navigation in insects: coupling of egocentric and geocentric information. *J Exp Biol 199, 129–140,* figure 11.

49. The egocentric storage and use of view-based landmark information is implicitly assumed in almost all recent papers on ant navigation, see, e.g., Graham P (2010): Insect navigation. In: Breed HD, Moore J, eds., Encyclopedia of Animal Behavior, Vol. 2, pp. 167–175, *Oxford, Academic Press.* Zeil J (2011): Visual homing: an insect perspective. *Curr Opin Neurobiol 22, 1–9.* Collett M, Chittka L, Collett TS (2013): Spatial memory in insect navigation. *Curr Biol 23, R789–R800.* Previously, it has been argued that bees attach home vectors to visual scenes experienced at particular food sites: Menzel R, Geiger K, Joerges J, Müller U, Chittka L (1998): Bees travel novel homeward routes by integrating separately acquired vector memories. *Anim Behav 55, 139–152.*

50. For the template-matching hypothesis accounting for olfactory nestmate recognition, see Chapter 1 and *n16* therein.

51. Vowles DM (1955): The structure and connections of the corpora pedunculata in bees and ants. *Quart J Microsc*

Sci 96, 239–255. Menzel R, Erber J, Masuhr T (1974): Learning and memory in the honeybee. In: Barton Browne L, ed., Experimental Analysis of Insect Behaviour, pp. 195–217. *Berlin, Springer.*—In cockroaches, lesions made by slivers of aluminum foil in the medial lobes of the mushroom bodies impaired place memory: Mizunami M, Weibrecht JM, Strausfeld NJ (1998): Mushroom bodies of the cockroach: their participation in place memory. *J Comp Neurol 402, 520–537.*

52. Seelig JD, Jayaraman V (2015), *n25.* Green J, Adachi A, Shah KK, Hirokawa JD, Magani PS, Maimon G (2017): A neural circuit architecture for angular triangulation in *Drosophila. Nature 546, 101–106.* Turner-Evans D, Wegener S, Rouault H, Franconville R, et al., Jayaraman V (2017): Angular velocity integration in a fly heading circuit. *eLife 6, e04577.*

53. It came as a real surprise when Daniel Kronauer published the first gene editing study via CRISPR/Cas in ants, soon followed by a second one put forth by Claude Desplan. Trible W, Olivos-Cisneros L, McKenzie SK, Saragosti J, et al., Kronauer DJC (2017): Orco mutagenesis causes loss of antennal lobe glomeruli and impaired social behavior in ants. *Cell 170, 727–735.* Yan H, Opachaloemphan C, Mancini G, Yang H, et al., Desplan C (2017): An engineered orco mutation produces aberrant social behavior and defective neural development in ants. *Cell 170, 736–747.*

54. Wehner R, Fukushi T, Isler K (2007): On being small: ant brain allometry. *Brain Behav Evol 69, 220–228.* Seid MA, Castillo A, Wcislo WT (2011): The allometry of brain miniaturization in ants. *Brain Behav Evol 77, 5–13.*

55. Ardin P, Peng F, Mangan M, Webb B (2016), *n26.* Stone T, Webb B, Adden A, Wedding NB, Honkanen A, et al., Heinze S (2017), *n25.* Hoinville T, Wehner R (2018), *n24.*—In this context it is also worth pointing out that several 'advanced' cognitive tasks that insects are able to perform, can be accomplished by artificial neural networks, which consist of an amazingly small number of neurons; see, e.g., Chittka L, Niven J (2009): Are bigger brains better? *Curr Biol 19, R995–R1008.* Chittka L, Rossiter SJ, Skorupski P, Fernando C (2012): What is comparable in comparative cognition? *Phil Trans R Soc B 367, 2677–2685.* Roper M, Fernando C, Chittka L (2017): Insect bio-inspired neural network provides new evidence on how single feature de-

tectors can enable complex visual generalization and stimulus location invariance in the miniature brain of honeybees. *PLoS Comput Biol 13, e1005333.*

56. For intricate discussions of the cost/benefit relations, potentialities and constraints of small (e.g., insect) brains, see especially Chittka L, Niven J (2009), *n55.* Chittka L, Skorupski P (2011): Information processing in miniature brains. *Proc R Soc B 278, 885–888.* Niven JE, Farris SM (2012): Miniaturization of nervous systems and neurons. *Curr Biol 22, R323–R329.*

57. Tolman EC (1948): Cognitive maps in rats and men. *Psychol Rev 55, 189–208, quots.* pp. 192 and 207.—Because of its wide intellectual scope, including concepts of space in the history of philosophy, O'Keefe's and Nadel's book, and the report of subsequent discussions about the book, make it highly recommended reading: O'Keefe J, Nadel L (1978): The Hippocampus as a Cognitive Map. *Oxford: Oxford University Press.* O'Keefe J, Nadel L (1979): Précis of O'Keefe and Nadel's The Hippocampus as a Cognitive Map. *Behav Brain Sci 2, 487–533.*

58. Place cells, grid cells, and head-direction cells have first been described in, respectively, O'Keefe J, Dostrovsky J (1971): The hippocampus as a spatial map. Preliminary evidence from unit activity in the freely-moving rat. *Brain Res 34, 171–175*; Hafting T, Fyhn M, Molden S, Moser MB, Moser EI (2005): Microstructure of a spatial map in the entorhinal cortex. *Nature 436, 801–806*; and Ranck JB (1985): Head direction cells in the deep cell layer of dorsal presubiculum in freely moving rats. In: Buzsáki G, Vanderwolf CH, eds., Electrical Activity of the Archicortex, pp. 217–220. *Budapest, Akademiai Kiado.* Taube JS, Muller RU, Ranck JB (1990): Head-direction cells recorded from the postsubiculum in freely moving rats. I. Description and quantitative analysis. *J Neurosci 10, 420–435.*—For reviews, see the articles in Hartley T, Lever C, Burgess N, O'Keefe J, eds. (2014): Space in the brain: cells, circuits, codes and cognition. *Phil Trans R Soc B 369, issue 1635.* Rowland DC, Roudi Y, Moser M-B, Moser EI (2016): Ten years of grid cells. *Annu Rev Neurosci 39, 19–40.*—For evidence of place and grid cells in humans, see Ekström AD, Kahana MJ, Caplan JB, Fields TA, et al., Fried I. (2003): Cellular networks underlying human spatial navigation. *Nature 425, 184–188.* Jacobs J, Weidemann CT, Miller JF, Solway A et al., Kahane MJ (2013): Di-

rect recordings of grid-like neuronal activity in human spatial navigation. *Nat Neurosci 16, 1188–1190.*

59. Manns JR, Howard MW, Eichenbaum H (2007): Gradual changes in hippocampal activity support remembering the order of events. *Neuron 56, 530–540.* Pfeiffer BE, Foster DJ (2013): Hippocampal place-cell sequences depict future paths to remembered goals. *Nature 497, 74–79.* MacDonald CJ, Carrow S, Place R, Eichenbaum H (2013): Distinct hippocampal time cell sequences represent odor memories in immobilized rats. *J Neurosci 33, 14607–14616.* Aronov D, Nevers R, Tank DW (2017): Mapping of a nonspatial dimension by the hippocampal-entorhinal circuit. *Nature 543, 719–722.* Ekstrom AD, Ranganath C (2017): Space, time, and episodic memory: the hippocampus is all over the cognitive map. *Hippocampus 28, 680–687.*

60. Tsoar A, Natham R, Bartan Y, Vyssotski A, Dell'Omo G, Ulansovsky N (2011): Large-scale navigational map in a mammal. *Proc Natl Acad Sci USA 108, E718–E724.* Geva-Sagiv M, Las L, Yovel Y, Ulanovsky N (2015): Spatial cognition in bats and rats: from sensory acquisition to multiscale maps and navigation. *Nature Rev Neurosci 16, 94–108.* Sarel A, Finkelstein A, Las L, Ulanovsky N (2017): Vectorial representation of spatial goals in the hippocampus of bats. *Science 355, 176–180.*

61. For the possible involvement of the insect's mushroom bodies in place learning, see Mizunami M, Weibrecht JM, Strausfeld NJ (1998), *n51.*—What would come as a real surprise is the discovery of a grid cell system in insects.

62. Calhoun JB (1963): The Ecology and Sociobiology of the Norway Rat. *Bethesda MD, U.S. Department of Health, Education, and Welfare.* Taylor KD (1978): Range of movement and activity of common rats *(Rattus norvegicus)* on agricultural land. *J Appl Ecol 15, 663–677.* Recht MA (1982): The fine structure of the home range and activity pattern of free-ranging telemetered urban Norway rats, *Rattus norvegicus. Bull Soc Vect Ecol 7, 29–35.* Russell JC, McMorland AJC, MacKay JWB (2010): Exploratory behaviour of colonizing rats in novel environments. *Anim Behav 79, 159–164.*

63. The quotations are, respectively, from Small WS (1899): Notes on the psychic development of the young white rat. *Amer J. Psychol 11, 80–100, quot.* p. 99. Locke NM (1936): A preliminary study of the social drive in the white rat. *J Psychol 1, 255–260, quot.* p. 259.—The term 'latent learning' was introduced by Blodgett HC (1929): The effect of the introduction of reward upon the maze performance of rats. *Univ Calif Publ Psychol 4, 113–134,* p. 122.—In rats, latent learning has mostly been studied by exposing adult rats to new maze environments. Barnett SA (1958): Exploratory behaviour. *Brit J Psychol 49, 289–310.*

64. Eilam D, Golani I (1989): Home base behavior of rats *(Rattus norvegicus)* exploring a novel environment. *Behav Brain Res 34, 199–211.* Golani I, Benjamini Y, Eilam D (1993): Stopping behavior: constraints on exploration in rats *(Rattus norvegicus). Behav Brain Res 53, 21–33.* Zadicario P, Avni R, Zadicario E, Eilam D (2005): 'Looping'—an exploration mechanism in a dark open field. *Behav Brain Res 159, 27–36.* Golani I, Benjamini Y, Dvorkin A, Lipkind D, Kafkafi N (2005): Locomotor and explorative behavior. In: Whishaw IQ, Kolb B, eds., The Behaviour of the Laboratory Rat: A Handbook with Tests. *New York, Oxford University Press,* pp. 171–182.

65. Etienne AS, Maurer R, Boulens V, Levy A, Rowe T (2004): Resetting the path integrator: a basic condition for route-based navigation. *J Exp Biol 207, 1491–1508.* Etienne AS, Jeffery K (2004), *n37.*

66. In fruit bats equipped with GPS loggers and foraging in their natural habitat, or moving around in large flight rooms decked out with cameras, experimental paradigms could be envisaged that resemble those tested in the cataglyphs. Tsoar A, Nathan R, Bartan Y, Vyssotski A, Dell'Omo G, Ulanovsky N (2011): Large-scale navigational map in a mammal. *Proc Natl Acad Sci USA 108, E718–E724.* Finkelstein A, Derdikman D, Rubin A, Foerster JN, Las L, Ulanovsky N (2015): Three-dimensional head-direction coding in the bat brain. *Nature 517, 159–164.*

67. In a stunning paper, Detlev Arendt and his colleagues at the European Molecular Biology Laboratory in Heidelberg, Germany, report that the developing mushroom body of an annelid worm and the developing pallium in the mouse brain exhibit high similarities in gene sequences and temporal sequence expression patterns. Tomer R, Denes AS, Tessmar-Raible K, Arendt D (2010): Profiling by image registration reveals common origin of annelid mushroom bodies and vertebrate pallium. *Cell 142, 800–809.*—Given this obviously deep homology between forebrain structures of invertebrates and vertebrates, the next question is what

happens later in ontogenetic development. Let us assume that these developmental processes finally ended up in similar neural circuitries. Then, in what ways and contexts do the cataglyphs and the rats take advantage of the computational capabilities of these circuitries in orchestrating their navigational tool set?

68. Interestingly, while researchers of insect navigation have developed a keen interest in the work done on spatial cognition in mammals, the reverse is not usually the case. Hence I hope that this book helps to draw attention to the multifaceted experimental approaches and emerging concepts that result from the work on insect navigators. After all, as central place foragers, the cataglyphs and their social insect companions share with rats, mice, and many other rodents the need to venture out in initially uncharted territory, and then to employ whatever spatial knowledge they have subsequently acquired to return to their home base and to other meaningful places in their environment.

Epilogue

Epigraph: Ramón y Cajal S (1937): Recollections of My Life. *Cambridge MA, MIT Press,* Reprint 1991, *quot.* p. 589.

1. Ramón y Cajal S (1921): Las sensaciones de las hormigas. *Arch Neurobiol Psicol Fisiol Neurol Psiquiat 2, 321–337.* Lubbock J (1882): Ants, Bees, and Wasps: A Record of Observations on the Habits of the Social Hymenoptera. *London, Kegan Paul, Trench, quots.* pp. v, vii. Hutchinson HG (1914): Life of Sir John Lubbock, Lord Avebury. *London, Macmillan.*—You may recall that we have met John Lubbock already in Chapter 1, together with Jean-Henri Fabre.

2. One could argue that within the experimental system of 'studying *Cataglyphis* in the test field' the ant's cockpit—the ant's art of wayfinding—has become what contemporary philosophers and historians of science call an 'epistemic thing,' a dynamic research body, which is not the final product but the driving force of research. Rheinberger HJ (1997): Toward a History of Epistemic Things: Synthesizing Proteins in the Test Tube. *Stanford CA, Stanford University Press.*—"Manipulating the researcher": Lenoir F (2010): Foreword to Rheinberger HJ, An Epistemology of the Concrete: Twentieth-Century Histories of Life. *Durham NC, Duke University Press, quot.* p. xvi.

3. In comparative animal cognition, researchers often tend to discover humanlike ways in how various animal species solve complex problems. They assume that general underlying concepts operate across taxa. As this expectation-driven top-down approach is easily constrained by particular concepts, it faces the danger of terminological ambiguities. For stringent critical arguments, see, e.g., de Waal FBM, Ferrari PF (2010): Towards a bottom-up perspective on animal and human cognition. *Trends Cogn Sci 14, 201–207.* Graham P (2010): Insect navigation. In: Breed HD, Moore J, eds., Encyclopedia of Animal Behavior, vol. 2, pp. 167–175. *London, Academic Press.* Döring TF, Chittka L (2011): How human are insects, and does it matter? *Formosan Entomol 31, 85–99.* Wystrach A, Graham P (2012): What can we learn from studies of insect navigation? *Anim Behav 84, 13–20.* Chittka L, Rossiter SJ, Skorupski P, Fernando C (2012): What is comparable in comparative cognition? *Phil Trans R Soc B 367, 2677–2685.*

Considering the insect navigator in the light of some higher-level cognitive concept, the argument would be that the cataglyphs develop a model of their outside world—their foraging environment—within which they plan actions and predict outcomes (action selection, see Chapter 1, *n22*). In this view, the ants' principal navigational routines are relegated to subordinate steering devices that the animals are free to flexibly use in one way or another to reach their goal. Of course, when starting an action—e.g., to leave the nest for a foraging journey to one of several familiar sites, or to return to the nest after food has been found—the ant must decide what to do. In the Navinet model proposed in Chapter 7 this decision process is represented by what we have called 'motivation units' (see Figure 7.10). Then, however, the ant is not assumed to consult an internal representation of its outside world, i.e., to spread out a map and mark out a preferred route, so to speak. Rather it is assumed to follow the instructions of interlinked navigational routines, which at any one time tell the animal what the most reliable direction is that it should steer. At present, even the high flexibility that we observe in the ants' navigational behavior is in accord with this view. Future experimental paradigms will decide whether this view is sufficient.

4. Tinbergen N (1951): The Study of Instinct. *New York, Oxford University Press, quot.* p. 12, *italics* by Niko Tinbergen.

5. For example, Risse B, Mangan M, Del Pero L, Webb B (2018): Visual tracking of small animals in cluttered natural environments using a freely moving camera. *Venice, Institute of Electrical and Electronics Engineers*, pp. 2840–2849.

6. Burger ML (1971): Zum Mechanismus der Gegenwendung nach mechanisch aufgezwungener Richtungsänderung bei *Schizophyllum sabulosum* (Julidae, Diplopoda). *Z Vergl Physiol 71, 219–254.*—The involvement of the central complex in steering undisturbed courses is borne out by lesion experiments in fruit flies and cockroaches: Strauss R (2002): The central complex and the genetic dissection of locomotor behaviour. *Curr Opin Neurobiol 12, 633–638.* Harley CM, Ritzmann RE (2010): Electrolytic lesions within central complex neuropils of the cockroach brain affect negotiation of barriers. *J Exp Biol 213, 2851–2864.*

7. Dethier VG (1976): The Hungry Fly. *Cambridge MA, Harvard University Press, pp. 27–30.* Kim IS, Dickinson MH (2017): Idiothetic path integration in the fruit fly *Drosophila melanogaster. Curr Biol 27, 2227–2238.* Brockmann A, Basu P, Shakeel M, Murata S, et al., Tanimura T (2018): Sugar intake elicits intelligent searching behavior in flies and honey bees. *Front Behav Neurosci 12, 280.*

8. Collett TS, Land MF (1975): Visual spatial memory in a hoverfly. *J Comp Physiol 100, 59–84.* Ofstad TA, Zuker CS, Reiser MB (2011): Visual place learning in *Drosophila melanogaster. Nature 474, 204–207.* Ostrowski D, Zars T (2014): Place learning. In: Dubnau J, ed., Behavioral Genetics of the Fly *(Drosophila melanogaster),* pp. 125–134. *Cambridge, Cambridge University Press.*—Often, rather small modifications of neural circuitries might open up new behavioral potentials; see, e.g., Chittka L, Niven J (2009): Are bigger brains better? *Curr Biol 19, R995–R1008.* Katz PS (2011): Neural mechanisms underlying the evolvability of behaviour. *Phil Trans R Soc B 366, 2086–2099.* Chittka L, Rossiter SJ, Skorupski P, Fernando C (2012), *n3.*

9. The "different story to tell" refers to Godfrey-Smith P (2016): Other Minds: The Octopus, the Sea, and the Deep Origins of Consciousness. *New York, Farrar, Straus and Giroux,* p. 13. For definitions of a "position sense" and "map sense," see Gallistel CR (1994): Space and time. In: Mackintosh N, ed., Handbook of Perception and Cognition. Vol. 9: Animal Learning and Cognition, pp. 221–253. *London, Academic Press, quot.* p. 222. Gould JL, Gould CG (2012): Nature's

Compass: The Mystery of Animal Navigation. *Princeton NJ, Princeton University Press, quot.* p. 185. A map sense forms the basis of "true navigation," as shown, e.g., in migrating birds. In insects, even long-distance migrators such as the North American monarch butterflies most certainly do not use maps *sensu* Gallistel and the Goulds. Mouritsen H, Derbyshire R, Stalleicken J, Mouritsen OØ, Frost BJ, Norris DR (2013): An experimental displacement and over 50 years of tag-recoveries show that monarch butterflies are not true navigators. *Proc Natl Acad Sci USA 110, 7348–7353.* See also rebuttals, *ibid., E3680, E3681.*

10. See Chapter 7, *n53,* for the first application of the CRISPR / Cas method in ants.—As outlined in Chapter 2, some *Cataglyphis* species, e.g., *C. cursor,* reproduce like *Ooceraea biroi*—the clonally reproducing ant used by David Kronauer—via unfertilized diploid eggs, i.e., exhibit so-called thelytokous parthenogenesis. However, there are also other means of gene editing in hymenopterans than relying on this kind of parthenogenetic reproduction.

11. For exemplary studies along these lines, see Seelig JD, Jayaraman V (2015): Neural dynamics for landmark orientation and angular path integration. *Nature 521, 186–191.* Stone T, Webb B, Adden A, Wedding NB, Honkanen A, et al., Heinze S (2017): An anatomically constrained model for path integration in the bee brain. *Curr Biol 27, 3069–3085.* Ardin P, Peng F, Mangan M, Lagogiannis K, Webb B (2016): Using an insect mushroom body circuit to encode route memory in complex natural environments. *PLoS Comput Biol 12, e1004683.*

12. Early papers on brain research in *Cataglyphis:* Kühn-Bühlmann S, Wehner R (2006): Age-dependent and task-related volume changes in the mushroom bodies of visually guided desert ants, *Cataglyphis bicolor. J Neurobiol 66, 511–521.* Seid MA, Wehner R (2008): Ultrastructure and synaptic differences of the boutons of the projection neurons between the lip and collar regions of the mushroom bodies in the ant *Cataglyphis albicans. J Comp Neurol 507, 1102–1108.* Seid MA, Wehner R (2009): Delayed axonal pruning in the ant brain: a study of developmental trajectories. *Dev Neurobiol 69, 350–364.* Stieb SM, Münz TS, Wehner R, Rössler W (2010): Visual experience and age affected synaptic organization in the mushroom bodies of the desert ant *Cataglyphis fortis. Dev Neurobiol 70, 408–423.* Grob R, Fleischmann PN,

Grübel K, Wehner R, Rössler W (2017): The role of celestial compass information in *Cataglyphis* ants during learning walks and for neuroplasticity in the central complex and mushroom bodies. *Front Behav Neurosci 11, 226.*

13. Attwell D, Laughlin SB (2001): An energy budget for signaling in the grey matter of the brain. *J Cereb Blood Flow Metab 21, 1133–1145.* Laughlin SB (2001): Energy as a constraint on the coding and processing of sensory information. *Curr Opin Neurobiol 11, 475–480.* Niven JE, Laughlin SB (2008): Energy limitation as a selective pressure on the evolution of sensory systems. *J Exp Biol 211, 1792–1804.*

14. Gronenberg W, Hölldobler B (1999): Morphologic representation of visual and antennal information in the ant brain. *J Comp Neurol 412, 229–240* (reference to *Cataglyphis*). For further allometric studies performed in ants by Wulfila Gronenberg, see Chapter 6, *n90.*—A first digital 3-D atlas of an ant brain has been provided by Bressan JMA, Benz M, Oettler J, Heinze J, Hartenstein V, Sprecher SG (2014): A map of brain neuropils and fiber systems in the ant *Cardiocondyla obscurior. Front Neuroanat 8, 166.*

15. Wehner R, Fukushi T, Isler K (2007): On being small: ant brain allometry. *Brain Behav Evol 69, 220–228.*

Acknowledgments

The work described in this book could not have been done without the cooperation, enthusiasm, and skill of a large number of graduate students whom over the course of several decades I had the good fortune to attract to work on the cataglyphs. I am greatly indebted to them and would like to thank them fullheartedly for the marvelous and inspiring times we spent together. In the pioneering days starting in the late 1960s new areas of research on compound eye vision, sky compass orientation, and landmark guidance were opened up by the work of Andreas Burkhalter, Peter Duelli, Immanuel Flatt, and Reto Weiler in the field, and by Paul Herrling, Robert Kretz, Eric Meyer, and Felix Räber in the laboratory. In the 1980s, Karl Fent, Bruno Lanfranconi, Ursula Menzi, Martin Müller, Paul Schmid-Hempel, and Christoph Zollikofer enriched the studies on vision, sky navigation, and path integration by advanced experimental paradigms, and started the work on the kinematics of locomotion and the ecology of space-use patterns. In the two decades around the turn of the millennium, the latter work was continued, and increasingly accompanied by studies on odometry and the combined use of path integration and landmark guidance routines, by Barbara and Per Antonsen-Michel, Barbara Dietrich, Franz-Xaver Dillier, Markus Knaden, Tobias Merkle, Rainer Neumeyer, Tobias Seidl, and Matthias Wittlinger. It was a source of great pleasure and personal enrichment to work with this community of students, who all contributed with their own expertise, attitudes, and approaches to our joint *Cataglyphis* endeavors.

Moreover, I am very happy and grateful indeed that from the 1990s onward I succeeded in luring colleagues from abroad to our North African field site at Mahrès. They were immediately captivated by *Cataglyphis*: Thomas Collett, later joined by Matthew Collett, was followed by Bernhard Ronacher, Harald Wolf, and finally Ken Cheng. Meanwhile, these colleagues and friends have attracted new generations of graduate students to work on various aspects of ant navigation. Special thanks are due to Wolfgang Rössler, who, by inviting me as an Alexander von Humboldt guest professor, established *Cataglyphis* work at the University of Würzburg. It was wonderful to co-supervise with him the graduate studies of Sara Stieb, Franziska Schmitt, Pauline Fleischmann, and Robin Grob.

In the early 1980s, when my wife Sibylle and I started to investigate *Ocymyrmex* in the Namibian gravel plains and the sandflats at Gobabeb, Mary Seely supported our stay, and all subsequent ones, in every conceivable way. Ten years later, in Australia, Steve Morton, in Australia, kindly led me around in the arid Acacia shrubland near Alice Springs, the homeland of *Melophorus bagoti*. This introductory tour paved the way for later studies of these "red honey ants." They started in 2001 as a joint project with Ken Cheng, and since then have largely been expanded by Ken and his students, some of whom—e.g., Ajay Narendra and Antoine Wystrach—have already attracted further generations of graduate students to desert ant research. I am very grateful to Ken for his cooperation over the many years of my adjunct professorship at Macquarie University, Sydney.

At home in our Zürich laboratory, my sincere thanks go to my longtime collaborators Thomas Labhart and Eric Meyer, who forcefully promoted work on the neuroanatomy and neurophysiology of the cataglyphs' visual system, as well as to the late Miriam Lehrer and to Ursula Menzi for many years of scientific cooperation. Thomas was also the first to record from a polarization-sensitive interneuron—in crickets, though, but later also in *Cataglyphis*. I am also indebted to Samuel Rossel, whom unfortunately I could not get enthusiastic enough to work on desert ants, but who instead became a superb honeybee researcher in projects on polarization vision and sky compass orientation, which we performed in parallel to those in the cataglyphs. It was a further delight to stay in

contact with Martin Müller and Stefan Sommer in discussing problems of ant navigation long after our scientific paths had parted. Both former graduate students, as well as Thomas Norgaard, also contributed recordings that had not been published previously. Among the many collaborators who assisted me during many years of research in various ways, I also owe special debts of gratitude to Karen Gossel, Claudia Fischer, and Hanna Michel, as well as to Valerie Domanico for her excellent technical assistance and to Michel Nakano for his great multifarious support in recent times.

Very special thanks are reserved for my dear friends Gary Bernard and Tsukasa Fukushi. During several sabbatical visits by them to Zürich we worked on quite a number of exciting projects ranging from optical physiology to brain allometry, and shared wonderful times together with their families in and around Zürich.

In the same vein, I am grateful to Bernhard Ronacher, Holk Cruse, and Thierry Hoinville for their longstanding cooperation, friendship, and innumerable discussions on several topics treated in the book, as well as for reading and commenting on parts (Holk and Thierry) or all (Bernhard) of the manuscript. I am also indebted to many colleagues who were patient enough to correspond with me on several aspects of *Cataglyphis* biology. Space limitations prevent me from fairly recognizing all of them, but *pars pro toto*, and from a rather wide range of disciplines, I would like to mention Christof Aegerter, Serge Aron, Johan Billen, Reinhard Blickhan, Ansgar Büschges, Xim Cerdá, Fred Dyer, Walter Federle, Robert Full, Wulfila Gronenberg, Bernd Heinrich, Andreas Herz, Stefan Hetz, Uwe Homberg, Robert Johnson, John Lighton, Ralf Möller, Edvard

Moser, Henrik Mouritsen, Ajay Narendra, Christian Peeters, Flavio Roces, Mandyam Srinivasan, Nicholas Strausfeld, Alberto Tinaut, Nachum Ulanovsky, Eric Warrant, Antoine Wystrach, and Jochen Zeil.

Looking to the future, I am pleased and grateful that many colleagues and friends are devoted to further promoting *Cataglyphis* research at our Mahrès study site: Bernhard Ronacher and Harald Wolf, who have greatly advanced our understanding of the ants' odometer and path integration strategies; Markus Knaden, who has opened up exciting new ways of investigating the olfactory world of the 'visual' desert navigator; and Matthias Wittlinger, who together with Hansjürgen Dahmen has recently developed and successfully applied the first *Cataglyphis* trackball device. This setup is now ready to further a wide range of studies in tetheredly navigating desert ants. Beyond Mahrès, I am deeply indebted to my colleagues and the entire staff of the Zürich Brain Research Institute, who with their kindness and friendship have provided me with a wonderful emeritus home.

Thanks in great measure go to Eva Weber for painting the beautiful portraits of our desert ants and for handing over to me her sketchbook, a gorgeous present. Rainer Foelix graciously produced several scanning electron micrographs portraying structural details of the cataglyphs. I also thank Karin Niffeler and Sarah Steinbacher for their kind cooperation in graphic design, and especially Thomas Heinemann for bringing all my graphs into their beautiful final shape. It was a great personal pleasure to work with him and to enjoy his professional expertise, unlimited help, kind understanding, and patience with my ever-growing suggestions. Finally, but not less cordially,

I wish to extend my sincere thanks to Janice Audet and her team at Harvard University Press for their manifold dedicated efforts to publish *Desert Navigator* in such an appealing form, and to Melody Negron from Westchester Publishing Services for her kind and effective help during the final stages of the production process, and to Melody Negron from Westchester Publishing Services for her kind and effective help during the final stages of the production process..

I gratefully acknowledge the constant readiness of the Tunisian government to provide us, year after year, with the necessary research permits, and the frequent help we enjoyed within the Mahrès community. Last but not least I am very appreciative of the generous financial support that I received from various organizations, most importantly the Swiss National Science Foundation, but also the Human Frontier Science Program, and most recently the Alexander von Humboldt Foundation.

My deepest gratitude is due to my wife, Sibylle, also a biologist, who accompanied me on almost all research trips to *Cataglyphis*, *Ocymyrmex*, and *Melophorus* land. Endowed with a clear grasp of the subtleties of the ants' behavior, she joined me in many pilot experiments that in one way or another should later lead to research projects of graduate students. Her broad knowledge and interest in all aspects of desert life, her patience and endurance, and not the least her language skills made our many journeys to the desert navigators unforgettable events in our lives. Without her love, steady support, and understanding I could neither have gotten so deeply immersed in a kind of research that might not be every scientist's first choice nor could I have written this book.

Illustration Credits

If not stated otherwise, photographs and illustrations are by the author.

Figure 1.3. Adapted from Wehner R and Wehner S. 1990. Insect navigation: use of maps or Ariadne's thread, *Ethology, Ecology, and Evolution* 2: 27–48, fig. 4, with permission from Taylor and Francis.

Inset: Courtesy of Tobias Seidl.

Figure 1.5c. Adapted from Dahmen H, Wahl VL, Pfeffer SE, Mallot HA, and Wittlinger M. 2017. Naturalistic path integration of *Cataglyphis* desert ants on an air-cushioned lightweight spherical treadmill, *The Journal of Experimental Biology* 220: 634–644, fig. 3a, with permission from the the Company of Biologists.

Figure 1.7. Courtesy of Matthias Wittlinger.

Figure 1.9. *(a)* Adapted from Wehner R. 1983. Taxonomie, Funktionsmorphologie und Zoogeographie der saharischen Wüstenameise *Cataglyphis fortis* (Forel 1902) stat. nov. (Insecta: Hymenoptera: Formicidae), *Senckenbergiana biologica* 64: 89–132, fig. 24, with permission from Senckenberg Gesellschaft für Naturforschung.

(b, c) Adapted from Wehner R. 1982. Himmelsnavigation bei Insekten. Neurophysiologie und Verhalten, *Neujahrsblatt der Naturforschenden Gesellschaft Zürich* 184: 1–132, figs. 23b and 24a, with permission from Naturforschende Gesellschaft Zürich.

Figure 1.10. Adapted from Wehner R and Rössler W. 2013. Bounded plasticity in the desert ant's navigational tool kit, in: Invertebrate Learning and Memory, edited by Menzel R and Benjamin PR, *London, Academic Press*, pp. 514–529, fig. 39.1, with permission from Elsevier.

Figure on page 23. Courtesy of Eva Weber. Drawing of a *Cataglyphis fortis* worker (based on a photograph by the author).

Figure 2.2. Reproduced from Description de l'Égypte, ou Recueil des observations et des recherches qui ont été faites en Égypte pendant l'expéditions de l'armée franççaise, 2nd edition (1826). *Paris: L'Imprimerie l'impériale de C.L.F. Panchoucke.*

Figure 2.3. *(a)* line drawing and *(b, c)* color pictures: courtesy of Eva Weber

Figure 2.4. Courtesy of Rainer F. Foelix.

Figure 2.5. *(a, b)* Courtesy of Eva Weber, *(c)* Courtesy of Rainer F. Foelix.

Figure 2.6. *(c, inset)* Courtesy of Donat Agosti.

Figure 2.7. Adapted from Wehner R. 1987. Spatial organization of foraging behavior in individually searching desert ants, *Cataglyphis* (Sahara Desert) and *Ocymyrmex* (Namib Desert), *Experientia Supplement* 54: 15–42, fig. 3, with permission from Birkhäuser.

Figure 2.8. *(a, b)* Adapted from Wenseleers T, Schoeters E, Billen J, and Wehner R. 1998. Distribution and comparative morphology of the cloacal gland in ants (Hymenoptera: Formicidae), *International Journal of Morphology and Embryology* 27: 121–128, fig. 1, with permission from Elsevier.

(c) Courtesy of Rainer F. Foelix.

Figure 2.9. *(a)* Courtesy of Cornelia Bühlmann.

Figure 2.11. *(left)* Adapted from Wehner R, Wehner S, and Agosti D. 1994. Patterns of biogeographic distribution within the *bicolor* species group of the North African desert ant *Cataglyphis*, Foerster 1850, *Senckenbergiana biologica* 74: 163–191, figs. 2, 7 and 8, with permission from Senckenberg Gesellschaft für Naturforschung.

(right) Courtesy of Eva Weber.

Figure 2.12. *(left)* Data source: Aron S, Mardulyn P, and Leniaud L. 2016. Evolution of reproductive traits in *Cataglyphis* desert ants, *Behavioral Ecology and Sociobiology* 7: 1367–1379, fig. 1.

(right) Courtesy of Sabine Hofkunst Schroer.

Figure 2.13. *(a)* Reproduced from Wheeler WM. 1915. The ants of the Baltic amber, *Schriften der physikalisch-ökonomischen Gesellschaft zu Königsberg (Leipzig, B.G. Teubner),* fig. 61.

(b,c) Adapted from Wehner R, Rabenstein R, and Habersetzer J. 2019. Long-leggedness in cataglyphoid Baltic amber ants, *Palaeobiodiversity and Palaeoenvironments,* in press, figs. 1a,c and 4a,c, with permission from Springer Nature.

Figure 2.14. Courtesy of Eva Weber.

Figure 2.17. Courtesy of Eva Weber.

Figure 2.21. Adapted from Wehner R and Wehner S. 2011. Parallel evolution of thermophilia: daily and seasonal foraging patterns in heat-adapted desert ants: *Cataglyphis* and *Ocymyrmex* species, *Physiological Entomology* 36: 271–281, fig. 1, with permission from John Wiley and Sons.

Figure 2.22. *(b, left)* Data source: Cerdá X, Retana J, and Cros S. 1998. Critical thermal limits in Mediterranean ant species: trade-off between mortality risk and foraging performance, *Functional Ecology* 12: 45–55, table 2.

(b, right) Cerdá X and Retana J. 2000. Alternative strategies by thermophilic ants to cope with extreme heat: individual versus colony level traits, *Oikos* 89: 155–163, fig. 3, with permission from John Wiley and Sons.

(c,d) Adapted from Marsh AC. 1985. Thermal responses and temperature tolerance in a diurnal desert ant, *Oxymyrmex barbinger, Physiological Zoology* 58: 629–636, figs. 2 and 3, with permission from University of Chicago Press Journals.

Figure 2.23. *(a,b)* Courtesy of Rainer F. Foelix.

(c) Adapted from Shi NN, Tsai CC, Camino F, Bernard GD, Yu N, and Wehner R. 2015. Keeping cool: enhanced optical reflection and radiative heat dissipation in Saharan silver ants, *Science* 349: 298–301, fig. 1D, with permission from the American Association for the Advancement of Science.

Figure 2.24. *(top)* Courtesy of Nanfang Yu.

(bottom) Adapted from Shi NN, Tsai CC, Camino F, Bernard GD, Yu N, and Wehner R. 2015. Keeping cool: enhanced optical reflection and radiative heat dissipation in Saharan silver ants, *Science* 349: 298–301, fig. 2, with permission from the American Association for the Advancement of Science.

Figure 2.26. Excerpted from Wehner R and Wehner S. 2011. Parallel evolution of thermophilia: daily and seasonal foraging patterns in heat-adapted desert ants: *Cataglyphis* and *Ocymyrmex* species, *Physiological Entomology* 36: 271–281, fig. 6, with permission from John Wiley and Sons.

Figure 2.27. *(inset)* Adapted from Wehner R. 1989. Strategien gegen den Hitzetod. Thermophilie und Thermoregulation bei Wüstenameisen *(Cataglyphis bombycina), Akademie der Wissenschaften und der Literatur Mainz, 1949–1989,* pp. 101–112, fig. 2, with permission from Akademie der Wissenschaften und der Literatur Mainz.

Figure 2.28. *(a,b)* Courtesy of Eric P. Meyer. *(c)* Courtesy of Stefan Hetz.

Figure 2.29. Adapted from Sommer S and Wehner R. 2012. Leg allometry in ants: extreme long-leggedness in thermophilic species, *Arthropod Structure and Development* 41: 71–77, figs. 2 and 3, with permission from Elsevier, supplemented in *(a)* by data from Wehner R, Rabenstein R and Habersetzer J. 2019. Long-leggedness in cataglyphoid Baltic amber ants, *Palaeobiodiversity and Palaeoenvironments,* doi.org/10.1007/s12549-019-00372-9, in press, and in *(b)* by data from Matthias Wittlinger.

Figure 2.30. Courtesy of Eva Weber.

Figure 2.31. *(inset)* Adapted from McMeeking RM, Arzt E, and Wehner R. 2011. *Cataglyphis* desert ants improve their mobility by raising the gaster, *Journal of Theoretical Biology* 297, 17–25, fig. 3, with permission from Elsevier.

Figure 2.32. *(a)* Data source: Seidl T and Wehner R. 2008. Walking on inclines: how do desert ants monitor slope and step length, *Frontiers in Zoology* 5, fig. 3. License BioMed Central Ltd.

Figure 2.33. *(a)* Data source: Wahl V, Pfeffer SE, and Wittlinger M. 2015. Walking and running in the desert ant *Cataglyphis fortis, Journal of Comparative Physiology A* 201: 645–656, figs. 1a,b and 4c,d.

(b) Adapted from Wahl V, Pfeffer SE, and Wittlinger M. 2015. Walking and running in the desert ant *Cataglyphis fortis, Journal of Comparative Physiology A* 201: 645–656, figs. 2a–c, distributed under the terms of the Creative Commons Attribution License.

Figure 2.34. *(b,c)* Courtesy of Kathrin Steck.

Figure 2.35. *(a–c)* Courtesy of Rainer F. Foelix, *(d)* Courtesy of Bruno Erb.

Figure 2.36. *(a)* Adapted from Lighton JRB and Wehner R. 1993. Ventilation and respiratory metabolism in the thermophilic desert ant, *Cataglyphis bicolor* (Hymenoptera, Formicidae), *Journal of Comparative Physiology B* 163: 11–17, fig. 8, with permission from Springer-Verlag GmbH Germany.

(b) Adapted from Niven JE and Scharlemann JPW. 2005. Do insect metabolic rates at rest and during flight scale with body mass?, *Biology Letters* 1: 346–349, fig. 1, with permission from the Royal Society London. Original data kindly provided by Jeremy Niven.

Figure 2.38. Data source for *Messor arenarius:* Neumeyer R. 1994. Strategie der Nahrungsbeschaffung syntoper Arten der

Ernteameisengattung *Messor* im mitteltunesischen Steppengebiet, *PhD Thesis, University of Zürich*, fig. 18.

Figure on page 89. Courtesy of Eva Weber. Drawing of a *Cataglyphis fortis* worker (based on a photograph by the author).

Figure 3.1. Adapted from Pomozi I, Horváth G, and Wehner R. 2001. How the clear-sky angle of polarization pattern continues underneath clouds: full-sky measurements and implications for animal orientation, *The Journal of Experimental Biology* 204: 2933–2941, figs. 1 and 2.

Figure 3.3. *(leftmost figure)* Courtesy of Eric P. Meyer.

Figure 3.6. *(left top and bottom)* Courtesy of Per Antonsen.

Figure 3.7. Courtesy of Martin Müller.

Figure 3.9. *(a,b)* Data source: Fent K. 1985. Himmelsorientierung bei der Wüstenameise *Cataglyphis bicolor:* Bedeutung von Komplexaugen und Ocellen, *PhD Thesis, University of Zürich,* figs. 8 and 9.

(c,d,h) Data source: Duelli P and Wehner R. 1973. The spectral sensitivity of polarized light orientation in *Cataglyphis bicolor* (Formicidae, Hymenoptera), *Journal of Comparative Physiology* 86: 37–53, figs. 1, 7, and 9.

(g) Adapted from Wehner R. 1982. Himmelsnavigation bei Insekten. Neurophysiologie und Verhalten, *Neujahrsblatt der Naturforschenden Gesellschaft Zürich* 184: 1–132, figs. 25a,b, with permission from Naturforschende Gesellschaft Zürich.

Figure 3.11. Adapted from Duelli P. 1974. Polarisationsmusterorientierung bei der Wüstenameise *Cataglyphis bicolor* (Formicidae, Hymenoptera), *PhD Thesis, University of Zürich,* fig. 26.

Figure 3.13. *(a, bottom)* Adapted from Wehner R and Müller M. 2006. The significance of direct sunlight and polarized skylight in the ant's celestial system of navigation, *Proceedings of the National Academy of Sciences USA* 103: 12575–12579, fig.2, © 2006 National Academy of Sciences

(b, bottom) Wehner R and Rossel S. 1985. The bee's celestial compass—a case study in behavioural neurobiology, *Fortschritte der Zoologie* 31: 11–53, fig. 7a, with permission from Oxford University Press, supplemented by data from Fent K, 1985. Himmelsorientierung bei der Wüstenameise *Cataglyphis bicolor:* Bedeutung von Komplexaugen und Ocellen, *PhD Thesis, University of Zürich,* fig. 58.

Figure 3.14. Courtesy of Rainer F. Foelix.

Figure 3.15. *(a)* Adapted from Mote MI and Wehner R. 1980. Functional characteristics of photoreceptors in the compound eye and ocellus of the desert ant, *Cataglyphis bicolor, Journal of Comparative Physiology* 137: 63–71, fig. 5, with permission from Springer-Verlag GmbH Germany.

(c,d) Adapted from Fent K and Wehner R. 1985. Ocelli: a celestial compass in the desert ant *Cataglyphis, Science* 228:

192–194, figs. 1a,b, with permission from the American Association for the Advancement of Science.

Figure 3.16. *(a–c)* Adapted from Wehner R. 1997.The ant's celestial compass system: spectral and polarization channels, in: Orientation and Communication in Insects, edited by Miriam Lehrer, *Basel, Birkhäuser,* pp. 145–185, figs. 13a, 15a, and 18, with permission from Birkhäuser.

(d) Adapted from Fent K. 1985. Himmelsorientierung bei der Wüstenameise *Cataglyphis bicolor:* Bedeutung von Komplexaugen und Ocellen, *PhD Thesis, University of Zürich,* fig. 36b.

Figure 3.17. *(right)* Data source: Coulson KL, Dave JV, and Sekera Z. 1960. Tables Related to Radiation Emerging from a Planetary Atmosphere with Rayleigh Scattering, *Berkeley, University of California Press.*

Figure 3.18. *(d)* Adapted from Lanfranconi BC. 1982. Kompassorientierung nach dem rotierenden Himmelsmuser bei der Wüstenameise *Cataglyphis bicolor, PhD Thesis, University of Zürich,* fig. 14.

Figure 3.19. *(right)* Adapted from Lebhardt F and Ronacher B. 2014. Interactions of the polarization and the sun compass in path integration of desert ants. *Journal of Comparative Physiology A* 200: 711–720, fig. 3, with permission from Springer-Verlag GmbH Germany.

Figure 3.21. *(a)* Adapted from Wehner R and Lanfranconi, B. 1981.What do the ants know about the rotation of the sky?, *Nature* 293: 731–733, fig. 2, with permission from Springer Nature.

Figure 3.22. Adapted from Dyer FE and Dickinson J. 1994. Development of sun compensation by honeybees: how partially experienced bees estimate the sun's course, *Proceedings of the National Academy of Sciences USA* 91: 4471–4474, fig. 1, © 1994 National Academy of Sciences.

Figure 3.24. Adapted from Fleischmann PN, Grob R, Müller VL, Wehner R, and Rössler W. 2018. The geomagnetic field is a compass cue in *Cataglyphis* ant navigation, *Current Biology* 28: 1440–1444, fig. 2, with permission from Elsevier.

Figure 3.25. Adapted from Fleischmann PN, Grob R, Müller VL, Wehner R, and Rössler W. 2018. The geomagnetic field is a compass cue in *Cataglyphis* ant navigation, *Current Biology* 28: 1440–1444, fig. 4, with permission from Elsevier.

Figure 3.26. *(b)* Adapted from Wehner R and Duelli P. 1971. The spatial orientation of desert ants, *Cataglyphis bicolor,* before sunrise and after sunset, *Experientia* 27: 1364–1366, fig. 2, with permission from Springer Nature.

Figure 3.27. Adapted from Duelli P. 1971. Die Konkurrenz von Polarisationsmuster des Himmels, Mondazimut und Windrichtung bei der Orientierung der Wüstenameise *Cataglyphis bicolor* im

Zeitintervall zwischen Sonnenunter- und Sonnenaufgang, *Diploma Thesis, University of Zürich*, fig. 22.

Figure 3.28. *(a)* Adapted from Wehner R, Bernard GD, and Geiger E. 1975. Twisted and non-twisted rhabdoms and their significance for polarization detection in the bee, *Journal of Comparative Physiology* 104: 225–245, fig. 4, with permission from Springer-Verlag GmbH Germany. Wagner-Boller EC. 1987. Ontogenese des peripheren visuellen Systems der Honigbiene *(Apis mellifera), PhD thesis, University Zurich*, fig. 53.

(b) Data source: Meyer EP and Domanico V. 1999. Microvillar orientation in the photoreceptors of the ant *Cataglyphis bicolor, Cell Tissue Research* 295: 355–361, fig. 3c.

Figure 3.29. *(a)* Adapted from Meyer EP. 1984. Retrograde labelling of photoreceptors in different regions of the compound eyes of bees and ants, *Journal of Neurocytology* 13: 825–836, fig. 4b, with permission from Springer Nature.

(b) Adapted from Meyer EP. 1976. Strukturanalyse der Neuronen I. und II. Ordnung im Sehsystem der Ameise *Cataglyphis bicolor* (Formicidae, Hymenoptera), *PhD Thesis, University of Zürich*, fig. 7. Meyer EP and Nässel DR. 1986. Terminations of photoreceptor axons from different regions of the compound eye of the desert ant *Cataglyphis bicolor, Proceedings of the Royal Society of London B* 228: 59–69, fig. 1b, with permission from the Royal Society London.

(c) Adapted from Lambrinos D, Möller R, Labhart T, Pfeifer R, and Wehner R. 2000. A mobile robot employing insect strategies for navigation, *Robotics and Autonomous Systems* 30: 39–64, fig. 3b, with permission from Elsevier.

Figure 3.30. *(a)* Adapted from Wehner R and Labhart T. 2006. Polarization vision, in: Invertebrate Vision, edited by Warrant E and Nilsson D-E. *Cambridge UK, Cambridge University Press*, pp. 291–348, fig. 8.15a, with permission from Cambridge University Press.

(b) Adapted from Petzold J. 2001. Polarisationsempfindliche Neuronen im Sehsystem der Feldgrille, *Gryllus campestris*: Elektrophysiologie, Anatomie und Modellrechnungen, *PhD Thesis, University of Zürich*, figs. 29a,b.

(c) Adapted from Labhart T and Petzold P. 1993. Processing of polarized light information in the visual system of crickets, in: Sensory System of Arthropods, edited by Wiese K, Gribakin FG, Popov AV, and Renninger G, pp. 158–168, fig. 2b, *Basel, Birkhäuser*, with permission from Birkhäuser.

Figure 3.31. *(a)* Adapted from Grob R. 2016. Experience-dependent behavioral changes and neuronal plasticity in visual pathways of *Cataglyphis* desert ants, *MSc Thesis, University of Würzburg*, figs. 12b,c,d.

(b) Adapted from Stieb SM, Münz TS, Wehner R, and Rössler W. 2010. Visual experience and age affect synaptic organization in the mushroom bodies of the desert ant *Cataglyphis fortis, Developmental Neurobiology* 70: 408–423, fig. 1a, with permission from John Wiley and Sons.

(c) Adapted from Schmitt F, Stieb SM, Wehner R, and Rössler W. 2016. Experience-related reorganization of giant synapses in the lateral complex: potential role in plasticity of the sky-compass pathway in the desert ant *Cataglyphis fortis, Developmental Neurobiology* 76: 390–404, fig. 3, with permission from John Wiley and Sons.

Figure 3.32. *(a)* Courtesy of Sara M. Stieb.

(b,c) Adapted from Schmitt F, Stieb SM, Wehner R, and Rössler W. 2016. Experience-related reorganization of giant synapses in the lateral complex: potential role in plasticity of the sky-compass pathway in the desert ant *Cataglyphis fortis, Developmental Neurobiology* 76: 390–404, figs. 2b and 4c, with permission from John Wiley and Sons.

Figure 3.33. *(a,b)* Adapted from Heinze S and Homberg U. 2007. Maplike representation of celestial e-vector orientations in the brain of an insect, *Science* 315: 995–997, figs. 1e and 2g, with permission from the American Association for the Advancement of Science.

(c) Adapted from Heinze S and Homberg U. 2008. Neuroarchitecture of the central complex of the desert locust: intrinsic and columnar neurons, *Journal of Comparative Neurology* 511: 454–478, fig. 11b, with permission from John Wiley and Sons.

Figure 4.2. *(b)* Data source: Heusser D and Wehner R. 2002. The visual centering response in desert ants, *Cataglyphis fortis, The Journal of Experimental Biology* 205: 585–590.

Figure 4.3. Courtesy of Matthias Wittlinger.

Figure 4.4. *(a)* Adapted from Wittlinger M, Wehner R, and Wolf H. 2007. The desert ant odometer: a stride integrator that accounts for stride length and walking speed, *The Journal of Experimental Biology* 210: 198–207, fig. 4.

(b) Courtesy of Matthias Wittlinger.

(c) Adapted from Wittlinger M, Wehner R, and Wolf H. 2006. The ant odometer: stepping on stilts and stumps, *Science* 312: 1965–1967, fig. 2, with permission from the American Association for the Advancement of Science.

Figure 4.5. *(b,c)* Adapted from Ronacher B and Wehner R.1995. Desert ants *Cataglyphis fortis* use self-induced optic flow to measure distances travelled, *Journal of Comparative Physiology A* 177: 21–27, figs. 1 and 2, with permission from Springer-Verlag GmbH Germany.

Figure 4.6. *(b)* Courtesy of Eva Weber.

Figure 4.7. *(a–c, bottom)* Adapted from Pfeffer SE and Wittlinger M. 2016. Optic flow odometry operates independently of stride integration in carried ants, *Science* 353: 1155–1157, figs. 2b and 3b, with permission from the American Association for the Advancement of Science.

Figure 4.8. Adapted from Seidl, T Knaden M, and Wehner R. 2006. Desert ants: is active locomotion a prerequisite for path integration?, *Journal of Comparative Physiology A* 192: 1125–1131, figs. 3 and 4d, with permission from Springer-Verlag GmbH Germany.

Figure 4.9. *(bottom)* Data source: Srinivasan MV, Zhang S-W, and Bidwell NJ.1997. Visually mediated odometry in honeybees, *The Journal of Experimental Biology* 200: 2513–2522, figs. 3b and c.

Figure 4.11. Adapted from Wohlgemuth S, Ronacher B, and Wehner R. 2001. Ant odometry in the third dimension, *Nature* 411: 795–798, fig. 1.

Figure 4.12. (a) Courtesy of Bernhard Ronacher.

(b) Data source: Maronde K, Wohlgemuth S, Ronacher B, and Wehner R. 2003. Ground instead of walking distances determine the direction of the home vector in 3-D path integration of desert ants, *Proceedings of the Neurobiology Conference Göttingen* 29: 548–549.

Figure 5.2. Adapted from Wehner R. 2003. Desert ant navigation: how miniature brains solve complex tasks. *Journal of Comparative Physiology A* 189: 579–588, fig. 2, with permission from Springer-Verlag GmbH Germany.

Figure 5.3. Adapted from Merkle TFC. 2007. Orientation and search strategies of desert arthropods: path integration models and experiments with desert ants, *Cataglyphis fortis* (Forel 1902). *PhD Thesis, University of Bonn*, figs. II.1 and II.2.

Figure 5.4. Data source: Mittelstaedt H and Mittelstaedt M-L. 1982. Homing by path integration, in Aviation Navigation, edited by Papi F and Wallraff HG, *Berlin, Springer-Verlag*, pp. 290–297, fig. 5.

Figure 5.6. Adapted from Wehner R, Gallizzi K, Frei C, and Vesely M. 2002. Calibration processes in desert ant navigation: vector courses and systematic search, *Journal of Comparative Physiology A* 188: 683–693, figs. 1 and 6, with permission from Springer-Verlag GmbH Germany.

Figure 5.8. Data source: Vögeli B. 2006. Spatial shortcutting in desert ants, *Cataglyphis fortis, Diploma Thesis, University of Zürich,* figs. 12 and 17.

Figure 5.9. *(top)* Courtesy of Markus Knaden.

(a,b) Adapted from Knaden M and Wehner R.2006. Ant navigation: resetting the path integrator, *The Journal of Experimental Biology* 209: 26–31, figs. 1a and 2a.

Figure 5.10. *(a)* Adapted from Wehner R and Srinivasan MV. 1981. Searching behavior of desert ants, genus *Cataglyphis* (Formicidae, Hymenoptera), *Journal of Comparative Physiology* 142: 315–338, fig. 4, with permission from Springer-Verlag GmbH Germany.

(b,c) Adapted from Müller M and Wehner R. 1994. The hidden spiral: systematic search and path integration in desert ants, *Cataglyphis fortis, Journal of Comparative Physiology A* 175: 525–530, fig. 2, with permission from Springer-Verlag GmbH Germany.

Figure 5.11. *(a)* Adapted from Merkle T, Knaden M, and Wehner R. 2006. Uncertainty about nest position influences systematic search strategies in desert ants, *The Journal of Experimental Biology* 209: 3545–3549, figs. 2a,c.

(b, left) Adapted from Merkle T, Knaden M, and Wehner R. 2006. Uncertainty about nest position influences systematic search strategies in desert ants, *The Journal of Experimental Biology* 209: 3545–3549, fig. 3b.

(b, right) Adapted from Vickerstaff RJ and Merkle T. 2012. Path integration mediated systematic search: a Bayesian model, *Journal of Theoretical Biology* 307: 1–19, fig. 3, with permission from Elsevier.

Figure 5.12 *(a)* Adapted from Wehner R and Wehner S. 1986. Path integration in desert ants: approaching a long-standing puzzle in insect navigation, *Monitore Zoologico Italiano* 20: 309–331, fig. 5, with permission from Taylor and Francis.

(b) Adapted from Wehner R and Wehner S. 1986. Path integration in desert ants: approaching a long-standing puzzle in insect navigation, *Monitore Zoologico Italiano* 20: 309–331, fig. 8a, with permission from Taylor and Francis.

(c) Data sources: Wehner R and Srinavasan MV. 1981. Searching behavior of desert ants, genus *Cataglyphis* (Formicidae, Hymenoptera), *Journal of Comparative Physiology* 142: 315–338, fig. 11, supplemented (red curve) by data source: Heinze S, Narendra A, and Cheung A. 2018. Principles of insect path integration, *Current Biology* 28: R1043–R1058, Box1 / II.

Figure 5.13. Data source: Bernasconi L. 1989. Das Nestsuchverhalten der Wüstenameise *Cataglyphis fortis* (Formicidae, Hymenoptera), *Diploma Thesis, University of Zürich,* fig. 12. Nieuwlands Y. 1989. Einfluss künstlicher Landmarken auf das Nestsuchverhalten der Wüstenameise *Cataglyphis fortis* (Formicidae, Hymenoptera), *Diploma Thesis; University of Zürich,* fig. 10.

Figure 5.15. *(b)* Adapted from Müller M and Wehner R. 1988. Path integration in desert ants, *Cataglyphis fortis, Proceedings of the National Academy of Sciences USA* 85: 5287–5290, fig. 2b.

Figure 5.16. Adapted from Müller M. 1989. Mechanismus der Wegintegration bei *Cataglyphis fortis* (Hymenoptera, Formicidae), *PhD Thesis, University of Zürich,* fig. 71.

Figure 5.17. *(a, bottom)* Adapted from Sommer S and Wehner R. 2005. Vector navigation in desert ants, *Cataglyphis fortis:* celestial compass cues are essential for the proper use of distance information, *Naturwissenschaften* 92: 468–471, fig. 1, with permission from Springer Nature.

(b, right) Adapted from Ronacher B, Westwig E, and Wehner R. 2006. Integrating two-dimensional paths: do desert ants process distance information in the absence of celestial compass cues?, *The Journal of Experimental Biology* 209: 3301–3308, fig. 4.

Figure 5.18. *(a)* Adapted from Grob R. 2016. Experience-dependent behavioral changes and neuronal plasticity in visual pathways of *Cataglyphis* desert ants, *MSc Thesis, University of Würzburg,* fig. 12d.

(b) Adapted from Stone T, Webb B, Adden A, Weddig NB, Honkanen A, Templin R, Wcislo W, Scimeca L, Warrant E, and Heinze S. 2017. An anatomically constrained model for path integration in the bee brain, *Current Biology* 27: 3069–3085, fig. 5g, with permission from Elsevier.

Figure 6.1. Adapted from Fleischmann PN, Grob R, Wehner R, and Rössler W. 2017. Species-specific differences in the fine structure of learning walk elements in *Cataglyphis* ants, *The Journal of Experimental Biology* 220: 2426–2435, figs. 1a,b,d.

Figure 6.2. *(right)* Adapted from Wehner R and Müller M. 2010. Piloting in desert ants: pinpointing the goal by discrete landmarks, *The Journal of Experimental Biology* 213: 4174–4179, figs. 1b,d.

Figure 6.4. *(a)* Adapted from Möller R. 2001. Do insects use templates or parameters for landmark navigation?, *Journal of Theoretical Biology* 210: 33–45, fig. 1, with permission from Elsevier.

(b) Adapted from Wehner R, Michel B, and Antonsen P. 1996. Visual navigation in insects: coupling of egocentric and geocentric information, *The Journal of Experimental Biology* 199: 129–140, fig. 5.

Figure 6.5. *(right)* Adapted from Lent DD, Graham P, and Collett TS. 2010. Image-matching during ant navigation occurs through saccade-like body turns controlled by learned visual features, *Proceedings of the National Academy of Sciences USA* 107: 16348–16353, fig. 2e, with permission from the National Academy of Sciences USA.

Figure 6.6. *(a)* Data source: Lambrinos D, Möller R, Labhart T, Pfeifer R, and Wehner R. 2000. A mobile robot employing insect strategies for navigation, *Robotics and Autonomous Systems* 30: 39–64, figs. 12 and 19.

(b) Adapted from Zeil J. 2012. Visual Homing: an insect perspective, *Current Opinion in Neurobiology* 22: 285–293, fig. 1, with permission from Elsevier.

Figure 6.7. *(a)* Courtesy of Per Antonsen.

(b) Adapted from Wehner R, Michel B, and Antonsen P. 1996. Visual navigation in insects: coupling of egocentric and geocentric information, *The Journal of Experimental Biology* 199: 129–140, fig. 8.

Figure 6.8. Adapted from Åkesson S and Wehner R. 2002. Visual navigation in desert ants *Cataglyphis fortis*: are snapshots coupled to a celestial system of reference?, *The Journal of Experimental Biology* 205: 1971–1978: figs. 1 and 2.

Figure 6.9. *(c)* Adapted from Lambrinos D, Möller R, Labhart T, Pfeifer R, and Wehner R. 2000. A mobile robot employing insect strategies for navigation, *Robotics and Autonomous Systems* 30: 39–64, fig. 17, with permission from Elsevier.

Figure 6.10. Adapted from Wystrach A, Beugnon G, and Cheng K. 2011. Landmarks or panoramas: what do navigating ants attend to for guidance. *Frontiers in Zoology* 8:21, figs. 1, 2, and 4, open access article distributed under the terms of the Creative Commons Attribution License.

Figure 6.11. Courtesy of Jochen Zeil.

Figure 6.12. *(bottom)* Adapted from Steck K, Hansson BS, and Knaden M. 2009. Smells like home: desert ants, *Cataglyphis fortis*, use olfactory landmarks to pinpoint the nest, *Frontiers in Zoology* 6:5, fig. 4, open access article distributed under the terms of the Creative Commons Attribution License.

Figure 6.13. *(a, right)* Adapted from Bisch-Knaden S and Wehner R. 2001. Egocentric information helps desert ants to navigate around familiar objects, *The Journal of Experimental Biology* 204: 4177–4184, fig. 4.

(b, bottom) Adapted from Collett M, Collett TS, Bisch S, and Wehner R. 1998. Local and global vectors in desert ant navigation, *Nature* 394: 269–272, figs. 1 and 2a,g,h.

(c, bottom) Adapted from Knaden M, Lange CHJ, and Wehner R. 2006. The importance of procedural knowledge in desert ant navigation, *Current Biology* 16: 916–917, fig. 1, with permission from Elsevier.

Figure 6.15. *(a)* Adapted from Wehner R, Michel B, and Antonsen P. 1996. Visual navigation in insects: coupling of egocentric and geocentric information, *The Journal of Experimental Biology* 199: 129–140, fig. 9.

(b) Adapted from Wehner R. 2003. Desert ant navigation: how miniature brains solve complex tasks. *Journal of Comparative Physiology A* 189: 579–588, fig. 7b, with permission from Springer-Verlag GmbH Germany.

(c) Adapted from Wehner R, Boyer M, Loertscher F, Sommer S, and Menzi U. 2006. Ant navigation: one-way routes rather than maps, *Current Biology* 16: 75–79, figs. 1a,b and 2a,b, with permission from Elsevier.

Figure 6.16. *(b)* Adapted from Mangan M and Webb B. 2012. Spontaneous formation of multiple routes in individual desert ants *(Cataglyphis velox)*, *Behavioral Ecology* 23: 944–954, fig. 7, with permission from Oxford University Press.

(d) Adapted from Sommer S, von Beeren C, and Wehner R. 2008. Multiroute memories in desert ants, *Proceedings of the National Academy of Sciences USA* 105: 317–322, fig. 4.

Figure 6.17 *(a)* Courtesy of Rainer F. Foelix.

(b) Adapted from Macquart D, Garnier L, Combe M, and Beugnon G. 2006. Ant navigation en route to the goal: signature routes facilitate way-finding of *Gigantiops destructor, Journal of Comparative Physiology A* 192: 221–234, figs. 4d and 6c, with permission from Springer-Verlag GmbH Germany.

Figure 6.18. Adapted from Baddeley B, Graham P, Husbands P, and Philippides A. 2012. A model of ant route navigation driven by scene familiarity, *PLoS Computational Biology* 8: e1002336, fig. 5, distributed in accordance with the Creative Commons Attribution (CC BY) licence.

Figure 6.19. Adapted from Fleischmann PN, Grob R, Wehner R, and Rössler W. 2017. Species-specific differences in the fine structure of learning walk elements in *Cataglyphis* ants, *The Journal of Experimental Biology* 220: 2426–2435, figs. 2d,e,f.

Figure 6.20. Data source: Müller M and Wehner R. 2010. Path integration provides a scaffold for landmark learning in desert ants, *Current Biology* 20: 1368–1371, fig. 1.

Figure 6.21. Adapted from Fleischmann PN, Grob R, Wehner R, and Rössler W. 2017. Species-specific differences in the fine structure of learning walk elements in *Cataglyphis ants, The Journal of Experimental Biology* 220: 2426–2435, figs. 2A and 4B.

Figure 6.22. Adapted from Fleischmann PN, Christian M, Müller VL, Rössler W, and Wehner R. 2016. Ontogeny of learning walks and the acquisition of landmark information in desert ants, *Cataglyphis fortis, The Journal of Experimental Biology* 219: 3137–3145, fig. 3.

Figure 6.23. Adapted from Fleischmann PN, Rössler W, and Wehner R. 2018. Early foraging life: spatial and temporal aspects of landmark learning in the ant *Cataglyphis noda, Journal of Comparative Physiology A* 204:579–592, figs. 5a,c, with permission from Springer-Verlag GmbH Germany.

Figure 6.24. *(a)* Courtesy of Ajay Narendra.

(b,c) Adapted from Zeil J, Narendra A, and Stürzl W. 2014. Looking and homing: how displaced ants decide where to go, *Philosophical Transactions of the Royal Society B* 369: 20130034, figs. 1 and 2c, with permission from the Royal Society London.

Figure 6.25. *(a)* Adapted from Wehner R. 2003. Desert ant navigation: how miniature brains solve complex tasks. *Journal of Comparative Physiology A* 189: 579–588, fig. 1, with permission from Springer-Verlag GmbH Germany.

(b) Courtesy of Robin Grob.

Figure 6.26. *(a,b)* Courtesy of Sara M. Stieb.

(c) Adapted from Wehner R and Rössler W. 2013. Bounded plasticity in the desert ant's navigational tool kit, in: Invertebrate Learning and Memory, edited by Menzel R and Benjamin PR, pp. 514–529, *London, Academic Press*, fig. 39.6c, with permission from Elsevier.

Figure 6.27. Adapted from Steiger U. 1967. Über den Feinbau des Neuropils im Corpus pedunculatum der Waldameise, *Zeitschrift für Zellforschung* 81: 511–563, fig. 6, with permission from Springer Nature.

Figure 6.28. *(b)* Adapted from Goll W. 1967. Strukturuntersuchungen am Gehirn von *Formica, Zeitschrift für Morphologie und Ökologie der Tiere* 59: 143–210, fig. 13, with permission from Springer Nature.

(c) From Strausfeld NJ. 2012. Arthropod Brains, *Harvard University Press, Cambridge MA*, fig. 6.16, middle figure, with kind permission from Nicholas J. Strausfeld.

Figure 6.29. Adapted from Kühn-Bühlmann S and Wehner R. 2006. Age-dependent and task-related volume changes in the mushroom bodies of visually guided desert ants, *Cataglyphis bicolor, Developmental Neurobiology* 66: 511–521, fig.3a, with permission from John Wiley and Sons.

Figure 6.30. Adapted from Wehner R and Rössler W. 2013. Bounded plasticity in the desert ant's navigational tool kit, in: Invertebrate Learning and Memory, edited by Menzel R and Benjamin PR, pp. 514–529, *London, Academic Press*, fig. 39.7, with permission from Elsevier.

Figure 7.1. Design by Eric P. Meyer and Karin Niffeler

Figure 7.2. *(a)* Adapted from Bregy P, Sommer S, and Wehner R. 2008. Nest-mark orientation versus vector navigation in desert ants, *The Journal of Experimental Biology* 211: 1868–1873, figs. 2a,c.

(b) Adapted from Collett M. 2012. How navigational guidance systems are combined in a desert ant, *Current Biology* 22: 927–932, figs. 1a,b and 2a, with permission from Elsevier.

Figure 7.3. *(left)* Adapted from Wystrach A, Mangan M, and Webb B. 2015. Optimal cue integration in ants, *Proceedings of the Royal Society B* 282: 20151484, fig. 1a, with permission from the Royal Society London.

(right) Data source: Wystrach A, Mangan M, and Webb B. 2015. Optimal cue integration in ants, *Proceedings of the Royal Society B* 282: 20151484, fig. 2a.

Figure 7.4. *(bottom)* Adapted from Bühlmann C, Cheng K, Wehner R. 2011. Vector-based and landmark-guided navigation in desert ants inhabiting landmark-free and landmark-rich environments, *The Journal of Experimental Biology* 214: 2845–2853, fig. 2.

Figure 7.5. Data source: Andel D and Wehner R. 2004. Path integration in desert ants, *Cataglyphis:* how to make a homing ant run away from home, *Proceedings of the Royal Society B* 271: 1485–1489, fig. 4b.

Figure 7.6. Adapted from Wolf and Wehner R. 2000. Pinpointing food sources: olfactory and anemotactic orientation in desert ants, *Cataglyphis fortis, The Journal of Experimental Biology* 203: 857–868, figs. 3a,b, 6a, and 11a,b.

Figure 7.7. Courtesy of Harald Wolf.

Figure 7.8. *(a*)* Adapted from Wolf H and Wehner R. 2005. Desert ants compensate for navigation uncertainty, *The Journal of Experimental Biology* 208: 4223–4230, fig. 5a.

(b*) Adapted from Sommer S and Wehner R. 2004. The ant's estimation of distance travelled: experiments with desert ants, *Cataglyphis fortis, Journal of Comparative Physiology A* 190: 1–6, fig. 2, with permission from Springer-Verlag GmbH Germany.

Figure 7.9. Data source: Schwarz S, Mangan M, Zeil J, Webb B, and Wystrach A. 2017. How ants use vision when homing backward, *Current Biology* 27: 401–407.

Figure 7.10. Adapted from Hoinville T and Wehner R. 2018. Optimal multiguidance integration in insect navigation, *Proceedings of the National Academy of Sciences USA* 115: 2824–2829, fig. 1a.

Figure 7.11 Adapted from Hoinville T and Wehner R. 2018. Optimal multiguidance integration in insect navigation, *Proceedings of the National Academy of Sciences USA* 115: 2824–2829, fig. 3b.

Figure 7.12. Adapted from Meyer EP. 1971. Die Ausbildung von Ovar, Fettkörper und Labialdrüsen bei verschiedenen Funktionstypen, Grössenklassen und Altersstufen von *Cataglyphis bicolor* (Formicidae, Hymenoptera), *Diploma Thesis, University of Zürich*, fig. 10.

Figure 7.13. Adapted from Wehner R, Meier C, and Zollikofer C. 2004. The ontogeny of foraging behaviour in desert ants, *Cataglyphis bicolor, Ecological Entomology* 29: 240–250, fig. 6a, with permission from John Wiley and Sons.

Figure 7.15. *(a)* Data source: Sassi S and Wehner R. 1997. Dead reckoning in desert ants, *Cataglyphis fortis:* can homeward vectors be reactivated by familiar landmark configurations?, *Proceedings of the Neurobiology Conference Göttingen* 25: 484.

(b) Data source: Collett M, Collett TS, Chameron S, and Wehner R. 2003. Do familiar landmarks reset the global path integration system of desert ants?, *The Journal of Experimental Biology* 206: 877–882.

Figure 7.16. *(a)* Adapted from Narendra A. 2007. Homing strategies of the Australian desert ant *Melophorus bagoti.* II. Interaction of the path integrator with visual cue information, *The Journal of Experimental Biology* 210: 1804–1812, figs. 4a and 6a, with permission from the Company of Biologists.

(b) Adapted from Narendra A, Gourmaud S, and Zeil J. 2013. Mapping the navigational knowledge of individually foraging ants, *Myrmecia croslandi. Proceedings of the Royal Society B* 280: 20130683, fig. 5c, with permission from the Royal Society London.

Figure 7.17. *(inset)* Adapted from Grob R, Fleischmann PN, Gübel K, Wehner R, and Rössler W. 2017. The role of celestial compass information in *Cataglyphis* ants during learning walks and for neuroplasticity in the central complex and mushroom bodies, *Frontiers in Behavioral Neuroscience* 11: 226, fig. 3B.

Figure 7.18. Adapted from Wehner R, Fukushi T, and Isler K. 2007. On being small: brain allometry in ants, *Brain, Behavior, and Evolution* 69: 220–228, figs. 1 and 2, with permission from S. Karger AG.

Figure 7.19. Courtesy of Edvard I. Moser and Vadim Frolov.

Index